CREATIVITY IN ENGINEERING

EXPLORATIONS IN CREATIVITY RESEARCH

Series Editor

James C. Kaufman

CREATIVITY IN ENGINEERING

Novel Solutions to Complex Problems

DAVID H. CROPLEY
University of South Australia
Mawson Lakes, SA, Australia

ELSEVIER

AMSTERDAM • BOSTON • HEIDELBERG • LONDON
NEW YORK • OXFORD • PARIS • SAN DIEGO
SAN FRANCISCO • SINGAPORE • SYDNEY • TOKYO

Academic Press is an imprint of Elsevier

AP

Academic Press is an imprint of Elsevier
32 Jamestown Road, London NW1 7BY, UK
525 B Street, Suite 1800, San Diego, CA 92101-4495, USA
225 Wyman Street, Waltham, MA 02451, USA
The Boulevard, Langford Lane, Kidlington, Oxford OX5 1GB, UK

ISBN: 978-0-12-800225-4

British Library Cataloguing-in-Publication Data
A catalogue record for this book is available from the British Library

Library of Congress Cataloging-in-Publication Data
A catalog record for this book is available from the Library of Congress

For Information on all Academic Press publications
visit our website at http://store.elsevier.com/

Typeset by MPS Limited, Chennai, India
www.adi-mps.com

Dedication

"Invention, strictly speaking, is little more than a new combination of those images which have been previously gathered and deposited in the memory. Nothing can come of nothing. He who has laid up no materials can produce no combinations."

Sir Joshua Reynolds,[1] 1723–1792

"... innovation is the only mechanism that can actually change things in substantive ways. Innovation is where creative thinking and practical know-how meet to do new things in new ways, and old things in new ways."

Dr. Charles Vest,[2] 1941–2013

[1]A Discourse Delivered to the Students of the Royal Academy, on the Distribution of the Prizes, December 11, 1769, by the President. (See http://www.gutenberg.org/files/2176/2176-h/2176-h.htm).

[2]President, U.S. National Academy of Engineering, 2007–2013.

Contents

List of Figures

List of Tables

Foreword

There is a genuine risk in trying something new. As many have noted (such as Robert Sternberg), we say we want creative and different ideas ... but we generally don't actually want them. Studying creativity is itself a risky act; it is a topic of great general interest that nonetheless does not have a firm home in any one discipline. Psychology, education, and business probably make the most notable contributions, although they often use different terminology and publish in different journals. When scholars from different fields decide to study creativity, there is the temptation to not conduct due diligence. It is easy to assume that because no one in a field has delved deeply into creativity, there is nothing notable to learn. What often happens then is that people from disparate fields tend to try to research from scratch and end up reinventing the wheel.

David Cropley is a rare exception, which is why I am so delighted to have this book kick off the *Explorations in Creativity Research* book series for Academic Press. David has delved deeply into both classic and current creativity scholarship while simultaneously bringing his own nuanced and informed perspective to the debate. His engineering background and sharp, analytical mind have helped turn him into one of the most exciting current scholars in creativity studies.

I believe that *Creativity in Engineering: Novel Solutions to Complex Problems* is a major work in the field, and am thrilled to introduce it in this new series.

James C. Kaufman
Neag School of Education
University of Connecticut

Preface

THE APPROACH OF THIS BOOK

This book provides a set of research-based concepts, firmly grounded in the body of knowledge of creativity, which will help professional engineers, engineering managers, and engineering educators to understand *creativity* in a systematic way. The material in the book will help these stakeholders to identify its key aspects, develop it in themselves, foster its development in students and professionals, and acknowledge and reward it appropriately. The aim is

- to demystify the concept of creativity, i.e., help educators, managers, practitioners and students understand it in a practical, concrete way;
- to show that there is a common core to creativity in all disciplines;
- to help people acquire a foundation of creative skills, motives, attitudes, and values from the very beginning;
- to show educators and managers how to facilitate the development of these through their leadership;
- to show educators and managers how to evaluate other people's work in ways that foster creativity.

The approach to creativity in this book emphasizes three complementary facets:

1. the thinking skills and strategies people need for creativity (cognitive factors);
2. the personal and motivational properties that permit and activate these skills and strategies (noncognitive factors);
3. the characteristics of the environment (social and organizational factors) that influence the whole process.

WHY IS THIS BOOK DIFFERENT?

Much of what has been written about creativity in the domain of engineering is unsatisfactory. There is no shortage of books and articles that teach readers cognitive *tricks*. These approaches, however, are overly dependent on the reapplication of factual knowledge, often in the form of catchy, fixed techniques. They fail to encourage deep learning and fail to move beyond developing a knowledge of *what and how* by adding knowledge of *when and why*. This shallow approach to understanding leaves learners able to execute simple procedures and describe superficial concepts, but far less able to compare and contrast different approaches, analyze underlying causes, and reapply their knowledge to new and unfamiliar situations. By definition, creativity in engineering is about dealing with new and unprecedented problems. An overreliance on the reapplication of tried and trusted methods leaves learners unable to answer questions such as "why didn't this method work in this particular case?" or "what method is best suited to this particular group of people?" The shallow, factual-only approach has been characterized as "fast food" creativity (A. J. Cropley & Cropley, 2009). This book therefore advocates *spinach* creativity: a more complete approach that does not shy away from those aspects that are less glamorous and less readily digested.

Most people are familiar with Thomas Edison's famous quote: "Genius is one percent inspiration, ninety-nine percent perspiration" (published in Harper's magazine in 1932). The same, broadly speaking, may be said of creativity. Perhaps a better simile is that creativity is like an iceberg—the end result, in the form of a brilliant new idea, or a novel solution to an intractable problem, is only 10% the whole effort. The other 90% is the complex interplay of personal properties, feelings, motivation, cognitive processes, organizational and social factors that deliver the visible product. If we are to get the best possible result from our creative engineering efforts, we must understand not only the visible tip of the iceberg, but also everything that lies hidden beneath.

OUTLINE OF CHAPTERS

The focus of the following chapters of this book is therefore, fundamentally, to address the question of how to reconnect creativity and engineering. Before embarking on a detailed explanation of the 4Ps of creativity, what they are, and how they are measured, I will begin by fleshing out two issues already mentioned briefly as important for understanding creativity in engineering. **Chapter 2** first tackles the question of the importance of creativity in engineering. Why should engineering organizations, their leadership, and their professional practitioners, as well as educational institutions, be concerned with creativity in engineering? What is it that engineers do that requires creativity? What value does creativity add to products? Also, how does creativity feed the wider process of innovation?

In **Chapter 3**, I then begin the detailed development of the framework for understanding creativity (and innovation) in engineering by looking at the *Phases* involved in both engineering design and creativity. What are they, and how do they differ in terms of what is required? One consequence of this will be that creativity is recognized as *not* being a one-size-fits-all activity. This chapter will also show that engineering design and a generic characterization of the phases of creativity bear a close correspondence. The phase model developed also introduces the idea of paradoxes that will be examined in detail in each subsequent chapter.

Chapter 4 begins the process of examining the 4Ps in some detail. I will begin by looking at the desired outcome of the creative engineering activity—the *Product*. What are the things that are created (products, processes, systems, services), and what qualities do these possess (or need to possess) to be regarded as creative? What value do these qualities add to the product and how does that set a creative product apart from one that is not creative?

Chapter 5 studies *Process* in the sense of the cognitive processes that are used in the creation of technological solutions. In particular, this tackles how ideas are generated, and how both divergent and convergent thinking processes are achieved in the context of engineering creativity.

Chapter 6 tackles the personal, psychological aspects of creativity and innovation. What are the feelings, personal properties, and motivational factors that can foster or hinder creativity?

Chapter 7 addresses the role of the organizational and social environments and creativity. This *Press* can exert an influence on creativity in a variety of ways: it defines what is creative, who is creative, and the amount and kind of creativity that society can tolerate. The *Press* also plays an important role in assisting or resisting creativity.

Chapter 8 examines creativity in the wider context of innovation. How do they differ across the phases of the creativity and innovation? I will also discuss wider issues such as competition, and how this impacts on the transition from novel idea to acclaimed product. This chapter also examines ways that the innovative capacity of organizations can be *diagnosed*.

Chapter 9 starts the process of looking at educational issues and creativity by studying creativity training (as distinct from education in a university context). This chapter will look at the evidence for and against creativity training, and apply these particularly to the case of professional engineering activities.

Chapter 10 draws the book to a close by examining more specifically how creativity can, and should, be embedded in engineering education. Among the issues facing engineering educators is a general reluctance to give up anything in the engineering curriculum, meaning that creativity, if it is included at all, is usually tacked on in a piecemeal fashion, rather than being built into engineering programs in a systematic, top-down manner.

ACKNOWLEDGMENTS

Sir Isaac Newton once wrote to his fellow natural philosopher Robert Hooke "If I have seen further it is by standing on the shoulders of giants."[1] I am very fortunate, as a researcher and author in the field of creativity, to have been the beneficiary of a number of willing sets of shoulders. As you will discover in this book, the discipline of creativity has its home largely in the field of psychology, where it has grown to maturity over some 60 years. The shoulders that I have stood on, and the sights I have seen, have been supplied by eminent scholars in the field of psychology. Not only have they allowed me to stand on their shoulders, but also they have generously admitted me to their domain, and patiently tutored me, despite my non-psychological background. If I have contributed to the field, I hope it is, principally, to show that creativity plays an enormously important role in engineering, and that engineers must familiarize themselves with it. I also hope that I have shown that the psychological foundations of creativity are not mysterious and incomprehensible, but are open and accessible to engineers. Having said that, I think the best solution for engineers concerned with recognizing, developing, and fostering creativity is to work closely with psychologists, drawing on the complementary strengths of the two disciplines.

The most important giant I need to thank is Emeritus Professor Arthur Cropley of the University of Hamburg. Arthur is an educational psychologist and he began working in the field of creativity more than 50 years ago. He therefore joined the field in its pioneering days, and has an unrivaled knowledge and perspective of how it has developed from infancy to maturity. He, more than anyone, is responsible for my entry to the field, and this book reflects his knowledge and expertise as much as my own. When I was writing about the Sputnik Shock of October 1957 (see Chapter 1) and its impact on creativity, it was invaluable to be able to talk to someone like Arthur, who not only lived through that time, but also remembers standing on a street corner, listening to a portable radio tuned to Sputnik's beeps. He is also my father.

[1]Letter from Isaac Newton to Robert Hooke, February 5, 1676, as transcribed in Jean-Pierre Maury (1992) *Newton: Understanding the Cosmos*, New Horizons.

The next giant I need to acknowledge is Professor James Kaufman of the University of Connecticut. James first contacted me in about 2003, asking me to contribute a chapter on engineering creativity to a book he was editing. My first response was to email him asking if he had the right Cropley! I thought that perhaps he had confused me with the *real* creativity researching Cropley, namely Arthur! James, however, assured me that he had not made a mistake. It is thanks to James, and his willing collaboration, that I have had the opportunity to immerse myself in a creativity research environment, through many visits to his lab, first in California, and more recently in Connecticut. It is also through James that I have come to know, and collaborate with, the much wider community of creativity researchers in psychology.

The third giant I would like to thank is Professor Mark Runco of the University of Georgia. Mark is another preeminent scholar of creativity with a long and distinguished record of research in the field. I have been fortunate to be able to interact with him on many occasions and have benefited immensely from his knowledge and insight.

There are many others in this field—psychologists and engineers—with whom I have crossed paths, collaborated, and exchanged ideas. I don't know if it is a characteristic of creativity researchers, or of psychologists, or simply luck on my part, but all of the people I have interacted with in this field have been open, generous, helpful, and friendly. It is always a pleasure to work with them, and makes for a varied and stimulating career. I thank all those other researchers and thinkers who have given generously of their time and intelligence and look forward to many more years of fruitful collaboration in this fascinating area.

Finally, I would like to thank my wife, Melissa (another psychologist!), and my children, Matthew, Dana, and Daniel. They have all made contributions to my understanding of creativity, not only through their moral support, but also through their creative examples in acting, filmmaking, singing, writing, and art. They *live* creativity—I just write about it.

David H. Cropley
Adelaide, May 2014

CHAPTER

1

Introduction

"Creativity can solve almost any problem. The creative act, the defeat of habit by originality, overcomes everything." ***George Lois, 1931–, Art Director and Author***

THE SPUTNIK SHOCK

The next time you are driving to an unfamiliar location, listening to the synthetic tones of your smart phone telling you to "Take next exit onto I-10 East," spare a thought for the significance of the date *October 4, 1957*. This date saw the birth of a profound technological revolution, the outcomes of which affect each one of us, every day of our lives. It is the date on which the disparate fields of engineering, technology, psychology, management, and economics began to flow together to give to us an understanding of how and why we develop technological solutions to modern, complex problems, and the value that these solutions deliver to society. It is a date that has had a far-reaching impact on our modern, 21st century lifestyle, influencing our economic security, our physical security, our health, our education, and much more.

You may recognize this as the date of the launch, by the Soviet Union, of *Sputnik I*—the world's first artificial satellite. It was more than that, however, because it ushered in the Space Age. Although this may seem a somewhat distant and mundane milestone to Generation Y[1] and beyond, it is wise not to underestimate the profound impact that this event had on the Western psyche at the time (Dickson, 2001). Indeed, that impact has been described as the *Sputnik Shock* (A. J. Cropley, 2001; A. J. Cropley & Cropley, 2009), and Western

[1]Generally speaking, those people born after 1982.

newspapers at the time roared headlines such as "Red 'Moon' over London!" and "Space Age Is Here!" Set against the backdrop of the Cold War tension between the Soviet and Western blocs, both sides' preoccupation with nuclear weapons technology, and the recent conflict on the Korean peninsula, it is not difficult to imagine the fear and consternation that this technological trump card engendered in Western countries. Dickson (2001) reflects that "it was as if Sputnik was the starter's pistol in an exciting new race. I was electrified, delirious, as I witnessed the beginning of the Space Age" (p. 3).

As interesting as the history of the early Cold War years is, what is the connection between your navigational problems, the disembodied voice on your smart phone, and Sputnik I? For engineers, one obvious link is that this first artificial satellite opened up our minds to the possibilities of new application areas of engineering. Communications, for example, would no longer be bound by terrestrial constraints such as the curvature of the earth, the physical barriers presented by the earth's oceans, the vagaries of atmospheric conditions, and the like. The idea of bouncing radio signals off satellites to facilitate intercontinental communications has evolved, over the decades since Sputnik, into the Global Positioning System (GPS) network of satellites that is helping you to find the way to your destination. However, the connection between your drive down the I-10 and Sputnik runs far deeper than just the technological possibilities that it opened up. It has much to do with the economic success, and the consequent impact on standards of living, that Western countries such as the United States, Canada, Great Britain, Germany, Australia, and others have enjoyed for more than 50 years. Indeed, the Sputnik Shock is responsible, in many ways, for our modern understanding of what Mokyr (1990) describes as "the lever of riches."

In fact, the Sputnik Shock of October 4, 1957, triggered a series of actions and outcomes that first linked creativity (in the sense of *the generation of effective novelty*), innovation (the *exploitation* of effective novelty), and engineering (the design and development of technological solutions to problems) together in a systematic and scientific way. It kick-started a rigorous examination of the association between the creation of new products, processes, systems, and services—*technological creativity*—and economic progress that has underpinned the development and success of nations for centuries. For the first time, however, it prompted *not* an economic explanation for this success, but a *psychological* one. The Sputnik Shock, in short, started a revolution in thinking that has attempted to explain not only the new technology itself, but also *who* develops the new technology, *how* and *why* they develop it, and *where* this development takes place.

As engineers, we are familiar with the technological consequences of the launch of Sputnik I. This event kicked off the Space Race that reached its zenith in the moon landing of July 1969. It stimulated a large number of novel technological spin-offs that trace their antecedence either directly, or indirectly, to the activities of NASA in the 1950s and 1960s. Memory foam, anti-corrosion coatings, cochlear implants, scratch-resistant eyeglass lenses, insulin pumps, and charge-coupled devices can all be seen as innovations that grew out of the catalyst of the U.S. Space Program,[2] itself jump-started by the Sputnik Shock of October 1957.

The Sputnik Shock also had other profound technological effects that have left important legacies in the 21st century. DARPA, the United States' Defense Advanced Research Projects Agency, was created in 1958 in direct response to the launch of Sputnik I, and its founding mission was "to prevent and create strategic surprise."[3] DARPA can count Unmanned Aerial Vehicles (UAVs), Micro-Electro-Mechanical Systems (MEMS), RISC computing, global positioning satellites, and the *Internet* among its technological achievements. However, while the success of your drive to that unfamiliar location owes a great deal to the impact of Sputnik I, there is, arguably, a more significant impact buried in DARPA's founding mission that links creativity to engineering—technological *surprise*.

Even as the direct technological impact of Sputnik I exerted its influence in the West, U.S. lawmakers began to look more deeply for the underlying causes of the Soviet Union's strategic achievement. The U.S. government realized, for example, that similar technological achievements could be made only by highly skilled people, and Congress sought to address this aspect of the problem through the National Defense Education Act (NDEA[4]) of 1958. This legislative solution was designed to address, among other things, a shortage of graduates in mathematics and engineering—a key resource in the development of new and superior technology. However, the final piece of the puzzle linking creativity, innovation, engineering and technology fell into place as experts began to hypothesize (rightly or wrongly) that the Soviet threat in space was not only a *quantitative* problem (a shortage of engineers in the United States, for example) but also a *qualitative* one. In particular, there was a sense that Soviet engineering achievements, and their success with Sputnik I, could be attributed to superior *creativity*

[2]http://www.howstuffworks.com/innovation/inventions/top-5-nasa-inventions.htm#page = 0

[3]http://www.darpa.mil/

[4]http://en.wikipedia.org/wiki/National_Defense_Education_Act

(A. J. Cropley & Cropley, 2009). For the first time, therefore, attention began to turn from economic issues that underpin the growth and development of technology—for example, investment and capital/labor ratios—and instead began to focus on the particular *qualities* of a product that make it creative—surprisingness, novelty, and effectiveness. Attention simultaneously turned to the *qualities* of the people and organizations that make the technology, and the processes by which they achieve the development of new and effective technological solutions to problems.

It turned out that a scientific foundation linking creativity, engineering, and technology—in a way that could help to explain the qualitative nature of the connection—already existed. The psychologist J. P. Guilford had delivered, back in 1950, a groundbreaking presidential address to the American Psychological Association's annual convention. In very simple terms, Guilford (1950) argued that human intellectual ability had been defined too narrowly in terms of factors such as speed, accuracy, and correctness—what he termed *convergent* thinking—and instead needed to be conceived of in a broader sense, to include factors such as generating alternatives, seeing multiple possibilities and so forth. In other words, he saw intellectual ability as encompassing both convergent (analytical) thinking *and* divergent thinking. The latter—divergent thinking—is seen frequently as a defining characteristic of creativity. Engineers, in fact, are no strangers to this duality of thinking styles—design, after all, is characterized by the fact that (Horenstein, 2002) "... if more than one solution exists, and if deciding upon a suitable path demands ... making choices, performing tests, iterating, and evaluating, then the activity is most certainly design. Design can include analysis, but it also must involve at least one of these latter elements" (p. 23). Engineering, in short, is all about creative problem solving.

The Sputnik Shock of October 4, 1957, therefore brought together, for the first time, the apparently disparate fields of creativity, with its psychological foundations, and engineering by stimulating recognition of the important role of divergent thinking in the process of designing technological solutions to complex problems. It provided the spark that has seen an explosion of research into what makes people able to devise new and effective solutions to problems—the psychology of creativity—and it lit the fuse of technological innovation that has given us GPS, the Internet, and many other systems and products that we now take for granted. There remains, however, one puzzle that is all the more surprising given the common interests of both creativity and engineering as they moved through the Space Age and into in the modern Digital, or Information Age. That puzzle is, simply, why is there not a *stronger* connection between creativity and engineering in the 21st century?

Creativity is concerned with the generation of effective and novel solutions to problems. Engineering is concerned more specifically with generating *technological* solutions to problems. Despite this, engineering is still frequently seen as predominantly analytical in nature—"a common misconception ... is that engineering is 'just' applied math and science" (Brockman, 2009, p. x). It stands to reason that successful engineering must focus *not only* on analysis and convergent thinking, *but also* on the vital role that synthesis and divergent thinking play in the creation of technology. Focusing on one at the expense of the other risks not only the integrity of the solutions themselves, but also the skill base of the people involved in the creation of these solutions. This book is concerned with reestablishing and rebalancing the link between creativity and engineering.

THE LINK BETWEEN CREATIVITY AND ENGINEERING

On the surface, that rebalancing should be straightforward. We all agree that creativity is an essential element of 21st century life. There is widespread agreement that creativity is a vital component in the success and prosperity of organizations. Yet, despite this, it is also clear that many leaders, managers, professional practitioners, and educators, not least in the field of engineering, are either apathetic to creativity or, while theoretically aware of its importance, uncertain of how to foster and exploit it in practice. This situation is by no means unique to engineering, and is typically simply the result of a lack of practical understanding of what creativity is, of how it can add value to the solution of real problems, and of what needs to be done to foster it. This in turn results from the fact that creativity is frequently conceptualized too broadly as a general, all-or-nothing property—"you either have it, or you don't"—and at the same time too narrowly, as mainly to do with aesthetics—"creativity is about art, isn't it?" Creativity is also regarded too narrowly as simply a matter of *thinking* and especially *free and unconstrained* thinking. Benson (2004), for example, describes anecdotal evidence suggesting that primary school teachers think that creativity is simply a matter of letting children "do their own thing" (p. 138) and that creativity, at its core, is "developed mainly through art and music" (p. 138). Other researchers have found similar conceptual roadblocks. Kawenski (1991), for example, writing about students in an apparel design course, found that "In the first place, their romantic notions led them to believe that creative thinking consisted of just letting their minds waft about dreamily, waiting for the muse to strike them." (p. 263).

The result of this is that creativity is often associated with lack of rigor, impulsive behavior, free expression of ideas without regard to quality, and similar "soft" factors. Things, also, which hard-nosed engineers disdain as "not *real* engineering." Nothing could be further from the truth, especially in the field of engineering, where the focus is on solving practical problems and satisfying customer needs.

In recent years, it seems as though there is no cross-fertilization and sharing of ideas taking place. What is even more interesting is that in the years immediately after the Sputnik Shock, there *was* a strong connection between creativity (and its psychological foundations) and engineering. Buhl (1960) epitomizes this, but also draws attention to the fact that this cross-fertilization seemed to fade away, so that from the 1970s onwards the connections between creativity and engineering were lost. It may be that engineering, in relation to creativity, was a victim of its own success. By the end of the 1960s, the achievements of the Apollo Space Program may have engendered a feeling among engineers in the United States, as well as other Western countries, that the issues identified by the Sputnik Shock, a decade earlier, had been solved. U.S. and Western engineers had comprehensively demonstrated their prowess, and we could stop worrying about creativity in engineering! Nevertheless, in early 21st century engineering, the challenge remains. We know creativity is important to engineering, but we struggle to understand why or how, and often therefore, the role of creativity is ignored.

At the same time that engineers forgot about creativity, another factor was conspiring to make it harder to reestablish the connection. As the study of creativity grew within the field of psychology, a gradual shift in our understanding of the term *creativity* took place. Creativity became tied strongly to the arts (D. H. Cropley & Cropley, 2013) in the public eye (pp. 12–13), and this contributed to the difficulty of reconnecting creativity to engineering. Any manager or teacher working in engineering, and interested in creativity, must now actively "unhook" creativity from the arts (McWilliam, Dawson, & Tan, 2011) before he or she can absorb the wealth of material that is available on the subject.

It seems that before any reconnecting and rebalancing of creativity and engineering can take place, it is first necessary to dispel some of the myths and misconceptions of creativity. What is creativity, and how should we understand it?

WHAT IS CREATIVITY?

Possibly the most significant factor that is holding back the development of creativity in engineering is the fact that, beyond the field of psychology, creativity is frequently poorly understood. Baillie (2002) illustrates this problem perfectly. She stated, "It is however not clear how creativity can be nurtured or fostered in students or how it can be assessed. What is creativity? What blocks it and what facilitates it?" (p. 185). These questions, in fact, have been studied extensively, in most cases for at least 50 years, and the results are widely available!

According to Florida (2002), creativity involves the production of "meaningful new forms." He went on to point out that such forms involve

- physical objects that can be made, sold, and used;
- theorems or strategies that can be applied in many situations;
- systems for understanding the world that are adopted by many people;
- music that can be performed again and again.

Important in this approach to creativity is the emphasis on products and the idea that the product must be, so to speak, public (other people come to know about it and find it useful in some way) and also enduring (its application or use persists for some time—in some cases for a very long time). This means that the creativity of, for example, a clever remark or a bright idea that achieves momentary recognition but leads to nothing and is soon forgotten is of lesser interest in this approach. The emphasis on *meaningful new forms* in discussing what constitutes a creative product is useful for practical settings such as engineering, but is equally applicable to artistic settings such as fine art, literature, or music. Even in intellectual and more abstract settings, such as theoretical science or philosophy, a creative *product* has to be good for something, and other people must be able to see that this is the case. Among other things, this emphasizes the importance of communication of products to other people, as has been discussed by, for example, Csikszentmihalyi (1999).

THE DEFINITION OF CREATIVITY

Two basic components are needed by engineers entering the field of creativity to answer the question *what is creativity*? These not only answer the fundamental question, and remove the basic blocks to reconnecting creativity with engineering, but also ensure that progress is made with a minimum of duplication. The first component is a clear, and widely accepted, definition representing the consensus that has emerged over decades of creativity research. Such a definition should be broad enough to satisfy the needs of any domain. Plucker, Beghetto, and Dow (2004) have captured all the essential ingredients in the following: Creativity is "the interaction among *aptitude, process and environment* by which an individual or group produces a *perceptible product* that is both *novel and useful* as defined within a *social context*" (p. 90).

THE FOUR Ps OF CREATIVITY

The second component needed by engineers for the reconnection with creativity is to recognize that the phenomenon is characterized in terms of 4Ps: Person, Product, Process, and Press (environment). This conceptual framework was first described by Rhodes (1961) and provides an excellent framework for understanding the *who, what, when, where*, and *how* of creativity in engineering. Each of these will be tackled in more detail in subsequent chapters; however, a brief summary of the 4Ps is given in the following paragraphs.

The *Person* addresses the factors relating to the psychology of the individual actor involved in the creation of the *perceptible product*. Research has shown that personal properties (e.g., optimism, openness, self-confidence), motivation (both intrinsic and extrinsic), and feelings (e.g., excitement, hope, fear) are distinct dimensions of the Person that each have a bearing on creativity (D. H. Cropley & Cropley, 2013). Furthermore, these dimensions of the Person interact with each other in a variety of ways such that different combinations have unique consequences for creativity.

The *Product* addresses the output of the creative endeavor. Although it is no surprise that psychologists are interested in the creative person, it is also widely accepted that an essential core of creativity, whether in music and poetry, or engineering and science, is the tangible artifact. In fact, this definition of *Product* can be extended—any product, process, system, or service that is *both novel and useful* qualifies as a creative *product*. Mackinnon (1978) concluded that "analysis of creative products" is "the bedrock of all studies of creativity," and indeed, Morgan (1953) came to a similar conclusion. While more recent definitions of the creative product debate the existence of higher order characteristics (D. H. Cropley & Cropley, 2005), the foundation of definitions as far back as Stein (1953) is a combination of novelty and usefulness. For some *thing* to be considered creative, it must be original and surprising, and it must address a real problem or need.

The *Process* typically addresses the styles of thinking that result in creative products. Although more complex and nuanced than is suggested here, two main thinking styles are commonly associated with creativity. It was Guilford (1950) again who laid the groundwork for understanding the roles that convergent and divergent thinking play in the production of creativity. While divergent thinking is often exclusively associated with creativity, it is important to recognize that convergent thinking also plays a critical role, especially when creativity is considered in the context of problem solving, engineering, and other STEM disciplines. Engineers will immediately recognize this as a feature of the design process. Horenstein (2002) explained the essence of design as follows: "... if more than one solution exists, and if deciding upon a suitable path demands being creative, making choices, performing tests, iterating and evaluating, then the activity is most certainly design. Design can include analysis, but it must also involve at least one of these latter elements" (p. 23). The core of engineering design therefore involves two stages: a stage of creative synthesis (i.e., divergent thinking), followed by a stage of logical analysis (i.e., convergent thinking).

It may be tempting to see *Process* in terms of activities, or *procedures*, such as brainstorming; however, these are simply a tool to help people recognize and tap into divergent and convergent thinking. In activities such as engineering, it is also possible to think of *Process* in terms of the stages that an individual or team undertake as they solve problems and satisfy needs, creatively or otherwise. However, this should not detract from understanding *Process* as the core cognitive activities that underpin creativity: divergent thinking and convergent thinking (A. J. Cropley & Cropley, 2009).

Finally, the *Press* examines the role of environmental and social factors on creativity. More specifically, *Press* can be considered to address both (a) how the "climate" can either facilitate or inhibit creativity, and (b) how the "environment" reacts to the production of creativity. *Press* therefore touches on not only factors such as management support for creativity (e.g., rewarding creativity, encouraging risk taking), and how the physical environment may foster creativity (e.g., through the provision of plants and adequate lighting in the workplace), but also on the way that society tolerates radical deviations from norms (are creative people ridiculed or hailed?), and even the rules and standards that govern professional activities such as engineering.

THE FIFTH P

Even divided into the 4Ps, this framework for understanding creativity may be still too diffuse to provide a concrete framework for recognizing and fostering creativity in engineering. The fact is that creativity in engineering is about solving problems, and the solutions engineers devise do not spring into being fully formed. As engineers, we understand that there is a sequence of stages that we go through as we first understand that there is a problem to be solved, dream up possible ways of solving that problem, narrow these down to one, or a few, probable solutions, before selecting the best option for development and implementation. If creativity is embedded in engineering, then it is spread across this sequence of stages in some way. If we are to understand fully creativity in engineering, we need to be able to understand how the 4Ps intersect with the stages that we know characterize the real world of engineering problem solving.

The answer to this issue is the fifth P: *Phases*. These are the steps involved in the generation of novel and effective engineering products. The approach that is taken here is based on Wallas's (1926) well-known four-phase model: In the phase of *Preparation*, a person becomes thoroughly familiar with a content area; in the *Incubation* phase, the person "churns through" or "stews over" the information obtained in the previous phase; in the phase of *Illumination*, a solution emerges, not infrequently seeming to the person involved to come like a bolt from the blue; and finally comes the phase of *Verification*, in which the person tests the solution thrown up in the phases of *Incubation* and *Illumination*. However, for reasons that are discussed more fully in Chapter 3, I further differentiate Wallas's system by adding three additional phases (*Activation*, *Communication*, *Validation*), thus conceptualizing creativity as involving *seven* consecutive *Phases* (Figure 1.1). The idea of Phase constitutes the fifth P of creativity.

The Phase concept will also tie strongly to the idea of Engineering Design as the mechanism by which products and systems are realized.

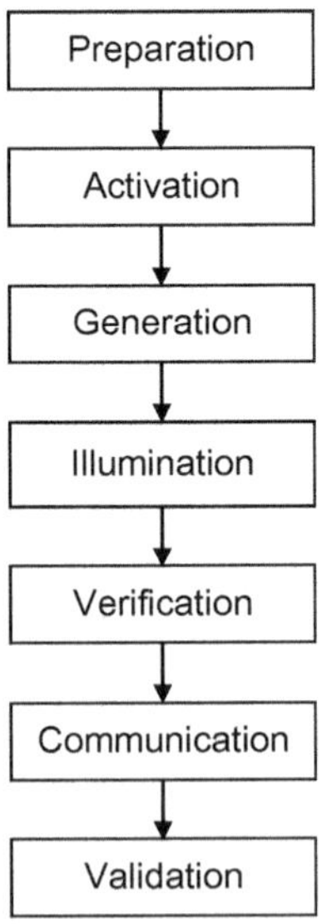

FIGURE 1.1 The extended phase model of the creative process.

PARADOXES OF CREATIVITY

There is one final puzzle, originating in the staged nature of the process of solving engineering problems, that must be tackled before we can successfully rebalance and reintegrate creativity and engineering. Horenstein's (2002) description of engineering, cited earlier, makes it clear that the steps involved in designing and developing an engineering solution involve different cognitive skills. Sometimes it is necessary, for example, to thinking analytically, sometimes synthetically. This seems to suggest a paradox in engineering creativity. Cognitive processes that appear, on the surface, to be mutually exclusive are both necessary for creativity. How can creativity in engineering be developed and fostered if it requires us simultaneously to think both convergently *and* divergently? For this reason, discussions of creativity are confronted by a number of apparent *paradoxes*: aspects of the processes of creativity, the personal properties associated with it, the conditions that foster its emergence, and the products it yields seem to be mutually incompatible. Further examples serve to illustrate this point. A lack of structure and management pressure in the environment may encourage creativity some of the time but inhibit it at other times. Properties of the individual—a willingness to take risks, for example—may be favorable to creativity at some points in the process, but unfavorable at other times. This paradoxical nature of creativity in engineering is resolved by the fifth P. By understanding engineering creativity as taking place across distinct phases, one is able to build a model of creativity in engineering that identifies more specifically the relationships between the Person, the Process, the Product, and the Press, at each Phase, and specifies exactly what conditions favor or inhibit creativity at each point in the problem-solving process.

SUMMARY

There is a tendency to speak of *creativity* in a global way. However, from early in the modern, post-Sputnik era, it was variously broken into the "Three Ps" (Barron, 1969): Person (*who*), Product (*what*), and Process (*how*), while Rhodes (1961) (p. 305) provided a fourth "P" (Press; i.e., the *press*ure exerted by the environment, or *where*). The *Four Ps* is now widely used as a framework for guiding the study and understanding of creativity, especially in the field of psychology (e.g., Kaufman [2009], A. J. Cropley & Cropley [2009] and Kozbelt, Beghetto, & Runco [2010]). As a result, creativity research variously addresses

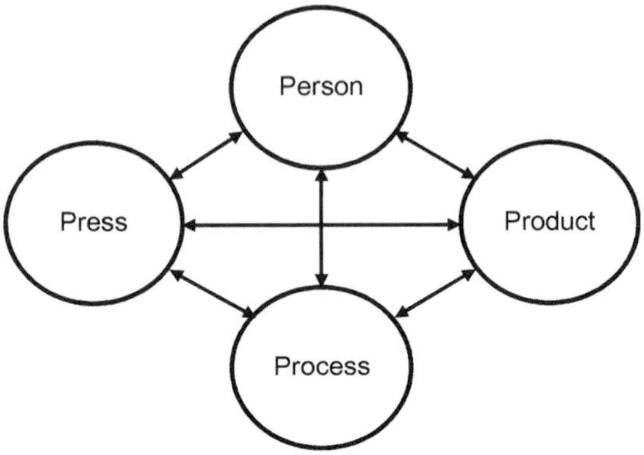

FIGURE 1.2 A "systems" concept of creativity.

- the **Process**, usually in the sense of the cognitive, or *thinking*, processes employed by creative individuals (as distinct from the steps involved in creativity);
- the **Person**, in the sense of the personal, psychological qualities of the individual engaged in creativity;
- the **Product**, meaning the outcome of creativity—the result. In engineering, in particular, we will think of the *product* as any of (Brockman, 2009, p. 4) or (A. J. Cropley & Cropley, 2009, p. 29):
  - tangible *artifacts* (e.g., chairs, cell phones);
  - complex *systems* (e.g., interactions of hardware, software and people, as found in airliners, ships);
  - *services* (organized, but usually intangible, systems of labor and material aids to meet particular needs);
  - *processes* (methods for achieving tangible or intangible outcomes);
- the **Press**, meaning the place in which creativity occurs. Most often, in an engineering context, this will be an organization, both in a physical, and a conceptual, sense.

It will be no surprise to the reader that the Four Ps do not operate in isolation from each other. In the same way that the value and purpose of a complex system is derived from the interaction of its parts, so *creativity* is a function of the interaction among the person, the process, the product and the press (Figure 1.2). Indeed, the definition of creativity given by Plucker et al. (2004) captures the *systems* nature of creativity perfectly: Creativity is "the interaction among aptitude, process, and environment by which an individual or group produces a perceptible product that is both novel and useful as defined within a social context" (p. 90).

CHAPTER

2

The Importance of Creativity in Engineering

"But out of limitations comes creativity." ***Debbie Allen, 1950–, Actress, Producer***

To reconnect creativity and engineering, we need to be clear on why the two should be connected in the first place. Why is creativity so important to engineering? If it is important, what value does it add? To understand why creativity is so important, we need to see engineering as an exercise in problem solving.

THE ECONOMICS OF CREATIVITY

Sputnik and the Apollo program aside, engineering rarely occurs in a vacuum. The development of *products* may, in ancient times, have served a narrower and more immediate purpose, but in the 21st century—indeed, since at least the time of the Industrial Revolution—engineering and technology have played a central role in the economic lifeblood of societies. In simple terms, societies that possessed the ability to create and exploit technology prospered, while those that did not have this capability struggled. Before I delve more deeply into the specific role that creativity plays in engineering, I want to discuss more generally why we are even having the conversation.

Mokyr (1990) attributes the enormous wealth and development of Western nations—"... the rise of the West"—to a foundation of *technological creativity*. He contrasted two generic types of person, stating that: "One [homo economicus] makes the most of what nature permits him to have. The other [homo creativus] rebels against nature's dictates.

Technological creativity, like all creativity, is an act of rebellion" (p. viii). Economic progress, built on the foundation of technological creativity, has resulted in a rise in living standards; improved nutrition, clothing, housing, health; reduced toil; and reduced disease. Furthermore, this *effect* of technological creativity has far outweighed the effort and cost.

Mokyr (1990) goes on to explain that economic growth and progress result from four processes: investment, commercial expansion, scale or size effects, and an increase in the stock of human knowledge. Economists naturally tend to concentrate on economic explanations, and even where they discuss the role of technological creativity, they tend to do so only in economic terms. Mokyr, however, shows that purely economic explanations may fail to explain adequately the role of technological creativity. Conventional economic wisdom, as embodied in Schumpeter (1942), holds that "capitalist expansion deriving from continuous, though fluctuating, technological change and innovation, [is] financed by the extension of credit" (Parker, 1984, p. 191). It is clear, however, from an engineering point of view, that improvements in production efficiency or the development of new and better products has always been possible without being "financed by the extension of credit." Throughout history, these effects have also been achieved simply as a result of human creativity and innovation. In other words, there is more to technological creativity as a driver of economic progress than just the economic relationship between supply and demand variables, research and development, and productivity growth.

Technological creativity is, without doubt, a driver of economic progress, but understanding how and why requires a multifaceted and multilevel approach. Economic growth is, by nature, an aggregate phenomenon, while creativity is concerned with smaller units of analysis (i.e., individuals, teams and organizations). Mokyr's argument is that, while not ignoring the aggregate-level *economics* of technological creativity, the appropriate focus is the "macro-foundation" of technological creativity—especially "what kind of social environment makes individuals innovative, what kind of stimuli, incentives and institutions create an economy that encourages technological creativity?" (Mokyr, 1990, p. 8). The point is this—paraphrasing Einstein's dictum—if the only tool we have is economics, then we tend to see every problem in terms of economics. Technological creativity, however, can only be analyzed fully as a social and an individual phenomenon.

The history of technology has always retained a component that could not be explained by economics. Mokyr (1990) argues that this component relates to luck and inspiration, and reminds us that those, in essence, are the domain of the individual. "Individuals matter" in the development of technology, so let us now look at *who* those individuals are and *what* they do.

ENGINEERING AND ENGINEERS

What is it that engineers do? I do not mean the obvious answers such as "they build stuff" or "they design computers," or other similar things. I mean, what is the single, common, unifying purpose behind all flavors of engineering, whether mechanical, electrical, civil, or any other subdiscipline? There is, in fact, a remarkably consistent answer to this question in introductory engineering texts. Jensen (2006), for example, says that "Engineers solve problems" (p. 17) and this process requires the ability to understand and define the problem, to apply standard approaches to solving the problem, and to "supplement the standard solution methods with creativity and insight" (p. 18). Burghardt (1995) describes the engineering profession as one "devoted to the creative solution of problems" (p. 2). Figure 2.1 shows this simple relationship in a form familiar to engineers—as a rudimentary flowchart.

Horenstein (2002), taking a slightly different tack, explains that "design" is what engineers do (p. 22), and that "design can be defined as any activity that results in the synthesis of something that meets a need" (p. 22). Brockman (2009) also connects engineering to needs-driven problem solving, reminding us that these problems arise from a desire to "satisfy mankind's complex needs and desires" (p. 3). Figure 2.2 illustrates this by modifying our flowchart of the process.

Buhl (1960), in his very prescient book *Creative Engineering Design*, stated that "a designer is one who satisfies mankind's needs through new answers to old problems" (p. 9). He went on to say, "The designer must deliberately create new products and processes which will fulfill mankind's needs. He must be creative in all stages of problem solution" (pp. 9–10).

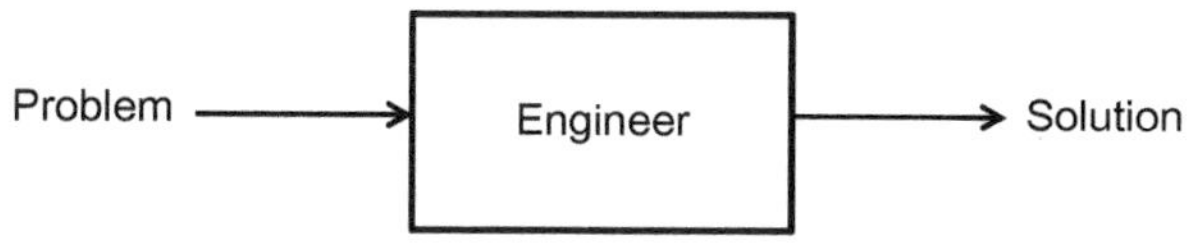

FIGURE 2.1 The basic process of engineering.

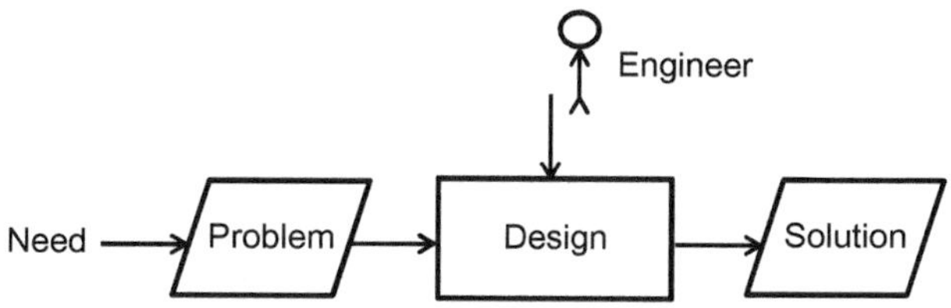

FIGURE 2.2 Engineering as needs-driven problem solving.

Therefore, the needs of society are the sources of problems. When engineers solve these problems, they do so specifically by developing *technological* solutions (as opposed to other kinds of solutions). However, this still has not answered the question about why creativity is a necessary part of the solution. To answer that, I first want to go back a step, and discuss what is driving *humankind's complex needs and desires.*

CHANGE

Pilzer (1990) builds a compelling, albeit unwitting, explanation for the importance of creativity in the context of engineering. Writing on the subject of economics, he describes the relationship between a society's wealth, its physical resources, and its technology. In simple terms, "technology determines what constitutes a physical resource" (p. 28) and "Technology determines ... both the efficiency with which we use resources and our ability to find, obtain, distribute, and store them" (p. 32). Without technology, a society's wealth is low because it is unable to make productive use of resources such as iron ore, oil, solar energy, fertile land, and even information. Without technology, it is unable to extract greater value from these raw materials or to turn them into valuable, tradable goods and services.

Since at least the time of the earliest modern people, humanity has made use of technology—sharpened sticks, shaped flints, animal skins, for example. Each of these represents the synthesis of something that meets a need—for hunting, for cutting, for warmth. However, the underlying pressure behind those needs has always been *change*. Burghardt (1995) suggests that there have been three "major revolutions in the way society has been organized" (p. 22) and that these came about as a result of major changes that stimulated humanity's needs and desires, resulting in problems that could be solved by engineers. These are changes at the highest level of abstraction, occurring over long periods of time. The first major change, which brought about a shift from hunter-gatherer to agrarian societies was environmental—the Ice Age. The change in climate meant that old methods of subsistence—hunting animals and gathering wild fruits, nuts, and vegetables—were no longer able to meet the needs of nomadic tribes. This change created new needs—for better food supply—and therefore problems that could be solved, potentially, by technology. Figure 2.3 adds to our previous flowchart to show how change provides the stimulus to needs, and therefore to the process of developing technological solutions to problems, i.e., engineering.

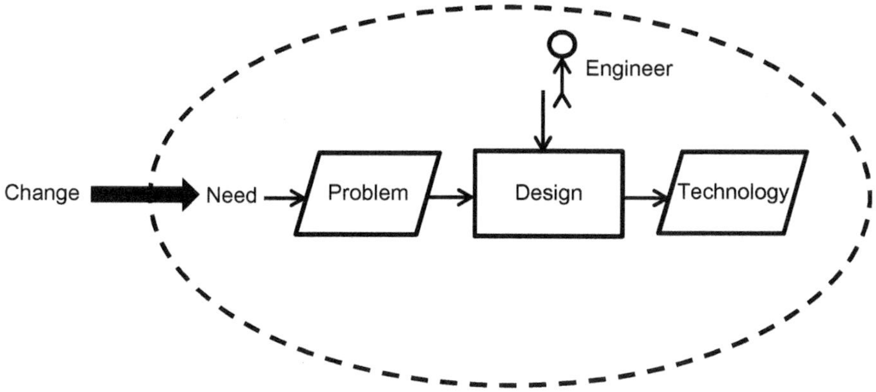

FIGURE 2.3 Change as the stimulus for engineering.

The second major change, which stimulated the shift from agrarian to industrial societies, resulted from a combination of famine and disease. As European societies had grown under the agrarian model, population growth outstripped food supply. Famine resulted, at the same time that more and more land was cleared, resulting in a shortage of the main fuel of the time, wood. The addition of the bubonic plague epidemic—the Black Death—that ravaged Europe in the 14th century left a shortage of people to work the land. The changes together provided the stimulus for new needs—for alternative sources of fuel, for less labor-intensive ways to work the land—that again were the source of problems with potential technological solutions.

The third major change, which has seen the shift to the modern, post-industrial information age, is underway, and therefore a little harder to characterize. However, we can say that changes to the earth's population—both its sheer size and its age profile—changes to the climate, and changes in the interconnectedness of people around the planet, i.e., globalization, are all generating needs that are the basis of problems ripe for technological solutions.

Even at a smaller scale, we see numerous examples from the history of humankind that illustrate *change* as a driver of society's needs and desires, and therefore as a source of engineering problems. It is also at this point that we can speculate that the effects of change manifested themselves in two different but related ways. The examples we have considered so far represent new needs generated by external changes. Thus, climate change generates a need for better food supply, and therefore problems like "how can we *grow* food to supplement our supplies?" However, it seems that a second pathway also occurs in parallel. External changes also generated *new solutions*. We will never know who first hit upon the idea that a stick used as a spear thrower (a *Woomera*

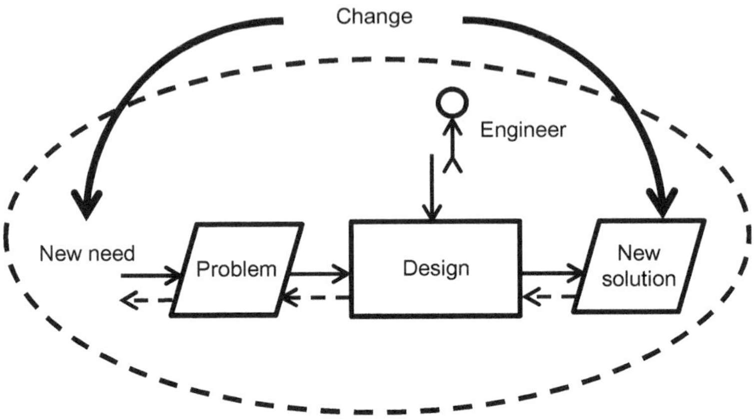

FIGURE 2.4 Change as a driver of needs and solutions.

in Australian Aboriginal culture) could improve the range and effectiveness of that weapon; however, it is plausible that the spear thrower was *not* developed *in response to* an identified need—"how can I throw my spear further?"—but was simply a lucky discovery—"look what happens when I do this!" Whether driven by the need or driven by the solution, the result is the same. New technologies—solutions—and new problems are linked together in a mutually beneficial way, with one or the other *driven by change*. Figure 2.4 extends our flowchart to show the role of change as stimulus for needs and solutions in the problem-solving process.

In the early days of agrarian society—in the Nile Valley, for example—we see more examples of change driving problems, resulting in technological solutions—in other words, *engineering*. Growing populations created new needs—for clothing, shelter, and food—that could not be met by the existing means. These new needs therefore drove technological solutions that could replace animal skins with woven fabrics and hand tilling of soil with ox-drawn plows. It is also likely that individuals developed new technologies—new solutions—either serendipitously or deliberately, that could then be matched to a need. In other words, once a society develops the wheel, its members suddenly realize how many needs they can satisfy with it. This *chicken and egg* situation does not, however, present us with any paradox because the end result is the same. Change generates both new needs and new solutions. The two pathways stemming from change (Figure 2.4) represent two general design paradigms that I will explore in the next chapter. One typifies a *top-down* design paradigm—new needs are satisfied by new solutions. The other describes a *bottom-up* approach—new solutions are able to satisfy new needs.

THE NEED FOR CREATIVITY

To understand the role that creativity plays in the engineering problem-solving process, I now draw your attention to a crucial feature of the process. The examples we have considered and the flowchart model of the process have both identified the fact that change stimulates *new* needs and *new* solutions. The changes that occurred to our ancestors were all unprecedented: there was no previous case or example to follow, and no prior solution that could be applied. Humans had never tackled the after-effects of an ice age. They had never before dealt with the problem of feeding large numbers of people, and they had never needed anything other than animal skins for warmth. New needs may look similar to past cases, but they are new precisely because they have unique characteristics and constraints that have never been encountered before. New solutions are new because they are identified as unique, original, and *never seen before*. It is the characteristic, indeed the *requirement*, of newness—in the needs and in the solutions—that is why creativity is central to engineering problem solving.

As engineers, however, we should note that not every need and not every solution is new, or needs to be new. There is also an important place for old solutions satisfying old needs, which I will discuss shortly.

It appears, therefore, that there are *six different pathways* through our problem-solving process. Some are driven by change, as I have already discussed, and some are driven by what I will call *stasis*—the maintenance of the status quo. It is important to note that there is nothing bad about stasis in the engineering context. Not every problem needs a creative solution. Buhl (1960) noted that if "we consciously try to solve each and every problem which arises as we drive from here to there we should soon lose our sanity" (p. 16). Buhl's point was that *precedented* problems, in other words, *old* or *existing* problems, can be addressed perfectly well with old or existing solutions. This is also important economically because if there is no requirement for creativity—because the need/problem and/or solution are routine—then costs can be lowered, risks reduced, and time saved. The urban myth of the *NASA Space Pen* illustrates this well. The popular story[1] tells us that NASA spent

[1]See, for example, http://www.truthorfiction.com/rumors/s/spacepen.htm#.U09JClWSzmd. In fact, the truth is that there were good reasons driving the development of a Space Pen to replace pencils, not least stemming, once again, from change. Greater attention to safety and risk reduction in the Space Program—a change in culture—stimulated new needs and therefore new problems. How can we reduce the risk of graphite dust in the closed environment of the capsule? How can we ensure that fragments of graphite will not cause a short circuit in the capsule's electronics? The answer—use a pen instead—was then confronted by a new problem. How can we make a pen that can operate in zero G?

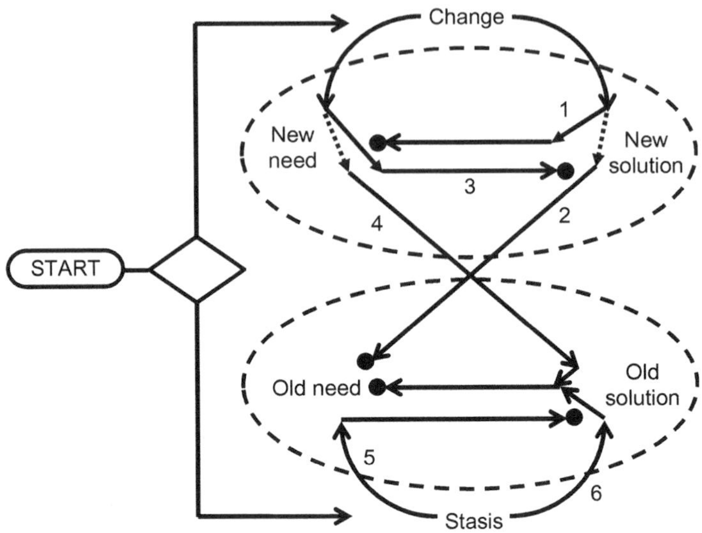

FIGURE 2.5 Six pathways in engineering problem solving.

10 years and $12 million developing a ballpoint pen that would work in zero gravity and extremes of temperature. Meanwhile, the Soviet cosmonauts simply used pencils. In other words, the myth is making the point that sometimes the *old*, routine solution is perfectly satisfactory.

Figure 2.5 illustrates the six pathways through our problem-solving process.

If the primary driver is **change**, the following pathways can be traced:

1. A new solution, e.g., an invention or discovery, creates a new, previously unrecognized, need. This is sometimes referred to as *technology push*, or a *supply-side problem*. Another label for this pathway is *Redirection*.[2] This type of problem driver also helps to explain why developing nations, for example, do not simply adopt old technologies as the solution to their needs. We do not see, for example, Afghanistan adopting the steam locomotive as its solution for mass and heavy transport. Nor do we see Afghanis installing a network of hard-wired telegraph stations to meet their

[2]I have drawn on the Propulsion Model of Kinds of Creative Contributions that is described by Sternberg, Kaufman, and Pretz (2002) for labels such as *Redirection*. However, I have not used all of the eight that they describe, and not necessarily with exactly the same meaning. The general spirit of their model, however, is at the core of my use of the terms.

communications needs. The availability of new technologies generates an up-to-date and modern need.

2. A new solution satisfies an old need. The wheel, for example, satisfied the need to be able to move large loads easily and quickly, for example for trade or food production. In engineering, this is quite common, and is described by the mantra "better, faster, cheaper." We can also call this *Forward Incrementation*.
3. A new need is satisfied by a new solution. This is often referred to as *market pull*, or a *demand-side problem*. We can also call this *Reinitiation*.
4. A new need is met with an old solution, which, unable to meet the new need, effectively drags the need back to what *can* be satisfied—an old need. This pathway has the effect of pulling change back to stasis and can be characterized as *Stagnation*. This is also what Albert Einstein was referring to when he said, "We can't solve problems [new needs] by using the same kind of thinking [old solutions] we used when we created them." In the end, the new need remains unfulfilled and the progress implicit in change is halted. The reader will also notice that creativity is absent, highlighting the fact that change can only be met successfully with the creativity that is implicit in novelty.

If the primary driver, by contrast, is **stasis**, then two pathways exist:

5. An old, or existing, need is satisfied by an old solution. This case is common to many engineering scenarios and is, in no way, inferior or deficient. If I need to hammer in a nail, then my trusty old hammer is a perfectly good solution. If I need to get traffic across a river, then a standard bridge design will almost certainly be satisfactory. This "more of the same" approach, or *Replication*, is the bread and butter of engineering. This is how engineers provide solutions to many of society's problems—and satisfy needs—in a cost-effective, timely, safe, and low-risk manner. While it involves no creativity, at least at the macro level, it should not be dismissed for at least two reasons. First, it comprises a large percentage of what engineers do, and second, because there may be a requirement for creativity buried deeper within it. For our current purposes, however, and for the case where old needs are satisfied by old solutions, we will set this aside as outside the scope of our discussion of creativity in engineering.
6. The last pathway is similar to the previous one. The driver is stasis, and the only difference is that the starting point is an old solution rather than an old need. You might wonder why someone would start with an old solution rather than an old need. This seems a little counterintuitive. Who starts by saying, "I have an old bridge here. Where is some traffic that needs to get across a river?" However, it is not as unlikely as you might think. Any time we salvage old parts

and reuse them—let us call it recycling—this is, in effect, what we are doing. Similarly, there are examples in the field of systems engineering where so-called *legacy items* may be incorporated into a new system. For reasons of cost reduction, increased effectiveness and developmental risk reduction, we may choose to reuse some existing subsystems as part of a larger complex system. In my hometown in Australia, for example, there was recently a project to upgrade the public train network. Old steel sleepers were replaced with concrete, and overhead power lines were installed so that new electric trains could replace the old diesel rolling stock. However, throughout this lengthy redevelopment, the same old steel railway tracks were used. The reason was that there was no fundamental change driving a new need as far as the interface between ground and train was concerned. The new trains still run steel wheels, and the new sleepers attach to the rails in the same old way. For this reason, the old solution—the steel rails—was able to satisfy the same old need—to provide a track for the trains to run on.

The need for creativity as a core component of engineering is now clear. Change—in the form of aging populations, carbon taxes, terrorism, globalization, food security, scientific discoveries, and more—drives new needs and new technologies. Engineering is a problem-solving process that connects new needs to new technologies. Creativity is concerned with the generation of effective, novel solutions; and therefore, creativity and engineering are, in essence, two sides of the same coin. In fact, we can characterize engineering as a special case of the more general process of generating effective, novel solutions to problems—i.e., creativity.

Gertner (2012) further illustrates the value of creativity in engineering, describing the driving force behind the activities of the famous Bell Labs: "... that the growth of the system [the U.S. telephone system run by AT&T] produced an unceasing stream of operational problems meant it had an unceasing need for inventive solutions" (p. 45). The impetus for creativity, moreover, was not simply technological in nature: "... the engineers weren't merely trying to improve the system functionally; their agreements with state and federal governments obliged them to improve it economically, too" (p. 45).

CREATIVE ENGINEERING PROBLEM SOLVING

Simplifying the previous discussion of needs, solutions, and driving factors, we can now separate out three generic types of creative engineering problem solving and one type of routine engineering problem

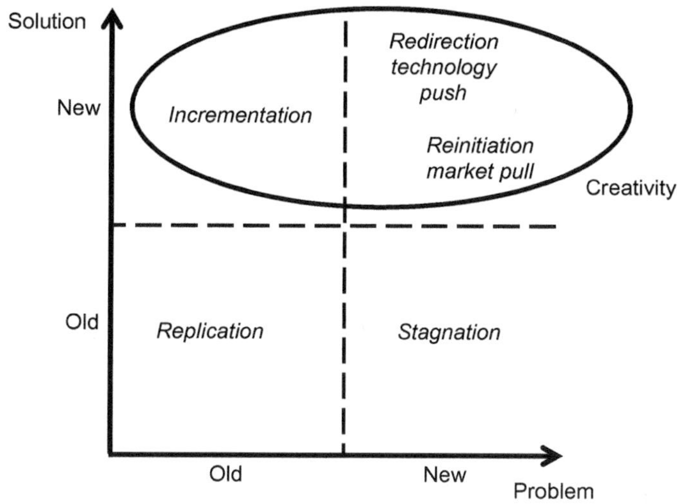

FIGURE 2.6 Types of creative engineering problem solving.

solving. Where problem and solution are *old* and the engineering process is one of matching these together, we are dealing with *routine*, but nevertheless important, engineering *replication*. Where *new* problems and *new* solutions are combined in some way, we are dealing with creative engineering problem solving that can be further characterized as *forward incrementation* (where new solution satisfies new problem), *redirection* (where new solution satisfies new problem), and *reinitiation* (where new problem is satisfied by new solution). Figure 2.6 illustrates.

The link between creativity and engineering is clear. Where new customer demands can be met by new technological solutions, we need engineers who are equipped—both technically and creatively—to generate those solutions. Where new technologies become available through scientific discovery, research, and development, we need engineers who are able to find new possibilities and new markets to exploit these possibilities. There will always remain a place for the application of engineering knowledge to the solution of routine (other quadrants)—well-understood, straightforward—problems. However, as the pace of change accelerates in the 21st century, we will see a growth in new problems that require creative—in other words, effective AND novel—technological solutions. To meet the challenge of these changes, we need engineers equipped with a blend of technical knowledge and creativity.

Another way to understand the need for creativity in technological problem solving is to imagine a world *without* novelty. Imagine what our world would look like, and how our society would function, if bridges were still made only of tree trunks, or if the only form of heating was from individuals burning wood. Imagine if the only form of augmented transport (i.e., something other than *by foot*) was the horse, or if the locomotive, to paraphrase Buhl (1960), had *not* been displaced by the automobile?

The value and importance of creativity—the *new* solutions—emerges in Buhl's discussion where he points out that "we expend a great deal of effort in modifying modification rather than attacking the problems at their core" and makes the vital point that "[i]ndustries are continually being supplanted, not by modifications but by innovations" (p.10). He gives a specific example that illustrates this point very well: "Locomotives were not displaced by modified locomotives but by a *new approach* [emphasis added] to transportation needs—the car" (p. 10).

At a very general level, Sternberg (2007) expresses a sentiment common in discussions of creativity and innovation, and the value that they bring to society: "The problems we confront, whether in our families, communities, or nations, are novel and difficult, and we need to think creatively and divergently to solve these problems" (p. 7). Creativity is of value because it tells us everything we need to know about generating the solutions to these novel and difficult problems—*how* to generate them, *who* can generate them, *how* to recognize them, and *how* to stimulate them.

THE OIL CRISIS OF 1973

I have already discussed how the Sputnik Shock of 1957 established a connection between creativity and engineering, so let us now turn our attention to another *shock*—the Oil Crisis of 1973—to develop a deeper understanding of exactly what creativity has to offer the business of engineering.

Prior to the first oil crisis in 1973, and also the so-called *energy crisis* of 1979, oil prices were historically comparatively low, and oil was plentiful. This meant that first-world economies and societies, such as that of the United States, developed around a fundamental dependence on oil. Cities evolved into extended suburban sprawls centered on the almost universal use of personal automobiles—for work, education, and recreation. Industry came to depend heavily on oil for the transportation of goods from factory to consumer. Farms and food production developed a dependency on oil, not only for the machinery used, but also for the production of ammonia for fertilizers. Homes depended on oil for heating, and to a lesser extent, for electricity. At the same time, developing nations benefited indirectly from cheap, plentiful oil because first-world countries found it cost-effective to shift manufacturing offshore—the higher transport costs were less than the savings obtained from cheaper labor. Additionally, highly efficient food production in developed nations, so dependent on oil, helped to feed growing populations in poorer, less advanced countries.

Between 1973 and 1979, however, the price of oil increased more than tenfold as a result of various political and economic activities, the net effects of which were to limit the supply of oil. One proximate effect of this was economic recession. In simple terms, the high cost of oil, and limits on its availability, damped down economic activity in countries like the United States. All of the parts of the economy dependent on oil saw costs increase, and as a result, people drove less, consumed less, and spent less.

However, one good thing did result from this state of affairs. It created a problem that had to be solved. The countries affected by the 1973 Oil Crisis, for the first time, had to tackle the question of how to reduce their dependency on oil. Even more important, from our point of view, was the fact that this problem was one that was amenable to technological solutions.

Oil Crisis—Solution Pathways

The economist Paul Pilzer (1990) argued, "technology determines what constitutes a physical resource" (p. 28). What this means is that if oil is in short supply, and the cost rises to problematic levels, there are two fundamental technological solution pathways to tackle this problem:

- Use technology to increase the actual amount of available oil.
  - Discover new reserves; extract previously inaccessible reserves; extract existing reserves faster and more efficiently; develop new, unconventional sources (e.g., oil sands, oil shale); develop synthetic oil technologies (e.g., CTL [coal-to-liquid]).
- Use technology to decrease the actual amount of oil used.
  - Replace oil with other forms of energy (e.g., solar, wind, hydroelectric, geothermal, natural gas); make cars more fuel-efficient; replace conventional automobiles with hybrid or fully electric vehicles; increase use of public transport; develop synthetic biofuels; telecommute to reduce personal travel; redesign cities to reduce dependence of cars; develop farming practices that are not dependent on synthetic fertilizers.

What is clear is that technology is pivotal to both of these solution pathways. Even more critical, for our purposes, is the fact that the technology inherent to these solution pathways is created by—**engineers**!

Oil Crisis—Technology Push and Market Pull

The two pathways identified remind us that there are two basic sources for the problems we encounter in engineering. The questions *How can we increase the amount of available oil?* and *How can we decrease the amount of oil used?* are, in essence, problems of technology push and market pull, or of *supply* and *demand*. In each case, the problem—and the solution—is technological in nature, and therefore the domain of engineers. In each case, we are faced with examples in which "we will not solve the problem by using the same kind of thinking that created it" to paraphrase Einstein. To increase the amount of available oil, for example, it is no good simply drilling deeper into an existing well. As William J. Cummings, an Exxon-Mobil spokesperson, said in December 2005,[3] "All the easy oil and gas in the world has pretty much been found. Now comes the harder work in finding and producing oil from more challenging environments and works areas." It was estimated, in 2005, that the technology available at the time was capable of extracting only 40% of the oil from existing wells. Getting the other 60% out and usable requires new technological solutions to tackle new problems such as extreme depth, extreme down-hole temperatures, environmental sensitivity, high oil viscosity, and high levels of sulfur and metals contamination.

In each case, the solution of the problem will require the design, development, and implementation of new technologies—new products, processes, systems, and services. These solutions will need to be not only effective, but also novel. In other words, the solution of these challenges is not simply the core function of *engineers*, but the core function of *creative* engineers.

As the case of oil and energy illustrates, a failure to do so is not merely an abstract inconvenience. The global economy depends on availability of affordable, plentiful energy—traditionally obtained from oil. If oil is too expensive or scarce to satisfy that need, then we must find ways to increase the supply of oil (if only to reserve it for those functions for which there may be no viable alternative—e.g., certain fertilizers, detergents, plastics, and solvents), and we must find ways to replace or reduce our dependency on oil. A failure to do so raises the specter of global economic recession, a sharp decline in standards of living, decreased food security and even famine, declining population growth, and a reversal of economic development in poorer countries.

[3]http://en.wikipedia.org/wiki/Peak_Oil

CASE STUDY: CREATIVITY AND INNOVATION IN AEROSPACE

In practical terms, how do we go about using technology to change the supply of a physical resource? How do we increase the amount of oil available and decrease the amount used? To illustrate, I have developed a case study based around creativity and engineering in commercial aviation.

The business of commercial air travel turned 100 years old in January 2014.[4] It began on January 1, 1914, with a single paying customer, the mayor of St. Petersburg, Florida, who paid US$400 for a 23-minute flight across Tampa Bay.[5] Over the 100 years of commercial aviation since that flight, passenger numbers have grown to staggering levels. It is estimated that on January 1, 2014, 8 million passengers flew on 100,000 flights globally, and in 2013, some 3.1 billion people flew on a commercial service. The cost of air travel nowadays varies greatly and is influenced by many factors; however, it is clear that the cost of fuel is one of the most significant, if not the single largest, operating costs for typical commercial airlines.[6]

Southwest Airlines is a good example of a commercial operator[7] with a reputation for innovation (Schrage, 2001). It is therefore instructive to look at the problem of reducing dependency on oil—in other words, reducing aviation fuel consumption—in an airline that is already a proactive cost-saver. In 2010, for example, Southwest Airlines' total operating expense was US$11.116 billion. Of this, US$3.62 billion (32.6%) was spent on fuel and oil[8]—far more than *wages and salaries* (US $2.752 billion or 24.8%). It should be clear, therefore, that even a low-cost, innovative airline is highly constrained by the cost that fuel places on its overall operations and profitability.

On the surface, it would seem that there is limited scope to make significant savings on fuel. As Porter (1996) explained, "Simply put, a trade-off means that more [or less] of one thing necessitates less [or more] of another. An airline can choose to serve meals—adding cost and slowing turnaround time at the gate—or it can choose not to, but it cannot do both without bearing major inefficiencies" (p. 68). The obvious trade-off in the case of fuel is that *more* fuel saved requires *fewer*

[4]http://flying100years.com/

[5]http://www.ibtimes.com/how-airline-industry-has-evolved-100-years-commercial-air-travel-1524238

[6]http://online.wsj.com/news/articles/SB10001424052702303296604577450581396602106

[7]http://www.southwest.com/

[8]http://www.airlinefinancials.com/airline_data_comparisons.html

passengers or *fewer* flights or *fewer* routes—all things that do not position the airline for effective competition. The problem, therefore, requires better definition. One way to save fuel would be to shut down operations, but that is clearly *not* what is intended. The real problem, therefore, is how to save fuel without affecting the core function of the airline.

Drawing on the definitions already presented, what scope is there for effective, novel solutions that may take the form of a product, process, system, or service (referred to in the introduction) in this particular case study domain?

Writing in 1981, not long after the 1979 energy crisis, van Dam, Holmes, and Pitts (1981) noted that a key driver of design innovation in aircraft was the fact that "[i]n an era of fuel shortages and soaring fuel prices, there is a growing interest in reducing fuel consumption of existing and future aircraft" (p. 587).

Many readers will already be familiar with a *product* solution that is already in widespread use: *winglets*. van Dam et al. (1981) describe the technological characteristics of winglets and the drag-reducing effect that they create. By reducing drag by some 6%,[9] the winglets allow aircraft to operate at the same speed while burning less fuel. To illustrate the impact that this has on Southwest Airlines, consider that the airline operated[10] 477 Boeing 737–700 and 737–800 aircraft equipped with winglets (as of December 2013). The airline's own estimate is that winglets save it *54,000,000 gallons* of fuel per annum. With aviation fuel costing US$2.89/gallon as of March 2014,[11] this represents a cost saving of approximately US$156,000,000 per annum!

This *product* solution, in fact, achieves other desirable secondary outcomes as well.[10] As a result of the reduced drag, an aircraft equipped with winglets has a reduced power requirement during takeoff and landing. Therefore, noise levels emitted by the aircraft during takeoff and landing are lower, making the aircraft 6.5% quieter. This is an important factor for people living close to airports. Additionally, the reduced fuel requirement for a given flight reduces undesirable emissions of NOx[12] by 5%, making the aircraft *cleaner*. In an era of levies and taxes imposed on carbon emissions, the winglet solution offers a

[9]http://www.b737.org.uk/winglets.htm#ProductionWinglets

[10]http://swamedia.com/channels/Corporate-Fact-Sheet/pages/corporate-fact-sheet#fleet

[11]http://www.iata.org/publications/economics/fuel-monitor/Pages/index.aspx

[12]NOx is the generic term for nitric oxide (NO) and nitrogen dioxide (NO_2). These gases form when combustion takes place in the presence of nitrogen, and they contribute to smog, acid rain, and the depletion of ozone.

further benefit. Less fuel burned means lower carbon emissions,[13] as well as lower taxes and penalties for the airline. Other solution options also emerge from this product. Airlines may choose to trade the fuel saving for the possibility of extended ranges. A saving of 6% fuel means that the aircraft can travel 6% farther for the same amount of fuel, opening up the possibility of changes and improvements to the service that the airline offers customers.

The fuel savings made possible by winglets, of course, have to be offset against the cost of fitting the winglets. The cost of the product is approximately US$725,000,[10] and fitting them to an existing aircraft can take that aircraft out of service for approximately one week, representing significant lost revenue. Another factor that must be included in calculations is that the winglets themselves, on a typical Boeing 737–800, weigh between 170 Kg and 235 Kg, partially offsetting the fuel saving by making the aircraft heavier. However, when all of these considerations are taken into account, the figures for Southwest Airlines show that enormous fuel savings are possible.

Engineering solutions are not confined to tangible artifacts (products). The *way* that the desired outcome is achieved—the strategy and tactics for achieving tangible and intangible technological outcomes—is vitally important too. In engineering, these *process* solutions require technological expertise in order to get the best result possible from the product. Returning to the problem of using less fuel, a technological *process* solution is found, for example, in the strategies that airlines employ in *how* they operate their aircraft.

The takeoff and climb segment of a typical medium- or long-range commercial flight comprises only approximately 8%–15% of the total time of the operation of the flight (Roberson & Johns, 2007). In previous eras of *low* fuel prices, airlines typically were not overly concerned with this segment as a potential source of fuel conservation (and therefore cost reduction). However, with steep increases in fuel prices, "airlines are reviewing all phases of flight to determine how fuel burn savings can be gained..." (p. 25). The key here is that the potential solution to the problem of reducing dependency on expensive, nonrenewable, oil-based fuels is now one of process—a *way* of achieving an outcome—as opposed to a tangible, engineered artifact or system. There is no physical change to the existing product, just a change to the way that it is used.

The key factor in this segment of the flight is the takeoff flap setting. Flap setting determines the lift that the aircraft is able to generate at a given speed. Flaps are extended and lowered on takeoff to increase the lift-generating capacity of the wing. This is necessary because of the finite length of runways and the need to balance the performance

[13]http://www.iata.org/whatwedo/ops-infra/Pages/fuel-efficiency.aspx

TABLE 2.1 Fuel Burn and Takeoff Profiles

T/O & climb profile	Takeoff weight (Kg)	T/O flap setting (Deg)	Fuel used (Kg)	Fuel differential (Kg)
1	72,575	15	2,392	–
2		5	2,299	93

characteristics of the wing at all stages of flight: takeoff, cruising, landing. In theory, therefore, a pilot would like to lower the flaps on takeoff to maximize the lift so that she can get the plane off the ground in the shortest distance and at the lowest speed. As the aircraft gains altitude, the flaps are retracted so that the aircraft can cruise at a higher speed. Like most engineering problems, however, there is a trade-off. Lowered flaps for takeoff achieves the desired goal of increased lift, but also means an increase in drag, which slows down the aircraft and requires more engine power to achieve takeoff speed. Finding the optimal combination of flap setting, drag, speed, engine power, and fuel consumption is a nontrivial problem with ample scope for finding solutions that favor low fuel consumption ahead of other factors.

To illustrate the scale of the problem, consider the figures in Table 2.1. A typical commercial jet aircraft is the 737–800. Even though this aircraft is already fitted with drag-reducing winglets, the flap-setting for takeoff and climb still has a substantial impact on fuel burn.

While a saving of 93 Kg of fuel may seem relatively insignificant, consider that a single aircraft of this type will perform that same activity thousands of time per year. In a single year of operation, a commercial aircraft typically operates 6 days per week. The aircraft might comfortably make four flights per day, for 48 weeks per year, i.e., $6 \times 4 \times 48 = 1{,}152$ cycles of takeoff and climb per annum. Under those conditions, it is possible to achieve a saving of 93 Kg × 1,152 = *107,136 Kg* of fuel on a single aircraft per annum. This translates to a cost saving of (US\$1.0/Kg = US \$107,136 pa).[14] Southwest Airlines operates 477 such aircraft to which these figures can be applied.[15] The potential fuel saving is therefore more than *51,000,000 Kg* of fuel (or US\$51million)—all from a simple *procedural* solution.

[14]http://www.iata.org/publications/economics/fuel-monitor/Pages/index.aspx. The Jet Fuel price is highly volatile (no pun intended); however, in early 2014, an average figure of about US\$1,000 per metric tonne is reasonable, translating to approximately US\$1/Kg.

[15]http://swamedia.com/channels/Corporate-Fact-Sheet/pages/corporate-fact-sheet#fleet. The website of Southwest Airlines gives figures typical of a large commercial operator. The airline, as of December 2013, was operating approximately 475 Boeing 737-700 and 737-800 aircraft.

Another solution pathway for this problem of saving fuel employs no actual physical change to the product—in this case the aircraft—but instead employs a solution that we consider at the *system* level. The critical factor in this solution is to see the problem in a more holistic sense. It is tempting, in considering the aircraft, and the question of reducing dependency on oil/fuel, to draw the system boundary only around the physical structure of the aircraft (Figure 2.7). In other words, a *failure* to adopt a systems-level perspective is to see the context of the problem too narrowly, as only involving the *product*.

When we employ systems thinking and redefine the system of interest, we can simultaneously redefine the problem. The problem is now no longer confined to the product, in the sense of the physical structure of the aircraft, but is extended to the entire system—that of operating a commercial airliner (Figure 2.8). Immediately, this changes our focus

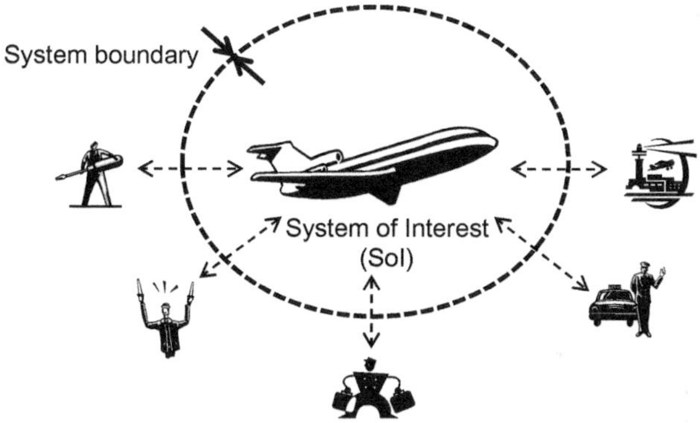

FIGURE 2.7 System boundary—fuel saving problem.

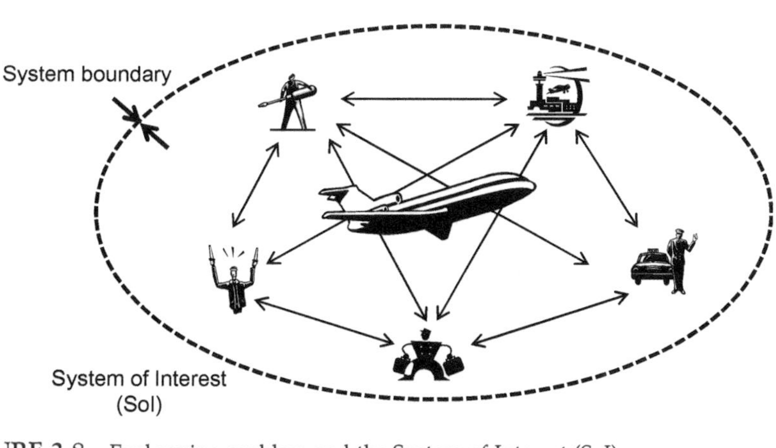

FIGURE 2.8 Fuel saving problem and the System of Interest (SoI).

from hardware only to the system as the interacting set of hardware, software, and people that deliver a certain outcome.

This then helps us to redefine the problem away from hardware-only solutions that change the product and to consider more diverse, systems-level solutions. We can consider the *Electronic Flight Bag* as one such system-level solution.

Commercial aircraft are bound by complex rules and operational procedures, one of which relates to the paperwork they are required to carry. Traditional *Flight Bags* could contain as much as 18 Kg of paperwork associated with a commercial flight. In recent times, airlines have been introducing "electronic" flight bags,[16] replacing the hard-copy paperwork with tablet computers, iPads, or similar devices. This change means that every flight has a reduced weight—even if only in the order of 16 or 17 Kg. However, for an operator like Southwest Airlines, operating 690 aircraft (as of December 2013), each of which could make approximately 1,152 flights per annum, the total weight saving is $690 \times 1{,}152\ \text{Kg} \times 16 = \mathit{12{,}718{,}080\ Kg}$. Now the additional fuel required for every extra kilogram depends on many factors, but a conservative and typical figure is that 135 Kg of additional fuel are required for every additional 1,000 Kg of weight on a typical commercial airliner. This means that the weight saved by the electronic flight bag represents a saving of 1,716,940 Kg of fuel per annum. Using our figure of US$1/Kg of fuel, this shows that the saving for the operator is still substantial. The fundamental nature of this solution was to conceive of the problem in a more holistic, *systems* sense.

The final solution pathway to the problem of reducing a commercial airline's fuel consumption is to consider *service* solutions. Recall (Chapter 1) that a service is *an organized system of labor and material aids used to supply the needs of the public.* Consider again Southwest Airlines as representative of the industry. In 2010, Southwest supplied its core service—flying people from destination to destination—through an organized system of labor (pilots, ground crew, flight attendants, baggage handlers, managers, cleaners) and material aids (principally, although not exclusively, aircraft). In doing so, the airline transported 88,191,000 passengers across 1,115,440 aircraft departures.[17] This was achieved with a *load factor*—the percentage of seats sold—of 79.4%. In other words, we can think of its *efficiency* as being 79.4%. What this means is that, on average, some 20% of seats on any given flight were unsold (and presumably empty). Because of the complex

[16] https://www.lhsystems.com/solutions-services/airline-solutions-services/flight-deck-solutions/lidoflightbag.html. This link describes the electronic flight bag system developed by Lufthansa.

[17] http://www.airlinefinancials.com/airline_data_comparisons.html

interrelationships between aircraft weight and fuel consumption, and the fact that some aspects of a flight are relatively fixed in terms of their fuel "cost" while others are variable, we can see that this efficiency of 79.4% is likely to have resulted in the wastage of a considerable amount of fuel. Put another way, the same number of passengers could, in theory, have been transported by Southwest Airlines using significantly fewer aircraft departures (and therefore, significantly less fuel burned). In fact, commercial airlines know this and strive to maximize the load factor figure—Southwest's performance is typical of the industry average. The issue for us, however, is that there is considerable scope to reduce the airline's dependency on oil, not through product, process, or system solutions, but simply through a *service* solution. In other words, a solution that, in this case, changes the way that the airline provides its core service and makes it more efficient, thus saving fuel. Examples of the ways that airlines may tackle this problem include canceling flights with low numbers of passengers, offering discounts to fill seats, and changing flight schedules to better match demand patterns.

Consider again only Southwest's fleet of Boeing 737–700 (143 seat) aircraft.[18] The 425 aircraft, assuming typical utilization patterns ($425 \times 1{,}152 = 489{,}600$ flights per annum) and a load factor of 100%, are capable of carrying some 70,000,000 passengers per annum. However, with a load factor of 79.3%, those 489,600 flights carried only 55,510,000 passengers. This means that the airline could have carried, *in theory*, the actual number of passengers it did on only approximately 388,000 flights. In other words, the same number of passengers could have been carried on 101,000 fewer flights, with no impact on the overall service outcome of transporting passengers. Given that a typical 737–700 burns on the order of 2,000 Kg of fuel per hour,[19] and assuming a typical flight is 2 hours (4,000 Kg of fuel), the potential saving to be achieved from a 100% load factor is as much as $101{,}000 \times 4{,}000$ Kg $= 404{,}000{,}000$ Kg of fuel (US$404,000,000).

Clearly, this figure would be reduced by an increase in fuel burn on the smaller number of more heavily loaded flights. However, even allowing for such factors, and assuming that the actual saving is only half of what is calculated, we can see that the potential to reduce fuel consumption with no basic impact on the service provided—through a change to *the way* the service is provided—still runs into hundreds of millions of kilograms of fuel and dollars.

[18]http://www.seatguru.com/airlines/Southwest_Airlines/fleetinfo.php

[19]http://www.b737.org.uk/techspecsdetailed.htm

TABLE 2.2 Total Weight and $ Savings—Southwest Airlines

Category	Solution	Fuel saved
Product	Winglets	156,000,000 Kg (477 Boeing 737–700 and 737–800 aircraft fitted with winglets)
Process	Takeoff Flap Setting & Climb Profile	51,000,000 Kg (477 Boeing 737–700 and 737–800 aircraft)
System	Electronic Flight Bags	1,716,940 Kg (690 Boeing 717 and 737 aircraft)
Service	Maximizing Load Factor	404,000,000 Kg (425 Boeing 737–700 aircraft)
	Total	**612,716,940 Kg**[a]

[a]*This is equivalent to approximately 735,000,000 liters or nearly 300 Olympic-sized swimming pools.*

If we combine all of these potential creative solutions together—product, process, system, and service—we see that there is enormous scope for airlines to reduce their dependence on oil (Table 2.2).

The purpose of this case study was to illustrate both the kinds of creative solutions—product, process, system, and service—that are possible in a technological problem domain, and also the scale of the impact of the problem-solving activity. Each of these solution pathways—winglets, better flight strategies, electronic flight bags, and a more efficient utilization of the existing solution—is based on technology, and the technology is created by engineers. In fact, over the past 40 years, aircraft fuel efficiency has improved by a staggering 70%, while in just the past 10 years, it has improved by 20%.[20] Coupled with the savings of this precious and nonrenewable resource are the substantial savings in the emission of greenhouse gasses and other pollutants, as well as substantial decreases in noise pollution. Where possible and available, I have used actual data for Southwest Airlines in order to illustrate the problem in a realistic manner. In many cases, however, assumptions need to be made because more detailed figures are not available. In some cases, these savings have already been implemented by Southwest Airlines, and other operators, while some of the examples discussed illustrate the potential for further savings.

[20]http://www.iata.org/whatwedo/ops-infra/Pages/fuel-efficiency.aspx

CHAPTER

3

Phases: Creativity and the Design Process

"OK, Houston, we've had a problem." **James Lovell, 1928–, Astronaut, Apollo 13**

The key to reconnecting creativity and engineering is that engineering appears to be a special case of creative problem solving. In Chapter 1, I introduced the idea of *Phases* and discussed the fact that engineering solutions come about as a result of a sequence of steps. In this chapter, I now explore these generic problem-solving steps, and I will show that the process of *engineering design* is the specific way that engineers enact that problem-solving process. Once that connection is made, it will then be possible to look at the impact of the *Person, Product, Process,* and *Press* on each stage of design.

PROBLEM SOLVING AND CREATIVITY

In the preceding two chapters, I developed the idea that engineers are concerned with solving problems, and that creativity, recognized in particular as a requirement for novelty, is central to this activity. I also used the term *creative problem solving* and suggested that engineering problem solving is really a special case of the former. Before I delve any further into *who* is being creative, *what* they produce, *where* these activities take place, and *how* they do it (the 4Ps)—all in the context of engineering—we need a model that describes the steps that take place in moving from recognizing that a problem exists to implementing a successful solution. I described this as the fifth P of creativity—*Phases*—in Chapter 1 and briefly outlined how this intersects with the 4Ps.

KNOWLEDGE AND PROBLEM SOLVING

It is tempting, as engineers, to think that we have our own special processes and methods for solving problems that are unique to the discipline. In engineering, as with many professions, we like to promote some degree of *mystique*. It is only through years of study and practical experience that a person can reach a position in which she is able to tackle the complex problems that arise through change and provide the technological solutions to address these problems. However, I argue that this is *not* the case, at least in one important sense! Of course, engineers need specialized technical knowledge and the experience of using that knowledge. In the same way that you probably *would not* seek to solve your medical, dental, legal, or psychological problems by visiting an untrained and unlicensed doctor, dentist, lawyer, or psychologist, you would not ask an untrained and unlicensed engineer to design and build you a new commercial airliner! My point is this—if you strip away the labels, the thing that makes each of these professions unique is not how they go about solving these problems, but the specialized knowledge that they apply within that process. In other words, whether the problem is *Why am I having trouble breathing?* or *How can I stop this pain in my upper jaw?* or *How can I stop my neighbor's dog from barking at night?* or *Why am I feeling anxious and stressed?* there is a commonality to *how* these different professions go about tackling these issues. That commonality is the *problem-solving process*. While the role and the value of creativity may differ across these professions—in each there will be situations in which old solutions satisfy old needs—there is also a class of activity in which redirection, forward incrementation, and reinitiation are both necessary and vital. For those cases, we can identify a generic creative problem-solving process.

The reason this is important to us is that it again emphasizes that creativity is important to engineering, and that it is accessible and transferable. In very simple terms, if lawyers and dentists can be creative problem solvers, then so too can engineers! This also means that we do not need to wait for someone to develop special, unique models of creativity for engineering. We can draw on those that already exist and immediately apply them to engineering. That, of course, is a major theme of this book.

Let us start, therefore, by surveying the different models of creative problem solving that exist. From these, we will see that there are certain common characteristics that describe how people, in any situation, go about solving problems creatively. I begin, once again, with Guilford, the initiator of the modern creativity era in psychological thinking

(1950). He described creativity as problem solving and defined it (Guilford, 1959) as having four stages:

- recognition that a problem exists;
- production of a variety of relevant ideas;
- evaluation of the various possibilities produced;
- drawing of appropriate conclusions that lead to the solution of the problem.

PROBLEM RECOGNITION

The first stage, problem recognition, highlights not only that there *is* a problem, but also that there are different *kinds* of problems. Kozbelt et al. (2010) note that, in general terms, problem solving has been studied in relation to so-called *puzzle problems*, typified by cryptarithmetic[1] problems (Newell & Simon, 1972). Michalewicz and Michalewicz (2008) give many examples of problems whose fundamental characteristic is that there is a single correct, and usually numerical, answer and that the task of the problem solver is to find it. While Kozbelt et al. (2010) question the relevance of such *convergent* problems to creativity, we will see that both well-defined, *single-answer* problems, and ill-defined, open-ended problems play an important role in creative problem solving. As Horenstein (2002) explains: "If only one answer to the problem exists, and finding it merely involves putting together the pieces of the puzzle, then the activity is probably analysis" (p. 23). In contrast, ill-defined, *divergent* problems have characteristics that are not predetermined. The problem may be imprecisely specified; there may be multiple solutions and solution pathways, any one of which is capable of satisfying the underlying need to greater or lesser degrees; and the criteria for recognizing a solution may be open.

The father of brainstorming, Alex Osborn (1953), made a very important point with regard to problem types. He pointed out that " . . . while every problem for creative attack is expressed as a question, not every question poses a problem for creative attack" (p. 97). This reiterates a point I have already made in previous chapters about problems that require creativity. Osborn went on to say, "One type of question calls for a factual answer, such as the question, 'How much money do we have to complete this project?' while '[another] type of question is one that calls for ideas, such as, 'In what ways can we profitably utilize

[1]A cryptarithmetic problem is one in which letters represent specific digits (e.g., X = 5; therefore XX = 55), and might take the form: SEND + MORE = MONEY. The object is to find the numerical value of each letter such that the letter problem is logically correct when converted to numbers.

available excess funds?' A question worded in this manner is a problem for creative attack."

Although Simon (1996) argued that ill-defined problems may be amenable to decomposition and solution of a set of well-defined sub-problems, there is a danger here that will be familiar to engineers. This *Cartesian*, or reductionist, approach overlooks the fact that in many complex systems, the desired behavior at the system level is *more than* the sum of the behaviors of the individual parts.[2] The most important aspect of this first stage of problem solving is to recognize that where a problem exists and where it cannot be solved simply by finding the single right answer, we are dealing with *creative problem solving*. Somewhat paradoxically, achieving that recognition—"is this problem convergent or divergent?" or indeed "what, exactly, is the problem?"—is itself a convergent problem! Thus, the first stage described by Guilford is convergent in nature.

This first phase has also been characterized as one of *finding* or *defining* the problem (Torrance, 1965). Creativity researchers speak of *problem awareness, problem recognition*, and the process of *problem finding* or *problem definition*. As Dillon (1982) pointed out, it is possible to distinguish between *recognizing* problems that are obvious to any qualified observer, *discovering* hidden problems, and finally, *inventing* problems. Merely *recognized* problems may well be solvable without creativity, or solving them may even be inhibited by creativity. In modern times, it has been recognized that invented problems have most to do with creativity (Jay & Perkins, 1997), and Mumford et al. (1996) identified "problem construction" as one of the main cognitive processes involved in creative problem solving.

Finding Good Problems

Simply *finding* any problem, while presumably better than nothing, is not enough. Even if we have determined that the problem is fundamentally divergent in nature, and therefore the focus of creative problem solving, there is more to do before we embark on the process of actually solving it. Higgins (1994) described this as "a vague feeling that something is wrong," while Luecke and Katz (2003), writing in the wider context of innovation, speak of the first stage of *Opportunity Recognition*. Tardif and Sternberg (1988) not only stressed the importance in creativity of *sensitivity to problems*, but went further and

[2]"Holism"—the whole is greater than the sum of the parts—is a fundamental characteristic of complex systems, and the same concept applies to problem solving. Many engineers have experienced frustration when they have tried to optimize the whole simply by optimizing the individual parts.

emphasized an additional element: finding *good* problems. Getzels and Csikszentmihalyi (1976) concluded that this is as much the case in artistic as in scientific creativity. A striking example of both these aspects of creative problem solving is Einstein's recognition that existing theories of electrodynamics were inadequate in dealing with moving bodies. He (a) *invented* a problem where many others saw none and (b) identified the *good* aspect of this problem. This quickly led to the special theory of relativity, revealed the need for a general theory of relativity, and ultimately resulted in lasting fame. *Good* problems are those that not only provoke a helpful answer to a specific situation, but also yield or even require elegant and generalizable solutions; i.e., they lead on to new things that go beyond the present situation.

Problem Awareness

Problem finding, however, is not as straightforward as might be thought. Sosa and Gero (2003) argued that many creative products are developed "to satisfy the needs of ... social groups" (p. 25). Those needs may be concrete and down-to-earth, such as cheaper power or a cure for a particular disease, but they may also be more general such as better educational methods or more beautiful ways of combining colors on canvas, or more abstract such as improved ways of expressing feelings through music. Generally, the social groups whose needs must be satisfied are (a) people who are knowledgeable in a domain—i.e., specialists or experts, or (b) users of the domain—people who are in some way affected by it. The people who are motivated to solve the problem are most commonly active in the domain as practitioners, experts, researchers, and so forth. In general, only such people are sufficiently immersed in a domain to notice problems. Those who have no contact with an area seldom (although perhaps not never) experience the need for solutions to problems or produce solutions in that domain. This brings us back to an important factor that I appeared to dismiss earlier: *domain knowledge*. Although the steps involved in problem solving may be common across different areas—domain-general, in other words—the special knowledge needed to both recognize problems in that domain and develop solutions is typically quite unique, or domain-specific. Thus, it is as unlikely that a nonengineer will recognize an engineering problem as it is that a nonengineer will solve an engineering problem—not for lack creativity, but for lack of technical expertise. This is, of course, good news for engineers, as creativity in engineering remains built on a foundation of solid technical knowledge and experience.

The Effect of the Problem on Creativity

One way of showing the relationship of creativity to problem solving is to focus not on creativity, but on the problems themselves. Sticking to the three dimensions introduced at the beginning, we can divide problems according to

- their degree of definition;
- the degree to which the solution pathway has already been defined;
- the clarity of the criteria for recognizing a solution.

Clearly defined problems that are solvable by means of standard techniques and for which there are obvious and well-known criteria identifying the solution constitute *routine* problems. They can often be solved without the need to generate novelty, although when existing knowledge is applied in settings where it has previously been treated as irrelevant, a certain technical or inventive creativity occurs. Nonetheless, creativity is not *absolutely necessary* and is probably not usual. By contrast, some problems require (a) becoming aware that there is a problem at all and finding a way of defining it, (b) working out techniques for solving the problem, and (c) developing criteria for recognizing a solution. Such *loosely defined* problems often demand a high level of creativity.

This raises the possibility that certain kinds of problems (routine problems) may actually inhibit creativity. It also seems that a too highly defined definition of the solution may hinder problem solving.

It is also conceivable that the reverse could occur: creativity could inhibit the solving of routine problems, for instance, by making the problem solver overlook perfectly effective and obvious (but not novel) solutions and look for obscure (novel) ones, or by encouraging the problem solver to go beyond the actual problem at hand and define it in an excessively complex fashion. In the case of loosely defined problems, on the other hand, creativity may be indispensable. However, although loose problem definition may facilitate or even demand creativity, it does not guarantee creative solutions (since it is possible that no effective solution may be achieved), but merely opens the doorway for them. Furthermore, since, as I have argued, creativity is not an all-or-nothing phenomenon, while there are different forms of creativity, loosely defined problems may elicit different amounts of creativity or creativity in different aspects of the solution, according to the aspects that are loose or the predilections of the problem solvers.

IDEA GENERATION

The second stage described by Guilford (1959)—the *production of a variety of relevant ideas*—assumes that the nature of the problem is now understood (recognized) and that for an ill-defined problem with no single correct answer, we must proceed with the process of creative problem solving. In the second stage, that process involves, in particular, the generation of many *possible* solutions. Why do we need a variety of possible solutions? That is the only way we can arrive at a point where we have something to compare. It goes without saying that comparisons require at least two things to compare. As soon as we are dealing with an ill-defined, open-ended problem for which there is no single correct answer, we are faced with *making a choice*. To be able to make a choice, we need some alternatives from which to make that choice. Generating those alternatives is the task in the second stage.

Whereas the fundamental characteristic of the first stage of problem solving was convergence, the basic feature of this second stage of creative problem solving is *divergence*. In other words, we branch outward from one problem or question to *many possible solutions*. At this stage, you may well have noticed that this sounds a lot like brainstorming or some other method. That is true, but I am avoiding talking about specific approaches, at least for the time being, precisely because I want to draw your attention only to the underlying concept and not to specific tools or techniques that can be used to execute the stage in question. I will discuss specific tools and techniques in later sections and chapters.

IDEA EVALUATION

The third stage described by Guilford (1959)—the *evaluation of the various possibilities produced*—operates on the set of possible solutions generated by the previous, divergent stage. It should be clear that this stage requires analysis, comparison, and convergence. An important point to remember here is that, although this stage is convergent in nature, it is not simply a case of reverting to discovering the right answer hidden among the set of possibilities that we generated in the previous stage. This stage is convergent because the goal is to converge toward an acceptable solution through a process of analyzing the candidates against a set of criteria that we have determined are appropriate. Thus, if the problem was *by what route should I go to work today?* you might have generated 20 alternatives in the second, divergent stage. Remembering that the overall goal of this process is to find *a* solution to our problem, we clearly need to eliminate 19 of the alternatives. Convergence and analysis here means that we must establish a set of criteria by which we can judge each of the alternative solutions. Those criteria must be meaningful and relevant to the problem. This is where constraints play an important role in creative problem solving. If we had no constraints whatsoever, then it would be difficult to find the best solution because *best* would be undefined. In fact, the first stage of our process—problem recognition—plays an important role in helping with this third stage because when the problem is clearly identified, the constraints are typically much clearer. Thus, *by what route should I go to work today* ***so that I get there by 9 a.m.?*** gives us a constraint that is now very helpful in judging the quality of the 20 candidate solutions. There is, however, another danger hidden in this process. Open-ended problems have no single right answer, and require us to generate a number of alternatives. Constraints give us the criteria against which we judge the alternatives and select the best solution. The more constraints we have, the easier it is to judge the alternatives, but the more we constrain the problem, the less freedom and novelty are possible. In other words, there is a delicate balance between too few constraints, making it impossible to choose among alternatives, and too many constraints, which can eliminate any creativity! It is a little like Henry Ford's statement that his customers could have any color car they wanted, as long as it was black. There would be little point in engaging in a creative problem-solving process if the constraints make only one answer possible.

SOLUTION VALIDATION

The final stage that Guilford (1959) described was that of *drawing of appropriate conclusions that lead to the solution of the problem*. This is, in essence, part of the stage of convergence described in stage three. It connects the chosen solution back to the original problem or need, and asks if this has actually been addressed. It avoids the danger that we identify a problem, generate alternatives, select among them in some way, but still fail to address the *real* problem. In simple terms, this stage maintains a holistic perspective across all stages of the process. The question is not *did we execute each stage correctly?* but *did we execute each stage correctly AND integrate these into an actual solution?* This stage, therefore, remains convergent and analytical in nature. Table 3.1 summarizes these generic problem-solving stages.

Guilford's (1959) model is a helpful foundation for understanding creative problem solving. Later in this chapter, I will show that there is a striking degree of similarity between this generic model and the way that engineers go about solving problems. However, before I do that, I would like to explore some other general models that have grown around Guilford's work.

TABLE 3.1 Stages of Creative Problem Solving

Stage	1	2	3	4
Description	Recognition that a problem exists	Production of a variety of relevant ideas	Evaluation of the various possibilities produced	Drawing of appropriate conclusions that lead to the solution of the problem
Summary	Problem Recognition	Idea Generation	Idea Evaluation	Solution Validation
Characteristic	Convergent	Divergent	Convergent	Convergent

(Guilford, 1959)

GENERAL MODELS OF CREATIVE PROBLEM SOLVING

The problem-solving approach just described (Guilford, 1959) has been the basis of a number of variations on the same theme over the years since Guilford (Newell, Shaw, & Simon, 1962) and has a well-established place in creativity research. It is attractive to us precisely because it is part of the body of knowledge that has developed over the past 60 years of creativity research. However, it is not the first such model, and it is useful for us to cast our net a little more widely to see who else has described this problem-solving process.

Even before the modern creativity era, i.e., post-Sputnik, some scholars had already described the process of *invention* in terms of a succession of steps. Prindle (1906), for example, studied inventors and concluded that an invention is the result of a series of small, interlinked steps. Each advances the development of the new invention by a small amount, generating an output (whether tangible or intangible) that serves as the input to the next step. However, the *classic* phase model, describing this problem-solving process, was first introduced about 90 years ago by Graham Wallas (1926). The Wallas approach, which has undergone some iteration, is important not so much for the number of steps involved, or the manner in which they are executed, but for his description of *what* those steps involve. The steps are understood in a more general manner as involving different *kinds of operations* rather than simply small gains in amount of content.

The Wallas (1926) model is often presented in terms four phases (Figure 3.1). In the phase of *Preparation*, a person becomes thoroughly familiar with a content area and defines the problem. In the phase of *Incubation*, a person "churns through" or "stews over" the information,

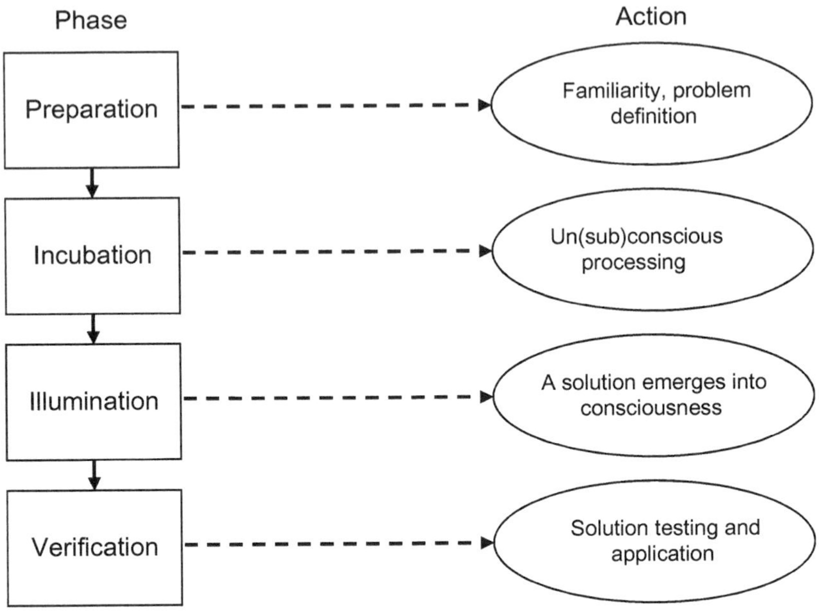

FIGURE 3.1 The phases of creative thinking, according to Wallas (1926).

typically *not consciously*. In the phase of *Illumination*, a solution emerges, often seeming to the person involved to come from nowhere. Finally, in the phase of *Verification*, a person tests and applies the solution that has resulted from the preceding phases.

It is worth noting, at this point, that we should not be tempted into believing that the process of invention, or creative problem solving, is strictly linear, as the models so far have implied. Empirical studies of people engaged in creative problem solving (Glover, Ronning, & Reynolds, 1989), as well as retrospective studies in which acknowledged creators described how they obtained new ideas (Csikszentmihalyi, 1996), have cast doubt on the validity of a phase model as an accurate description of the production of creative solutions in real-life settings. This is necessarily foreign to engineers—linear, sequential (Waterfall) models (Blanchard & Fabrycky, 2006; Stevens, Brook, Jackson, & Arnold, 1998) of engineering are now seen as too idealized a representation of how engineering actually happens. Nonetheless, the Wallas (1926) and Guilford (1959) models do offer helpful ways of looking at the production of effective novelty as a *process* rather than an event. In particular, they help us to understand the contributions of divergent and convergent thinking to this process as I indicated in Table 3.1.

Other Models of Problem Solving

As we scan the environment of approaches to creative problem solving, it is worth making one point of clarification. The phrase *creative problem solving* is used in two separate but related ways. When it is written in lowercase, I am using it in the general sense to mean "...efforts undertaken by individuals and teams to resolve open-ended problems through creative thinking" (Puccio & Cabra, 2010, p. 159), in contrast to its uppercase meaning, which specifically names the process devised by Alex Osborn, the father of brainstorming (Osborn, 1953).

Osborn, whose work was informed by both Guilford and Wallas, devised a comprehensive approach to the application of creativity in problem solving that he first described in terms of three *procedures*: (a) Fact-finding, (b) Idea-finding, and (c) Solution-finding. Creative Problem Solving (CPS) has undergone many refinements and changes in terminology. The current version is called the *thinking skills model* (Puccio & Cabra, 2010) and includes the following steps: exploring the vision, formulating challenges, exploring ideas, formulating solutions, exploring acceptance, and formulating a plan. Although the names are different, a core part of CPS remains recognition that divergence and convergence are essential elements of the process. The nonlinearity of the process is also addressed through an additional metacognitive step, assessing the situation, which is used to guide individuals and groups in the execution of the process.

Two other popular problem-solving processes, also summarized by Puccio and Cabra (2010), help to show a certain universality in the application of creativity. *Appreciative Inquiry* (AI) is an organizational development process, devised by Cooperrider and Srivastva (1987), that maps closely onto our generic problem-solving model. AI consists of four stages: *discovery* (identifying current, successful organizational processes and practices), *dream* (identifying ways to expand the use of successful processes and practices), *design* (constructing ideal *future* processes and practices), and *destiny* (identifying ways to implement the ideal processes and practices conceived in *design*). In this sense, AI describes a process that is broadly similar to that described by Wallas, Guilford, and Osborn.

Design Thinking (DT) is a user-centric approach to problem solving (Puccio & Cabra, 2010). This process involves careful observation of unmet needs (something familiar to engineers as *needs analysis*) prompted by gaps, frustrations, and other problems. The steps in Design Thinking are *understand* (problem-related information gathering); *observe* (watching and interviewing users to better understand the problem); *point of view* (analyzing the preceding observations); *visualize* (brainstorming outline solutions to identified the challenges); *prototype* (creating the promising physical solution or solutions), and *test and reiterate* (soliciting feedback on prototypes and modifying as required). Once again, we can see a pattern of convergent and divergent activities that together move the process from problem to solution.

The final model I will summarize, before I drill deeper into the process, is that described by Higgins (1994). He discusses four core stages of creative problem solving (which, just to confuse matters, he calls CPS) that reflect a similar progression to the preceding models. These stages cover:

- *Problem Identification*—This stage involves a focus on solving the real problem and not simply tackling or eliminating symptoms of the problem. To achieve this, we need to establish a set of criteria for evaluating the solution, to determine if the problem has been solved. This stage is largely analytical in nature.
- *Making Assumptions about Future*—This stage is concerned with identifying constraints, especially future conditions in the problem space, that may impact on the success of the problem-solving activity.
- *Generation of Alternatives*—This is the core divergent stage of the process.
- *Choice of Alternatives*—This stage uses analysis, drawing on the evaluation criteria established previously, to determine the best choice among the alternatives.

Table 3.2 now places these models alongside each other. One reason for doing this is that I want to establish a universality in the process. It does not matter if you are trying to take a creative approach to solving problems of an organizational, marketing and advertising, product design, or service-provision nature; certain fundamental steps are largely invariant across domains. The second reason for laboring this point is that I will shortly apply this same principle to solving engineering problems. In doing so, we will also see that the universality allows us to understand engineering problem solving using exactly the same constructs of creativity that apply in any other domain.

Although the terminology in Table 3.2 may differ, and emphasis is given different parts in each model, the fact remains: there is a common core that is perhaps best explained not by attempting to describe the specifics of the stage, but by stating the *character* of each stage. Problem solving moves through three key stages:

- *Convergence*—to recognize that there *is* a problem, and exactly *what* it is;
- *Divergence*—to develop a large set of *possible solutions* to the problem;
- *Convergence*—to *select the best* among those possible solutions.

TABLE 3.2 General Models of Problem Solving

Stage	1	2	3	4
Description	Recognition that a problem exists	Production of a variety of relevant ideas	Evaluation of the various possibilities produced	Drawing of appropriate conclusions that lead to the solution of the problem
Guilford	Problem Recognition	Idea Generation	Idea Evaluation	Solution Validation
Wallas	Preparation	Incubation	Illumination	Verification
Osborn CPS	Fact-Finding	Idea-Finding	Solution-Finding	
Design Thinking	Understand, Point of View, Observe	Visualize	Prototype	Test & Reiterate
Appreciative Inquiry	Discovery, Dream	Design	Destiny	
Higgins CPS	Problem Identification, Making Assumptions about Future	Generation of Alternatives	Choice of Alternatives	
Characteristic	**Convergent**	**Divergent**	**Convergent**	**Convergent**

The Extended Phase Model

Despite the underlying commonality of the models described—creative problem solving moves from convergence to divergence and back to convergence—there is probably a sense that we need to know a little more before we can apply this to engineering. As engineers, we are used to processes, but to be useful, a process model needs to be sufficiently detailed that it can directly drive and inform practical activities. The models that we have looked at so far make the point about the nature of creative problem solving but may be just a little too superficial to drive concrete action. It is a little like the instructions you sometimes find accompanying build-it-yourself products. When they follow a sequence that can be summarized as (a) open the box; (b) look at the picture on the piece of paper; (c) Congratulations! You have built your bunk-bed/TV cabinet/Swing set, then we can find that they are not terribly helpful. There seem to be a few steps missing! It will help us in our efforts to embed creativity in engineering if we can develop a slightly more detailed model of creative problem solving.

Wallas (1926), in fact, suggested that there were *seven* phases in the problem-solving process. As I described earlier, these are often presented as a core of only four (Kozbelt et al., 2010) or sometimes five (Kaufman, 2009) stages. I want to go back to the seven stages Wallas originally defined because they give us a better model in terms of correspondence to engineering, as we will soon see. Wallas' (1926) original model included

- *Encounter* (a problem or challenge is identified);
- *Preparation* (information is gathered);
- *Concentration* (an effort is made to solve the problem);
- *Incubation* (ideas churn over in the person's head);
- *Illumination* (a solution suddenly becomes apparent);
- *Verification* (the individual checks out the apparent solution);
- *Persuasion* (the individual attempts to convince others that the product really does solve the problem).

In some of my own earlier individual and collaborative work (A. J. Cropley & D. H. Cropley, 2008; D. H. Cropley, 2006; D. H. Cropley & Cropley, 2010b, 2011, 2012; D. H. Cropley, Cropley, Chiera, & Kaufman, 2013), we used the understanding that creativity is just one part of a larger process of innovation, to develop an extended phase model of innovation, based on Wallas's (1926) seven steps.

The four-phase model attributed to Wallas may be appropriate for discussions of generation of novelty. However, the phase approach must be extended if it is to encompass *functional* creativity (D. H. Cropley & Cropley, 2005), which requires that novelty must not merely be generated, but must also be useful for some purpose. Thus, functional creativity requires going beyond generation of novelty to encompass (a) a phase of *Communication* to users (in the case of innovation, this most commonly means customers), and (b) acceptance of the novelty by users, referred to as *Validation*. Communication and Validation are a necessity in innovation implementation, where both communication to customers and also acceptance by them are essential (see Christensen [1997] and Besemer [2006]).

One further differentiation of the Wallas approach is also needed. In a meta-analysis of research findings, Davis (2009) reviewed the extensive discussion of the nature of creativity and concluded that it can be viewed as having three "controlling components" (p. 26), the first of which is "problem finding." This conclusion is consistent with emphasis on sensitivity to problems and problem finding going back to Guilford (1950) and Torrance (1963). Mumford and Moertl (2003) described a case study of innovation in management practice, and concluded that innovation was activated by "intense dissatisfaction" (p. 262) with the status quo. In this chapter, recognition that there is a problem and a resulting urge to do something about it is referred to as involving *Activation*. However, problem awareness does not come from nowhere: you cannot see problems in and be dissatisfied with something that you do not know anything about. Thus, the generation of novelty commences with acquisition of knowledge in an area, in a phase that is referred to here as *Preparation*.

These considerations lead to an extended, seven-phase framework involving *Preparation, Activation, Generation, Illumination, Verification, Communication,* and *Validation*. The principal difference between this phase structure and the commonly cited four-phase Wallas (1926) approach is that the *expanded* phase model wraps the core creative process (i.e., Activation, Generation, Illumination, Verification) in a front and back end that acknowledges the more convergent elements central to *exploitation* and therefore innovation (i.e., Preparation, Communication, Validation). It is still important to note that the phases do not necessarily form a lock-step progression of completely distinct stages. There may well be interactions, false starts, restarts, early break-offs, and the like: Haner (2005, p. 289) summarized the relevant literature as showing that the phases of creativity are "iterative and non-sequential," and occur in a recurring "nonlinear cycle." Writing from the point of view of organizational psychology, Gupta, Smith, and Shalley (2006, p. 693) contrasted sequential or lock-step generation of effective novelty in an organization ("punctuated equilibrium") with nonlinear development ("institutional ambidexterity"). This extended model of creative problem solving (perhaps better thought of as *innovation*) is summarized in Figure 3.2. The general action involved in each phase is shown alongside.

As was the case with our earlier models, these phases can each be characterized in terms of a contrast between convergence and divergence. In Table 3.3, I now place the various models alongside each other.

ENGINEERING PROBLEM SOLVING: DESIGN

In Chapter 1, I characterized engineers as problem solvers and explained that the core activity of engineering is *design*. Design is the means by which engineers solve problems. I also suggested that creativity is inherent to that process of design. As a final step before examining the role that the 4Ps—Person, Product, Process, and Press—play across the creative problem-solving engineering design process, I want to establish that there is a clear and logical correspondence between creative problem solving (exemplified by the models we have just studied) and engineering design. I will do that by outlining some typical examples of how engineers model the design process, and then I will show that these are just creative problem solving by another name. The aim of this is to demonstrate, as I suggested in Chapter 1, that creative problem solving and engineering design are two sides of the same coin.

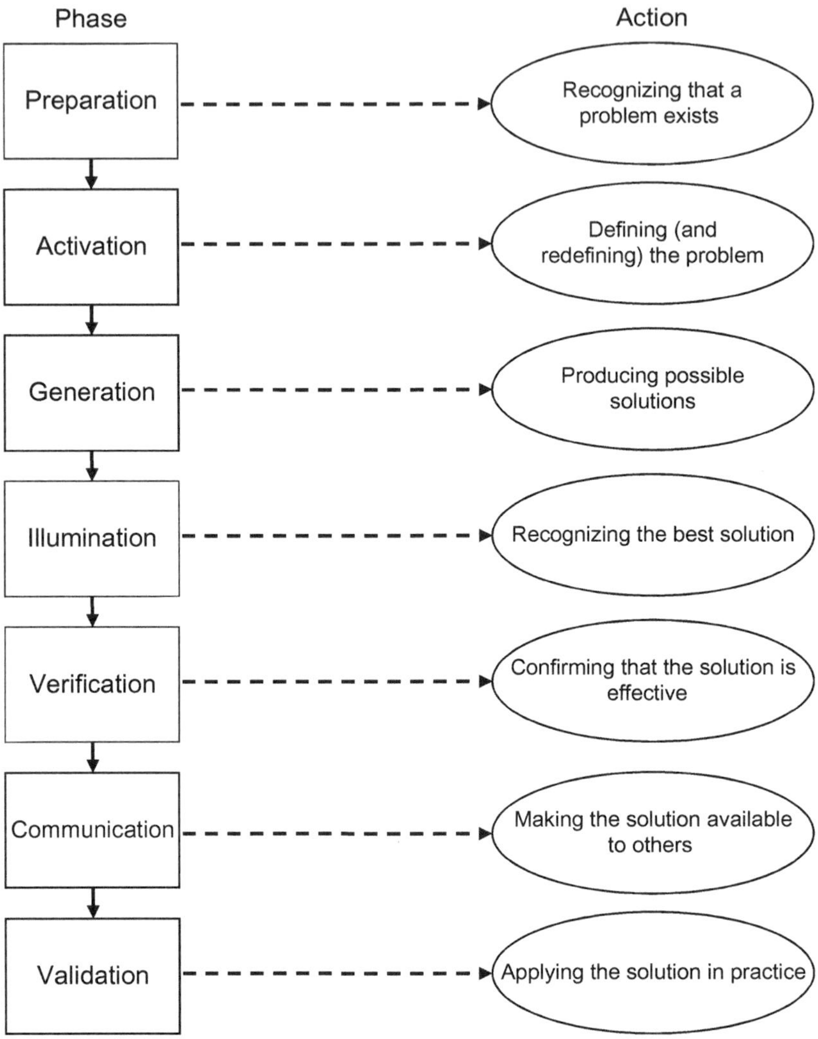

FIGURE 3.2 The extended phase model of creativity.

Once that is established, we are then in a position to import everything that the field of creativity research has learned about generating and exploiting effective novelty and apply it to engineering.

Engineering Design as Creativity

While it is true to say that engineering encompasses many varied activities, an essential core—indeed, a defining characteristic—of

TABLE 3.3 A Comparison of the Extended Phase Model and Other Problem-Solving Models

Generic phases (EPM)	**Preparation**	**Activation**	**Generation**	**Illumination**	**Verification**	**Communication**	**Validation**
***Super*-phases**	Invention					Exploitation	
Character	Convergent	Divergent	Divergent	Convergent	Convergent	Mixed	Convergent
Wallas	Encounter, Preparation		Concentration, Incubation	Illumination	Verification	Persuasion	
Osborn CPS	Problem Definition	Preparation	Idea Production	Idea Development		Solution Evaluation	Solution Adoption
Design Thinking	Understand	Observe/PoV	Visualize		Prototype	Test and Reiterate	
Appreciative Inquiry	Discovery	Dream	Design		Destiny	–	
Higgins CPS	Environmental Analysis, Problem Recognition	Problem Identification, Making Assumptions about Future	Generation of Alternatives	Evaluation & Choice of Alternatives		Implementation	Control

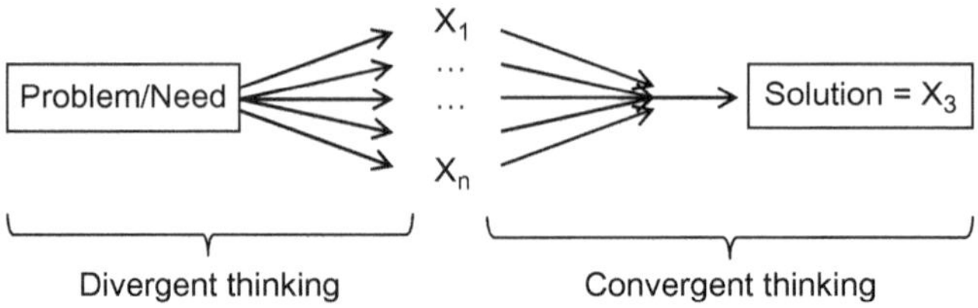

FIGURE 3.3 Convergence and divergence in problem solving.

engineering is design. Dieter and Schmidt (2012) remind us that "... it is true that the professional practice of engineering is largely concerned with design; it is often said that design is the essence of engineering" (p. 1). Citing Blumrich (1970), they characterize the process of design as "to pull together something new or to arrange existing things in a new way to satisfy a recognized need of society" (p. 1). Dieter and Schmidt (2012) describe the essence of design as *synthesis*.

Horenstein (2002) contrasted design with other essential activities in engineering by focusing on the process of solving problems. He stated that "If only one answer to a problem exists, and finding it merely involved putting together the pieces of the puzzle, then the activity is probably analysis ... if more than one solution exists, and if deciding upon a suitable path demands being creative, making choices, performing tests, iterating and evaluating, then the activity is most certainly design. Design can include analysis, but it must also involve at least one of these latter elements" (p. 23). The core of engineering practice is therefore design, but that design activity involves two stages: a stage of creative synthesis, followed by a stage of logical analysis. The first stage is synonymous with divergent thinking (Guilford, 1950), while the second is synonymous with convergent thinking. This may be illustrated as shown in Figure 3.3, and we usually think of this process proceeding, as illustrated, from left to right.

Another way to think of this needs-driven, problem-solving process is that it represents a *top-down* design paradigm. This approach is characteristic of the discipline of systems engineering (Blanchard & Fabrycky, 2006; Stevens et al., 1998) and is "[a] design methodology that proceeds from the highest level to the lowest and from the general to the particular..." (McGraw-Hill, 2003, p. 572).

Figure 3.4 illustrates the flow from User Need in the *functional* domain through to the System in the *physical* domain. This diagram also acknowledges that there is an important element of *traceability* between all parts of the process. The need is decomposed into a set of requirements which, in turn, are decomposed into a set of functions that describe what the system must do to meet the expressed need

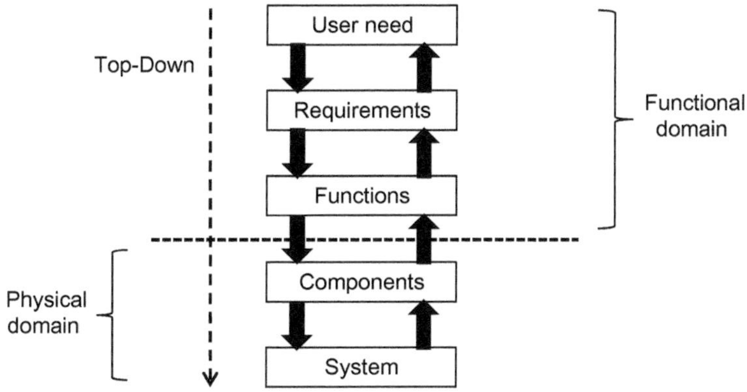

FIGURE 3.4 The *top-down* paradigm in engineering.

(i.e., to solve the problem). These functions are then allocated to physical components that combine into a system. It is in this step that we recognize the essential *divergent thinking* that we know characterizes creative problem solving—a given function, describing some aspect of what the system must do, is realized through a physical component that has been selected as the best solution, under the given set of circumstances. Each function is itself a problem that needs to be solved using divergent thinking.

Engineering Divergent Thinking

It is worth a momentary digression here. Figures 3.3 and 3.4 remind us that the key to engineering design is divergent thinking. A need—e.g., *I want to travel to my place of work faster than I can walk*—is decomposed into a set of requirements (e.g., "the system shall . . .") and then a set of corresponding functions that describe *what* the system must do to satisfy the given need. The core of each function is a verb-noun pair that expresses, in solution-free terms, what function is required. Thus, a function for our transport need might be *move person* or *carry passenger*. To connect the functional domain to the physical domain (Figure 3.4)—i.e., to solve the problem—we must explore a range of possible solutions for the problem *How to carry passengers?* and then select the most suitable of these solutions. For example, we might brainstorm this question and determine that options include car, train, bus, tram, bicycle, skateboard, hot-air balloon, jet-pack, slingshot, ferry, and so on. I will return to this example shortly, to discuss the next step, but for a moment, I would like to consider this as an example of divergent thinking.

Traditional definitions of divergent thinking—for example, "thinking . . . that generates a variety of ideas" (Russ & Fiorelli, 2010, p. 236)—as captured in the Torrance Tests of Creative Thinking (Torrance, 1966)—usually illustrate this process in the following way. As part of an Alternate Uses test, participants may be invited to think of as many uses as they can for an object, e.g., a tin can. Although there is no doubt that such a question does test divergent thinking (e.g., it can be used as a suit of armor for a mouse; a cup for drinking; one end of a communication device), divergent thinking in engineering design (and, arguably, in any practical problem-solving context) is manifest in a subtly different way. To illustrate the difference, consider the fact that engineers rarely select an object, for example, a balloon, and ask, *What are all the possible things I could do with this object?* Instead, the more typical design process, and that described in the preceding paragraph, is that a question is asked, or a problem posed—for example, *How can I carry passengers to work?* and a variety of possible solutions are proposed (car, bicycle, balloon, etc.). Both examples represent divergent thinking in the sense that a variety of ideas is generated in response to a question or problem. However, in the former case, the progression is really from *solution* (tin can) to *possible needs*, like the game-show *Jeopardy* in which there are many correct questions in response to the given answer. By contrast, in the engineering design example, the progression is from *need* (carry passengers) to *possible solutions* (including, among other things, a balloon). In engineering parlance, this is the difference between bottom-up design (*what need can I satisfy with this object?)* and top-down design (*how can I satisfy this need?*) (see Figure 3.4). You might wonder why we, as engineers, consider bottom-up as a possibility, especially in the context of creative problem solving. I explained the reason in the previous chapter when I talked about solution pathways and *legacy items* (see Chapter 2). Pure bottom-up design should be rare, but there is a third paradigm that we do encounter—one that incorporates legacy items into the normal, needs-driven, top-down paradigm. This is sometimes referred to as *middle-out design* (Stevens et al., 1998) and represents a fusion of top-down with some bottom-up. The important point is that this does not mean that the essential creative problem-solving process is bypassed. Legacy items, when they exist, still have to be matched to a function, and should be regarded simply as one of many possible ways to satisfy the function.

Constraints and Design

The differences between divergent thinking in the top-down sense and divergent thinking in the bottom-up sense may seem trivial, and indeed, this is of little consequence to the general concept of divergent thinking. However, in the context of the complete design (or creative problem-solving) process, where divergent synthesis is followed by convergent analysis, it highlights an element that is critical to any discussion of the interaction between engineering design and creativity. Divergent thinking, in isolation, is free to consider any possible solution that enters the mind of the designer. Thus, there is no limit on the *possible* uses of a tin can, and no limit to the number of ways passengers *could be* carried. However, divergent thinking, as the precursor to convergent thinking and taking place within a creative problem-solving (or top-down engineering design) process, does not have the same complete freedom. It is true that there are an infinite number of ways that passengers *could be* carried, but achieving a practical solution dictates that many of these will be rejected during the stage of convergent analysis that follows our divergent synthesis.

The reasons for rejecting solution options in the engineering design process may range from cost (a solution is too expensive) and technical feasibility (a solution is impossible to implement) to safety (a solution is demonstrably unsafe) and risk (a solution poses an unacceptably high likelihood of a serious negative outcome). Each of these parameters introduces *constraints* that, in effect, limit the available range of solutions from *all possible solutions* to a subset of feasible, practical solutions. However, knowing that these constraints exist should not cause us to bypass divergent thinking. Rather, these constraints are the basis of the analysis that follows our divergent thinking, and knowing what they are is a necessary part of the engineering problem-solving process.

Freedom versus Constraint

The preceding discussion highlights a paradox of creative problem solving and engineering design. At the heart of top-down creative engineering design is the ability to generate (Guilford, 1950, 1967) many different ideas (fluency), of different types (flexibility), that are unusual (originality) and to develop these ideas (elaboration). This divergent thinking characterizes creativity and depends for its success on *freedom from constraint*. The only way that we hope to identify and develop effective, competitive, technological solutions to engineering problems is to explore the largest possible design space—that is, to *maximize* fluency, flexibility, originality, and elaboration. However, in practice, we are bound by constraints that place limits on the design space. Gravity exists, objects cannot move faster than the speed of light, friction occurs between different objects, and so on. Successful design is contingent on maximizing the design space, while practical limitations act to minimize the design space. It would appear, therefore, that successful design may be impossible because constraints never permit the designer to explore the unfettered, maximum theoretical design space (Figure 3.5) in which

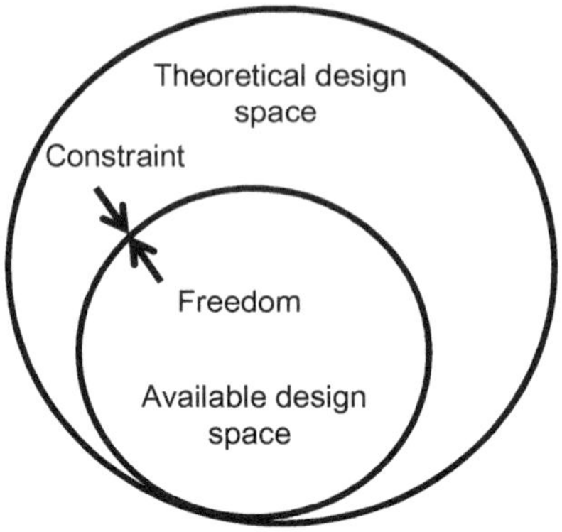

FIGURE 3.5 The theoretical and available design spaces.

reside all the highly effective and novel solutions that satisfy needs. The designer instead must settle for a limited and inherently less satisfactory *available* design space (Figure 3.5) that is more likely to be filled with routine, old solutions that, while probably effective, lack the novelty that opens up new markets and new possibilities.

The picture is not bleak, however, as it might first appear. As engineers, we know that we are constrained by fundamental physical laws and, more often than not by budgets, politics, and other factors. Engineering creativity is about finding ways to circumvent the constraints—to push the boundary of the available design space ever outward by exploring new possibilities, challenging norms, and taking calculated risks. Mokyr (1990) summed this up when he said, " ... technological change involves an attack by an individual on a constraint that everyone else takes as given" (p. 9).

ENGINEERING MODELS OF DESIGN

While you may now be convinced that engineering and creativity are closely linked, how does the engineering world describe the process used to solve problems? Blanchard and Fabrycky (2006) outline the role of the so-called Vee Model of the systems engineering process (Forsberg, Mooz, & Cotterman, 2000) as a model of engineering design. Figure 3.6 shows a standard version of the Vee Model, with the User Need driving the technical definition of the problem and design on the left, while the components of the solution are integrated and tested on the right. The Vee Model is particularly useful in emphasizing the impact that early problem identification has on the later process of evaluation. The Vee Model is very similar to the depiction of top-down

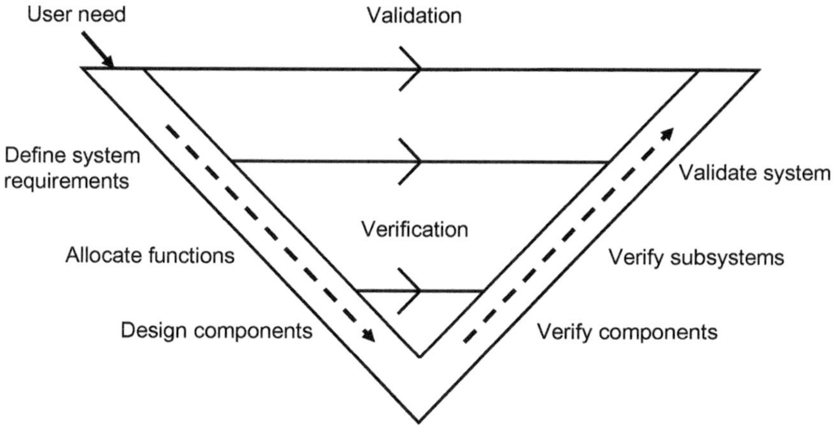

FIGURE 3.6 Vee Model.

process shown in Figure 3.4, if you imagine that the base of the "Vee" is the boundary between the functional and physical domains. Design proceeds from the top left, down to the base of the "Vee," and then back up to the top right. At each stage of the left-hand branch, we can look across to the equivalent point on the right-hand branch to understand that the information generated as we move down the *design arm* informs the process of validation that must take place as we move up the *integration arm*. This model emphasizes the importance of investing in the left-hand design activities—the problem definition, idea generation, and illumination—as a necessary prerequisite to successful verification and validation. In other words, the importance of the early stage of the problem-solving process described in Table 3.3.

Surveying a selection of other engineering design process models, we see that there is a high degree of commonality, not only between the models themselves, but also to the underlying *convergence—divergence—convergence* or *analysis—synthesis—analysis* that is the basis of creative problem solving. Pugh (1991), for instance, provides a good model of high-level, conceptual design. This begins with *Market/User Needs & Demands*. This *front end* of the design process is critical to the success of the latter stages, and Pugh gives examples of famous products and systems that experienced difficulties largely through a failure to execute this first stage well. The Sydney Opera House in my own country is an example of problems with the front end of the design process (Yeomans, 1968). Although it has become a cultural and national symbol, there were many problems and challenges during the design process, and it has been criticized, for example, as having poor acoustics.

The next stage that Pugh (1991) describes is *Product Design Specification (PDS)*. This is stage encompasses activities that we have already seen in the left-hand arm of the Vee Model (Figure 3.6). The PDS is both an activity and an output. It embodies the information gained through understanding the user need, and decomposing this into both *what* the system will do to meet the need and also *how well* it will need to do these things. Like the Vee Model, this looks ahead to the latter stages of design—evaluating our solutions—and also drives the next step of synthesis. *Generation of Solutions* therefore follows the definition of functions and constraints in the PDS. This is followed by *Evaluation of Solutions* in which we draw on the information in the PDS. Pugh (1981) makes a critical point that illustrates the importance of getting this front end of design right and especially of having good candidate solutions to choose from, as well as good criteria for evaluating those solutions—"the wrong choice of concept in a given design situation can rarely, if ever, be recouped by brilliant detail design." In other words, once this stage is complete, you tend to be stuck with what you have. The process concludes with a set of *Conceptual Phase outputs* that encapsulate the results of the preceding stages. Pugh (1991) also describes some of the specific tools that are used to generate ideas within the context of engineering design—brainstorming, for example—and stresses the importance of the constraints that play a role in engineering problem solving.

I have already discussed *systems* engineering and design (Blanchard & Fabrycky, 2006; Stevens et al., 1998), and an important point needs to be made here. Design happens at a number of possible levels. Systems engineering is a good discipline to understand if you wish to understand what these levels are and how they are interrelated, as well as understand that they all do much the same thing but at different levels of detail. Much of what I have described so far about engineering design is pitched at the level of *conceptual design*. The reason is that conceptual design is the highest level of abstraction of the design process, and it maintains a vital, direct link with the real source of the design problem—the user. It is important to understand that, no matter what level we are working at—*Conceptual, Preliminary*, or *Detail*—the core activities remain the same. That core is our process of *analysis—synthesis—analysis* repeating at increasing levels of detail. Thus, the output from a conceptual phase of design becomes the input to a more detailed level of preliminary design, and so on. We also recognize that in a complex system design environment, we are tackling as many different problems as there are functions, subfunctions, and so on. In other words, one basic user need may drive a set of system-level functions, which decompose into many subsystem-level functions, and so on. Each of these is a problem to which we apply our basic *analysis—synthesis—analysis* creative problem-solving process. In a paper on the relationship between creativity and systems engineering (D. H. Cropley & Cropley, 2000a), I examined the more detailed steps involved in conceptual design, described by the International Council on Systems Engineering (INCOSE) Systems Engineering Handbook.[3] These include

- list system elements;
- identify element options descriptors for system elements;
- define the design space envelope;
- define a process to generate a range of element options;
- select element options to populate the design space; and
- describe element choices in greater detail.

These steps outline a now familiar sequence of problem recognition and definition, constraints, idea generation, idea evaluation, and the development and communication of those ideas.

The same story is told repeatedly by engineers. Dieter and Schmidt (2012), for example, provide a recent comprehensive description of the engineering design process. This excellent book adds many details, tools, and techniques but still revolves around a core of (a) Define the Problem; (b) Gather Information; (c) Concept Generation; and (d) Evaluation and Selection of Concepts. By now, it should be clear that engineering design and creative problem solving are no different. As I stated earlier in the book, engineering design can be seen simply as a special case of creative problem solving.

Before I draw these engineering models together and compare them to the generic phase model developed earlier, I would like to finish with the oldest of the engineering design models that I consider. Buhl (1960) is important because his is the earliest work that I can find that explicitly describes engineering design in terms of creative problem solving. He characterized the process of solving engineering problems in terms of the following stages (Table 3.4):

- *Recognition*—in which the problem solver sees a mismatch between the world as it currently is, and some particular state of affairs.
- *Definition*—in which the stakeholder and the real problem are identified, and success criteria understood.
- *Preparation*—in which past and possible solutions are identified and a stock of relevant knowledge built.
- *Analysis*—in which the properties of past and possible solutions are determined.
- *Synthesis*—in which the analyzed information is compared, contrasted, and sorted in relation to the problem at hand, and where solutions with no merit are eliminated, while others are allowed to emerge from the data.

[3]See http://www.incose.org/ProductsPubs/products/sehandbook.aspx.

TABLE 3.4 Mapping Generic Phases to Engineering Design Models

Generic phases (EPM)	**Preparation**	**Activation**	**Generation**	**Illumination**	**Verification**	**Communication**	**Validation**
***Super*-phases**	Invention					Exploitation	
Character	Convergent	Divergent	Divergent	Convergent	Convergent	Mixed	Convergent
Pugh (1991)	Market/User Needs & Demands	Product Design Specification	Generation of solutions (conceptual design)	Evaluation of solution (conceptual design)		Conceptual Phase outputs	
Cropley & Cropley (2000a)	List system elements	Identify element option descriptors for system elements	Define the design space envelope	Define a process to generate a range of element options	Select element options to populate the design space	Describe element choices in greater detail	
Dieter & Schmidt (2012)—CD	Define Problem	Gather Information	Concept Generation	Evaluate & Select Concept			
Buhl (1960)	Recognition, Analysis	Definition, Preparation	Synthesis	Evaluation		Presentation	

- *Evaluation*—in which the merits of the emergent solution are examined.
- *Presentation*—in which the resultant solution is developed and implemented to the wider world.

Buhl (1960) notes many important and recurrent themes both in creative and engineering problem solving. They include the nonlinear progression that the process frequently follows. However, the most prescient of Buhl's (1960) points is that "[i]t is necessary to understand all the factors which tend to prohibit or retard the work at each phase, and to understand what things tend to increase the possibility of an unusual answer" (p. 15). The next four chapters address this very point. We now know that four factors—Person, Product, Process, and Press—either help or hinder creative problem solving (and therefore engineering problem solving) in each of the phases. How they do this is what follows next.

CHAPTER

4

Product: The Creativity of *Things*

"Great triumphs of engineering genius—the locomotive, the truss bridge, the steel rail— . . . are rather invention than engineering proper." ***Arthur Mellen Wellington, 1847–1895, Civil Engineer***

Creativity is not simply a matter of thinking or behaving in ways that differ from the customary—it involves the development of tangible solutions to practical problems, i.e., *Products*. Engineers, and engineering, are concerned fundamentally, with the application of knowledge and skills to solve problems. Those solutions typically take the form of tangible artifacts (products), more complex arrangements of physical elements (systems), ways of doing things (processes), or other intangible, value-adding solutions (services). For each type of solution, while there may be routine answers that merely address the problem, there are many more cases where, in the words of Hungarian-American scientist and engineer Theodore von Kármán (1881–1963), "... the engineer creates what never was" (Mackay, 1991). Products are what unite engineering and creativity. Understanding the role of creativity in engineering therefore means understanding the characteristics that make a product creative (e.g., novelty and effectiveness), and how these can be recognized and measured in practical settings.

The key theme of this book—reconnecting creativity and engineering—is concerned with understanding how novel and effective technological solutions to society's problems and needs are generated. We call this Engineering Design, and I have suggested that four factors impact on the way that we go about executing this creative problem-solving activity. These factors are referred to as the *4Ps*—Person, Product, Process, and Press. As the stages of engineering design unfold, the 4Ps, and our understanding of them, can act to either help, or hinder, process. We now need to examine each of the 4Ps in more detail so that we are aware of the role that each plays in the success of engineering design. As engineers, we understand that our purpose in life is to create *things*, and for this reason, we will begin our detailed journey into Engineering Creativity by looking more closely at the *Product*.

WHAT ARE PRODUCTS?

The most obvious result of any creative activity is its product—in other words, the output generated. Furthermore, it is the product that is exploited when it is inserted into a particular problem context and setting to achieve innovation (D. H. Cropley & Cropley, 2010a). The phases in Table 3.3—preparation, activation, and so on—trace the journey of the product from the initial driving need to the successful solution. MacKinnon (1978) concluded, "... analysis of creative products..." is "... the bedrock of all studies of creativity." (p. 187). Despite the fact that "Of the four P's, the creative product is probably the most widely studied and measured" (Kaufman, 2009, p. 23), less progress of practical use to engineering creativity has been made than might be thought. Kaufman (2009) acknowledges that, despite much research categorizing creative products, "[m]ore difficult, however, is figuring out how to measure them" (p. 30).

If we accept that creativity is vital to engineering, it is reasonable to ask what it is that makes a product creative and how to measure this objectively. William Thomson (Lord Kelvin) noted (1889) that "... when you can measure what you are speaking about, and express it in numbers, you know something about it; but when you cannot measure it, when you cannot express it in numbers, your knowledge is of a meager and unsatisfactory kind..." (p. 73). Unless we have the means to *quantify* the creative product, there may be little practical value in telling engineers that their products need to be more creative.

Some writers have concluded that it is too difficult to define creative products in a practical, objective way, because the concept is so subjective, and have instead recommended focusing on creative *processes* and characteristics of the creative *person*. Bailin (1988), however, was a strong advocate of the opposite view. She criticized the tendency to look at creativity purely in terms of psychological processes, and urged writers to focus on products, labeling efforts to foster creativity without reference to products "misleading" and "dangerous." Even where research has focused on the product, this has often been simply as a means to establish external criteria by which measures of the *Person* or *Process* could be compared and validated (Plucker & Renzulli, 1999). Although there are challenges to making objective measurements of product creativity that I will discuss later, the creativity of products is not as diffuse a concept as might at first appear to be the case.

One apparent advantage of the domain of engineering, in relation to discussions of product creativity, is that it is comparatively easy to define what we mean by a *product*. Tangible, functioning, physically useful *artifacts* such as tools and machines can be seen and touched. Similarly, *processes* such as the means for assembling an automobile in a factory or packaging food are readily understandable. In the domain of engineering, I have already introduced the idea of *solutions* to the needs of society as encompassing four types of product: artifacts, processes, systems, and services (Table 4.1).

Whether artifact, process, system, or service—all potential technological solutions to problems that arise in society—how do we begin to judge, in a formal and objective sense, what determines if these products are creative or not?

External Indicators of Creative Products

A step in this direction is to explore the different *kinds* of products that can be created. The "propulsion model" (Sternberg, 1999; Sternberg & Kaufman, 2012; Sternberg, Kaufman, & Pretz, 2002, 2003) does this by considering *external indicators* of products. A creative product achieves its external effect by *propelling* a field. This can occur, for example, through conceptual *replication* (the known is transferred to a new setting), *redefinition* (the known is seen in a new way), *redirection* (the known is extended in a new direction), and *reinitiation* (thinking begins at a radically different point from the current one and takes off in a new direction).

TABLE 4.1 Different Types of Creative Product

Product type	Product characteristics
Artifact	A manufactured object—tools, devices, consumer goods. E.g., hammer, cell phone, scissors.
Process	A method of doing or producing something—a production line, a procedure, a defined sequence of actions to achieve a particular outcome.
System	A combination of interacting elements forming a complex, unitary whole—systems have properties that emerge only at the level of the whole. E.g., an aircraft, a car, a transportation system, a communication system.
Service	An organized system of labor and material aids used to satisfy defined needs. Bank accounts, retirement plans, home delivery, babysitting, etc.

THE FUNDAMENTAL CRITERIA OF THE CREATIVITY OF PRODUCTS

Whether an artifact or a service and regardless of its external effect (e.g., redefinition), how do we characterize the creativity of a product? In domain-general discussions of creativity, it is more or less self-evident that the first characteristic of a creative product is *novelty*—creativity should always lead to something *new*. Paradoxically, however, there is a limit to the amount of deviation from the status quo (i.e., surprise/novelty) that an individual or a society can tolerate. When a person is confronted by excessive novelty, thinking may become disorganized, or the person may cling rigidly to what he already knows, and the reverse effect from the one we desire may result; in other words, too much novelty may shut down creativity. In a similar way, when a product is so surprising that it surpasses the society's capacity to tolerate novelty, the product may be rejected, or even worse, the person generating the surprise may be rejected or persecuted (see, for instance, the example of Ignaz Semmelweiss in Chapter 7). This point will be developed more fully in a later chapter, where creativity will be discussed in terms of society's ability to tolerate surprise, not as a property of products.

Leaving those concerns aside, novelty is an absolute *prerequisite* for creativity. This holds true in engineering as much as in other fields (Sprecher, 1959). It is not, however, sufficient on its own. If it were, every crazy idea or absurd suggestion would be creative. This makes intuitive sense to engineers. As a solution to our problem of *getting to work faster than walking*, being teleported there in a flux-capacitor-powered time machine is certainly novel, but may have some other limitations! In fact, Amabile and Tighe (1993, p. 9) emphasized that products must be "appropriate," "correct," "useful," or "valuable." Thus, creative products must be *not only* novel, but also socially tolerable and capable of doing what they were designed for: they must be *relevant and effective*. Cattell and Butcher (1968, p. 271) popularized the term *pseudo-creativity* to refer to variability whose novelty derives only from nonconformity, lack of discipline, blind rejection of what already exists and simply letting oneself go. In this sense, at least as far as engineers are concerned, *effectiveness* is perhaps the most basic constraint of all (see Chapter 3). We can design the most original, never-before-seen solution to a problem, but the requirement of effectiveness ensures that engineering creativity remains firmly grounded in reality, even if our job is to push the boundaries to their limits. "Quasi-creativity" (A. J. Cropley, 1997b, p. 89), in a translation of Heinelt (1974), also has many of the elements of genuine creativity—e.g., a high level of fantasy—but retains only a tenuous connection with reality. An example would be the novelty generated in daydreams. Once again, our problem-solving focus helps us to understand that, no matter how original and surprising, engineering creativity must remain purposeful and relevant. The product, in other words, *must* solve the problem it was created to solve.

Another side note here on the subject of definitions. A minor challenge that we face when discussing creativity is the fact that there seem to be certain entrenched *implicit beliefs* about the topic. Foremost among these may be an association between creativity and artistic pursuits. The issue that this raises relates to our understanding of relevance and effectiveness. Effectiveness seems very clear to an engineer but may be a more diffuse concept when talking about creativity in the context of art or poetry.

Another important consideration for engineers is that the order of these criteria—novelty and effectiveness—is not arbitrary. Although novelty might seem to take precedence over effectiveness, the preceding paragraphs suggest that, at the very least in engineering, there can be no discussion of creativity without *first* dealing with the issue of effectiveness. In other words, there is an implied order, or hierarchy, to these criteria. To take a simple example, engineers design bridges, primarily, to solve the problem of getting traffic across a river, under some given set of constraints. If the bridge does not satisfy that core function—the defining problem—then it is a bad product, no matter how unusual, surprising, or beautiful it might be. Higgins (1994) reiterates this point in a more general, business-oriented, sense when he states that "to be a true creative product it must have value and not just be original." It is conceivable that in other fields—music, art, and literature, for example—novelty may have primacy, but in engineering, *effectiveness is king*.

Further Criteria of the Creativity of Products

Two additional criteria help to define the creativity of products. The first additional criterion is *elegance*. Albert Einstein argued that it is not difficult to find novel solutions to problems: the difficult part is finding solutions that are *elegant* (Miller, 1992). Grudin (1990) reinforced this idea when he titled his book "The *grace* of great things [italics added]." Elegant solutions frequently cause a "shock of recognition" when they occur, and provoke a "Why didn't I think of that?" reaction. An elegant solution may look so simple and obvious—after the fact—that observers may underrate its creativity or denigrate it as "banal." Another relevant concept that emerges when we consider elegance is *aesthetics*. To a certain extent, elegance is concerned with aesthetic aspects of the product—style, for example—and I will consider this separately a little later when I talk about how the discipline of industrial design intersects with engineering. For the time being, it is sufficient to think of elegance in these terms: "a good solution usually looks nice" (Rechtin & Maier, 2000).

The second additional criterion is *genesis*. To explain this, consider that in 1605 Renaissance philosopher Francis Bacon developed a binary cipher using only "five-bit" combinations of the letters *a* and *b*, thereby showing that complex messages could be represented without loss of information, using only two discrete values. The 17th century mathematician Gottfried Leibnitz built on this concept to invent the binary number system in 1679. Although the two men could hardly have conceived of modern digital computers, they nevertheless laid the foundation for digital computing as we know it today. This is an excellent illustration of *genesis*.

The particular quality of genesis as a criterion of creativity is that a solution possessing this quality *not only* offers new possibilities for the situation for which the novelty was generated, but also

- is applicable in other apparently unrelated situations (i.e., the solution is transferable to other situations, whether or not this was intended or foreseen);
- introduces a new way of conceptualizing a whole area, or opens up new approaches to existing problems, possibly in many areas (i.e., the solution is *germinal*);
- demonstrates the existence of previously unnoticed problems and suggests the need for new work (i.e., the solution is *seminal*);
- lays a foundation for later innovations for which the original novelty is necessary, although the original innovator may have had no idea of the future innovation (i.e., the solution is *foundational*).

Genesis may be the most abstract of the four criteria of product creativity, but also the one that offers the greatest potential rewards. If *ordinary* novelty propels us further along a known path, genesis propels us along entirely new and unanticipated paths. Genesis is paradigm breaking. In terms of engineering products, genesis is perhaps characterized by situations in which we find ourselves solving problems that we were not aware existed and that could not be articulated by the user.

THE HIERARCHICAL ORGANIZATION OF CREATIVE PRODUCTS

Before I describe these four criteria of product creativity in more detail, I would like to outline some other ways that creative products can be classified, in a more general sense. One way to do this is to use the four dimensions just listed to arrange products in a hierarchy ranging from the "routine" product (characterized by effectiveness alone) at one pole to the "innovative" product (characterized by effectiveness, novelty, elegance, and genesis) at the other, with "original" and "elegant" products between these poles. This relationship, shown in Table 4.2, mirrors my discussion of *old* and *new* solutions in Chapter 2. In the table, a plus sign means that a criterion is associated with this kind of product, while a minus sign indicates that it is not. The associations in Table 4.2 can also be used to demonstrate the position of pseudo- and quasi-creativity, where the only *necessary* property of products seems to be novelty. The table shows that products *higher* in the hierarchy incorporate all of the properties of products at *lower* levels, but add something to them. According to this classification, *routine* products are not creative, because the second necessary criterion (novelty) is absent. As I indicated in Chapter 2, this does not mean that these products are useless or that they are not common. In engineering, a very large number of products perform important functions that benefit humankind and contribute to the advancement of society, yet are devoid of creativity. They are effective, and they solve an *old* problem in what I described as *replication,* but that is all. Improvements to these products, rather than representing creativity, are instead simply *evolutionary* changes that exploit existing technologies. When we move beyond routine products, however, then we enter the realm of *revolutionary* change thanks largely to the addition of novelty.

The hierarchical organization of products shown in Table 4.2 introduces an important principle into the discussion of creativity: creativity is not an all-or-nothing quality of a product; there are *levels* or *kinds* of creativity. It is not something that products either have or do not have. Different products can have creativity to greater or lesser degrees (as assessed against the four *criteria*), or they can display different kinds of it (i.e., *different combinations* of the four criteria). Table 4.2 suggests different labels for different *kinds* of creativity ("original," "elegant," "innovative"), while the hierarchical organization of these kinds of

TABLE 4.2 The Hierarchical Organization of Products

	Kind of product				
Criterion	**Routine**	**Original**	**Elegant**	**Innovative**	**Pseudo- or quasi-creativity**
Effectiveness	+	+	+	+	–
Novelty	–	+	+	+	+
Elegance	–	–	+	+	?
Genesis	–	–	–	+	?

creativity means that there are also *levels* of creativity (innovative is more creative than elegant, while elegant is more creative than original). I will shortly explore in more detail how we measure those levels, but before I do that, I will look at some other ways to understand the *kind* of creativity we encounter in products.

Situation versus Domain Relevance of Creative Products

Another way of distinguishing between *kinds* of creativity involves a distinction between what we can call *situation-relevant* and *domain-relevant* creativity. Situation-relevant creativity solves a specific, concrete problem, of the kind that we have already discussed—e.g., how to get traffic across a river. The effectiveness of the solution is judged according to specific criteria related to the particular situation. In the bridge example, these may include cost, speed with which traffic is able to move, durability, safety, style, environmental impact, and the like. A creative solution would satisfy these constraints in a novel and effective way, and the constraints would normally be specified by the customer. This kind of solution would therefore be classified as *original* (Table 4.2).

Domain-relevant creativity, by contrast,

- expands the way the domain is conceptualized;
- emphasizes new issues not previously noticed; and
- suggests new ways of solving problems in the area.

It involves, in other words, the presence of elegance and genesis (Table 4.2). Ideally, domain-relevant creativity also leads to solutions that are situation-relevant (i.e., the solution would also do what it was supposed to do, such as getting the traffic across the river). However, it is possible for the two kinds of creativity to exist separately: a product could involve situation-relevant creativity without domain-relevance, or domain-relevance without situational-relevance.

Domain-relevant creativity *unaccompanied* by situation-relevant creativity is probably more acceptable, and usual, in aesthetic and philosophical domains, where the development of new *approaches* to problems may be at least as important as practical issues such as getting traffic across a river. In engineering, however, situation-relevant creativity is a prerequisite to higher-order domain-relevance.

This question of situation and domain-relevance, and the roles of the different criteria of creativity, is illustrated by an example we have already considered. The Sydney Opera House in Australia, of course, was built as a venue for staging operas. The original intent was undoubtedly functional, or focused mainly on the situation-relevant criteria of effectiveness and novelty. The problems that occurred during building and the subsequent criticisms that have been leveled at the structure (e.g., poor acoustics and seating) have been subsumed, to a large degree, by the domain-relevance of the Opera House. It is now widely recognized that the building helped to transform construction techniques (e.g., using prefabrication methods) and opened up new perspectives on building design (e.g., the sail-like roof shape). The Opera House has also become a national icon and a highly recognizable symbol of Australia. In other words, it possesses higher-order qualities of elegance and genesis that we can characterize as domain-relevance, even if situation-relevant creativity (especially, questionable effectiveness) has left something to be desired.

This example also illustrates another feature of the higher-order criteria of creativity. Genesis, in particular, is often most recognizable after the fact. In other words, unlike novelty, effectiveness, and elegance, it is harder deliberately to build it into the product. Engineers can be reasonably sure, in advance, that their products will be effective, and that they are novel and elegant, but how do we build into a product the fact that it transforms a society? Genesis, therefore, may be the hardest of the criteria to design into the product, but is also the criterion that may carry the greatest rewards. Think of some examples of transformative *products* like the World Wide Web (WWW). Tim Berners-Lee did not set out to transform society when he devised a means for integrating hypertext, the Transmission Control Protocol (TCP) and the domain name system (DNS). He was interested in more mundane and practical questions of improving information management in his organization. However, it is undeniable that the Web, as a product, possesses enormous *genesis*, in the sense that it has resulted in changes to the way people communicate, are educated, conduct business, and are entertained.

For engineers, this discussion of *kinds* of creativity leads directly to the next issue. If our business is developing original, elegant, and/or innovative solutions, how do we recognize the criteria that define these qualities? Furthermore, if I were to tell you that your product needs to be *more innovative*—i.e., it needs to be *more* novel, *more* effective, *more* elegant, and/or have *more* genesis—how do you make sense of that?

PRODUCT CREATIVITY AS A SYSTEM

The most important aspect of the four-criterion model of creativity that I have presented is that the criteria, like the kinds of products, also form a hierarchy. In other words, as is evident from the discussion in the previous section, the order of effectiveness, novelty, elegance, and genesis *matters*. It is no accident that effectiveness is the basic, defining criterion of the different kinds of products, while genesis is only seen in a subset. In fact, if we ignore *routine* products as outside the scope of this book (refer back to Chapter 2's discussion on this), then both effectiveness *and* novelty are the necessary, or prerequisite, qualities for a product to be considered creative. Only when *both* are present is it possible to talk about creativity. Elegance and genesis sit further up in the hierarchy (Figure 4.1).

In engineering creativity, while it is possible to talk about creativity *without* novelty and effectiveness (i.e., creativity that is only domain-relevant, or quasi/pseudo-creativity), in the normal course of events, elegance and genesis are only interesting and significant when the first two criteria have been met.

Perhaps even more significantly, the relationship among the criteria is also *dynamic*. In other words, addition of the higher-order criteria *adds value* to those below them. What this means is that novelty *increases* effectiveness. Elegance increases both novelty and effectiveness, and genesis adds to all three of the lower-order criteria. Thus, although elegance and genesis are *not prerequisites* for creativity, they do add value to the overall creativity of a product when present. Importantly, a change in the context or the particular purpose to which the product is applied can also have the opposite effect, destroying the effectiveness of a product. For this reason, in a sense, creativity is not an aspect of the product at all, but of the context: *The context determines creativity* by defining a product's relevance and effectiveness. The role of the context as part of a system of creativity will be discussed in more detail in a later chapter.

The issue of the hierarchy of creativity criteria becomes particularly important in situations in which a product is competing with a *rival*—something that is frequently characteristic of situations involving engineering products. A product that is novel and effective—i.e., creative—may have its effectiveness and thus its creativity destroyed by a rival.

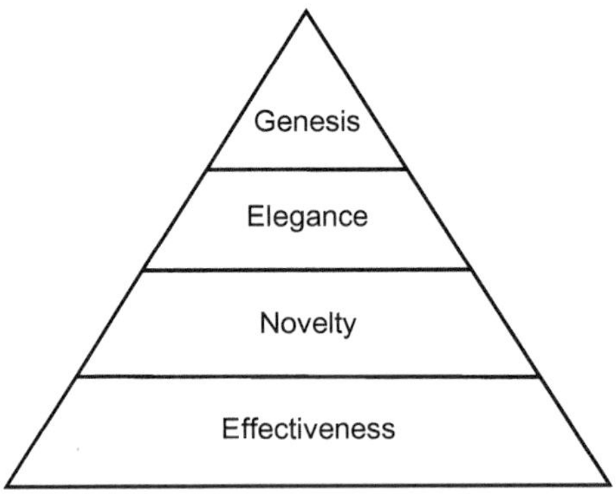

FIGURE 4.1 The hierarchy of creativity criteria.

Consider the relationship between the vacuum tube and the solid-state transistor. The relevance and effectiveness of the former was destroyed by the introduction of the latter. The vacuum tube was novel in its day, but was rendered obsolete not by the fact that it ceased to solve the problem for which it was designed, but simply because of the introduction of a more novel, and ultimately more effective, rival. Gertner (2012) details the fascinating history of these products, developed by Bell Labs, and his book is a must-read for engineers interested in creativity, innovation, and engineering.

The preceding discussion is one way of explaining the value of creativity in a competitive, commercial environment. For another explanation of this, see D.H. Cropley, Kaufman, and Cropley (2008). The dynamic hierarchy of creativity criteria in a product influences the way that products compete with each other. The need to make a product robust in the face of a rival product, or even capable of subtracting value from the rival, supports the importance of "loading" new products with novelty, and suggests several reasons for doing this:

- Novelty may add so much value to a product that it is immune to value subtractions resulting from a rival's novelty.
- The product's novelty may also give it the capacity to subtract value from a rival product (i.e., to nullify the rival's effectiveness).
- *Extra* novelty may add to a product's higher-order *genesis* value.

However, since the nature of the rival may be unknown at the time a product is being developed, the benefit resulting from novelty may initially be only *latent*.

LATENT FUNCTIONAL CREATIVITY

In Table 4.2, I touched on the possibility that some products, while displaying novelty, may not involve other criteria such as effectiveness. I suggested that these were of less interest to engineers, addressing only domain-relevance. However, the hierarchical, value-adding nature of novelty, elegance, and genesis means that we should not dismiss this apparent pseudo/quasi-creativity too quickly. Despite my earlier insistence on the primacy of effectiveness in engineering creativity, there may be another kind of novelty that is worth taking seriously, even when it occurs in isolation. This other kind of novelty is one that *has not yet* been seen to contribute to the effective solution of a problem. However, it may eventually do so, if and when an appropriate context is encountered. It is an *abstract* novelty that yields potential, or *latent*, creativity. For this reason, we need to be cautious about dismissing products as lacking creativity simply because they cannot be used to solve a particular problem in a given situation. Their latent creativity may simply mean that the technology is ahead of its time, and that the usefulness of the novelty is not yet apparent. Even seemingly *ineffective* novelty could possess "hidden" or "latent" effectiveness, and the creative engineer, under the pressure of delivering solutions to pressing needs, should not forget the *potential value* of the new solutions generated in the design process. The Bell Labs is again an excellent example of this concept. It is true that the bulk of the company's effort, in terms of both expenditure and people, was concerned with *development*. However, the engine room of the organization was the *applied research*—what Stokes calls the *Pasteur Quadrant* (1997)—that delivered some of its most significant technological contributions to modern society.

MEASURING THE CREATIVITY OF PRODUCTS

As helpful and enlightening as it is to understand what characteristics make a product creative, and as helpful as these are in categorizing different kinds of products, if we are to extract the promised value from creativity in the domain of engineering, we need to be able to answer not just qualitative questions about creativity, but also quantitative ones. In other words, when you bring your product to me, and I tell you that it is *not* creative, your natural reaction might be to ask, "OK, but what do you mean?" If I responded that it needs to be, for example, more effective or more novel, you might still ask either "What does *that* mean?" or "Yes, but *how much* more?" As engineers, we are used to tackling quantitative questions, and there is no reason why we should not have the same expectations when we deal with creativity.

The measurement of product creativity is by no means a new problem in creativity research. There have been extensive studies on how to measure it in its broadest sense. Besemer and O'Quin (1999) describe three common approaches used to measure product creativity: indirect measurement, global judgment and criterion-based measurement. These approaches have been developed in both a domain-general and a domain-specific context (for a more detailed discussion of domains and creativity, see Baer [2010]). Some of the possible solutions that span these different approaches include the use of expert raters (Amabile, 1996), divergent thinking-based scoring of creative products for originality or fluency (Reiter-Palmon et al., 2009), or assessment of a product's historical impact (Simonton, 2009).

Horn and Salvendy (2006) compared various product creativity measurement tools, including rating scales and subjective assessments. The former include Besemer and O'Quin's (1987, 1999) *Creative Product Semantic Scale* (CPSS) and Reis and Renzulli's (1991) *Student Product*

Assessment Form, and the latter concentrated on Amabile's (1983, 1996) *Consensual Assessment Technique* (CAT). Horn and Salvendy (2006) also reported on the range of different domains to which a rating scale have been applied, including artwork, cartoons, chairs, advertisements, scientific and creative writing, audio-visual products, and social studies. The CAT, by contrast, has been applied to stories, art, poetry, and other aesthetic products. As is evident, most of this research has been focused on the evaluation of either aesthetic or organizational products. The assessment of aesthetic works (such as paintings or poems) has, in fact, been extensively investigated for nearly a century (Baer, Kaufman, & Gentile, 2004; Cattell, Glascock, & Washburn, 1918; Child & Iwao, 1968).

There are, however, surprisingly few studies aimed at assessing the creativity of products in the sense of tangible, scientific, or technological products—that is, engineered artifacts or manufactured consumer goods. Where studies do examine engineering products, it is primarily in connection with related concepts, such as "usability" (see, for example, Han, Hwan Yun, Kim, & Kwahk [2000]). In one such domain (mathematics), Mann (2009) argues that many of the current assessments are time-consuming to score; they also tend to be separate instruments designed to measure the specific domains. As a result, most of the work on mathematical creativity assessment cannot be applied easily to related domains (such as engineering). What is needed is a *universal aesthetic* of creativity—a set of indicators that "can be recognized with a substantial level of agreement by different observers, and can be used to judge both amount and kind of creativity" (D. H. Cropley & Cropley, 2008, p. 155).

Consensual Assessment

It is tempting to dismiss some of the techniques as unsuitable for engineering creativity. However, there are important lessons to be learned about product creativity measurement. The most straightforward way of determining the creativity of a product is to ask people who know about such things whether it is creative. This sensible idea is at the heart of the method of *consensual assessment* (for a summary, see Hennessey & Amabile [1999]). Amabile and her colleagues have developed and refined this approach, and the *Consensual Assessment Technique* (CAT) is now relatively well known among creativity researchers. The method frequently involves recruiting a panel of judges to rate the creativity of a product, often experts in the field to which the product belongs. There is evidence, however, that even people without deep, expert knowledge of a field are capable of identifying creativity when they see it (Kaufman et al., 2013). Judges' ratings, whether or not they are experts, seem to relate to genuine differences between products (i.e., they are valid), and they are reliable in a statistical sense.

Although it involves assessing the creativity of *products*, paradoxically, the CAT is most widely used as an instrument for identifying creative *people*. The CAT is often used by giving all members of a group, such as a class of students, the same standardized task, leading to a closely specified product (such as a collage made from an egg carton, a sheet of writing paper, a paper clip, and as much string as they like). Judges then rate each person's product in order to identify the more (or less) creative members of the group. Subsequently, the people may be divided into subgroups on the basis of the score received by their product (e.g., the most creative third, the least creative third, and the people in the middle, or something similar). Thus, the CAT is often used more as a test for assessing the creativity of *people* rather than products. Nonetheless, the CAT popularized an important principle for assessing the creativity of products: *agreement among observers*. It is not hard to see that there are practical limitations to the use of such a method in engineering design: it is time-consuming and expensive to assemble a panel of experts every time we need to rate the creativity of our designs.

Rating Scales

A more attractive proposition for measuring creativity in engineering design is the rating scale. Psychologists have developed instruments based on observers' ratings for systematically determining the creativity of products. An early example is Taylor's (1975) *Creative Product Inventory*, which measures the dimensions Generation, Reformulation, Originality, Relevancy, Hedonics, Complexity, and Condensation. The criterion of hedonics raises an interesting issue: it is reminiscent of Jackson and Messick's (1965) very early distinction between *external* criteria of the effectiveness of a novel product (i.e., does it work?) and *internal* criteria such as logic, harmony among the elements of the product, and pleasingness (i.e., is it beautiful?). Taylor thus reinforces the importance of both *functional* criteria and *aesthetic* criteria in the measurement of product creativity.

More recently, Besemer and O'Quin (1987) developed the *Creative Product Semantic Scale*, based on three dimensions: *Novelty* (the product is original, surprising, and germinal), *Resolution* (the product is valuable, logical, useful, and understandable), and *Elaboration and Synthesis* (the product is organic, elegant, complex, and well-crafted). A later version of the Besemer and O'Quin scale has 43 items (Besemer & O'Quin, 1999). Besemer (1998) confirmed empirically that the scale measures three dimensions, and demonstrated its ability to distinguish consistently among products (three chairs of quite different design, for example). Reliabilities of the three dimensions ranged from .69 to .87 (alpha coefficients), with the majority of coefficients being in excess of .80 (which is widely accepted as highly satisfactory). In the *Creative Product Semantic Scale*, these criteria are assessed by asking raters to rate a product on bipolar dimensions (e.g., "surprising—unsurprising"; "logical—illogical"; "elegant—inelegant"). The raters' task is to indicate how close the object being rated is to one or the other pole of each bipolar dimension.

The criteria used in the CPSS—for example, *surprisingness*, *complexity*, or *germinality*—may seem to be highly subjective, but psychological research has shown that even untrained judges, working without knowledge of what other judges are saying, can reach much the same conclusions as the other judges about the prominence of many of the criteria in a solution. This means that the method has good inter-rater reliability and consistency (i.e., scale reliability) and also satisfactory test-retest reliability (similar ratings are given if raters are asked to rerate the same products at a later date).[1]

Hennessey (1994) reported inter-rater agreement ranging up to .93 (excellent) even among untrained undergraduates who rated geometric designs or Picasso drawings in terms of both the *Creativity of Product* and *Creativity of Process*, simply applying their own subjective understanding of these qualities. Internal reliabilities of individual people's ratings of creativity ranged from .73 to .93. Vosburg (1998) also reported that untrained judges who rated products on 7-point scales such as "Very complex—Not at all complex" or "Very understandable—Not at all understandable" achieved inter-rater reliabilities of about .90. In other words, different people seem to have a common and reliable understanding of novelty, complexity, elegance, and the like; can recognize them when they see them; and can express their judgments of the level of the characteristics in a quantifiable way. This strongly suggests that trained engineers could be expected to be able to make the same consistent ratings of creativity in their domain using this type of rating scale.

Baer, Kaufman, and Gentile (2004) extended such findings to some degree by looking at products that were not based on a narrowly defined task for each person. They gave 13 raters personal narratives, stories, and poems written by schoolchildren in different classrooms, under different conditions, and with varying instructions, i.e., under diverse conditions. The raters achieved very high levels of agreement on the creativity of the products: .87 for the poems, .94 for the stories,

[1]A general word about measurement: engineers may be less familiar with the methods and statistical techniques I am describing here, being more firmly grounded in physical measurement science. For a good overview of the concepts of psychometric approaches to measurement and the statistical techniques used, see DeVellis (2012).

and .96 for the personal narratives. However, it is important to note that the raters were all experts in the field of creative writing. It seems that people who have experience and knowledge of a field of activity—presumably this would include professional engineers—can agree on the creativity of domain-relevant products of different kinds and produced under varying conditions. The question is—can we improve on this? Can we develop a rating scale that does not require deep domain expertise, or deep knowledge of creativity, but yields valid and reliable measures of product creativity? Can this then be used to diagnose and improve creativity in engineering design?

The Creative Solution Diagnosis Scale (CSDS)

By combining the indicators of creativity of the type used in the scales presented in the preceding section (see Table 4.3) with the four criteria of functional creativity already outlined (relevance and effectiveness, novelty, elegance, genesis), D. H. Cropley and Cropley (2005) created a more detailed scale for the measurement of product creativity: the *Creative Solution Diagnosis Scale* (CSDS).

The Creative Solution Diagnosis Scale (CSDS) combines:

- *principles of creativity* (relevance and effectiveness, novelty, elegance, genesis);
- *criteria* of the principles (possession and use of knowledge, problematization, adding to existing knowledge, going beyond existing knowledge, external elegance, internal elegance, going beyond the immediate problem); and
- *indicators* of the presence of the criteria (e.g., diagnosis, prescription, redefinition, reconstruction, convincingness, completeness, germinality, seminality).

TABLE 4.3 Criteria of Creativity in a Solution

Kind of criterion	Level of creativity	Kind of creativity
External	• differs from what already exists • leads to surprise • is generalizable • is seminal • is germinal	• relevant • valuable • effective • useful
Internal	• generates many ideas • leads to substantial reformulation of ideas • opens up new principles	• logical • elegant • understandable • well-crafted • harmonious • complex

The original 30-item scale (Table 4.4) was developed to facilitate the assessment (or *diagnosis*) both of the *amount* of creativity and the *kind* of creativity of products, including but not limited to artifacts, systems, processes, and services. The scale expands on the four basic criteria of creativity with *indicators* that represent *observable characteristics* of a creative product. It is intended as a general, diagnostic instrument for engineers to use in the evaluation products (and product concepts) and as a tool to drive the enhancement of creativity in those products.

The observable characteristics represent the operationalization of the abstract criteria of creativity. In other words, whereas novelty is an abstract concept that we think characterizes a creative product, *diagnosis, prescription, prognosis,* and so on can be observed in the product and *quantified* by a judge.

Subsequent research (D. H. Cropley & Cropley, 2008) proposed a refinement of the original CSDS with the addition of an intermediate layer of descriptors that help to link the four criteria to the 30 indicators. These descriptors provide a better operationalization of novelty and elegance (Table 4.5). As a result, the CSDS that was proposed as a measure of product creativity could be said to consist of a total of seven factors, or dimensions, each of which was made measurable through observable indicators. While this is a positive step for measuring product creativity in engineering, it remains a theoretical construct. To be convinced that the CSDS measures what we claim, that it does so accurately and consistently, and to show that it can be used by people other than deep experts requires empirical evidence and statistical analysis.

Recent empirical research with this scale (D. H. Cropley & Kaufman, 2012, 2013; D. H. Cropley, Kaufman, & Cropley, 2011; Kaufman, et al., 2013) has demonstrated that the CSDS can be used by engineers (both students and professionals) and nonengineers with high reliability to rate the creativity of manufactured objects. Analysis of the data collected in these empirical studies did suggest, however, a modification to the original structure of criteria and indicators. Factor analysis of two different datasets using the scale shown in Table 4.5 suggested that raters differentiate between only five *factors* of functional, or product, creativity (not the seven shown). These factors are broadly similar to the original four *criteria* (effectiveness, novelty, elegance, and genesis) with the exception that novelty is better represented by two distinct factors: *problematization* and *propulsion*. In addition, three of the original 30 indicators proved to be unconnected to any of the factors/criteria. Perhaps not surprisingly, the three indicators that were eliminated were those that characterized *existing knowledge*. This revised structure is shown in Table 4.6.

TABLE 4.4 The Original 30-Item Creative Solution Diagnosis Scale (CSDS)

Criterion of creativity	Indicator
Relevance & Effectiveness	CORRECTNESS (the solution accurately reflects conventional knowledge and/or techniques)
	PERFORMANCE (the solution does what it is supposed to do)
	APPROPRIATENESS (the solution fits within task constraints)
	OPERABILITY (the solution is easy to use)
	SAFETY (the solution is safe to use)
	DURABILITY (the solution is reasonably strong)
Novelty	DIAGNOSIS (the solution draws attention to shortcomings in other existing solutions)
	PRESCRIPTION (the solution shows how existing solutions could be improved)
	PROGNOSIS (the solution helps the beholder to anticipate likely effects of changes)
	REPLICATION (the solution uses existing knowledge to generate novelty)
	COMBINATION (the solution makes use of new mixture[s] of existing elements)
	INCREMENTATION (the solution extends the known in an existing direction)
	REDIRECTION (the solution shows how to extend the known in a new direction)
	RECONSTRUCTION (the solution shows that an approach previously abandoned is still useful)
	REINITIATION (the solution indicates a radically new approach)
	REDEFINITION (the solution helps the beholder see new & different ways of using the solution)
	GENERATION (the solution offers a fundamentally new perspective on possible solutions)
Elegance	RECOGNITION (the beholder sees at once that the solution "makes sense")
	CONVINCINGNESS (the beholder sees the solution as skillfully executed, well-finished)
	PLEASINGNESS (the beholder finds the solution neat, well done)
	COMPLETENESS (the solution is well worked out and "rounded")
	GRACEFULNESS (the solution is well proportioned, nicely formed)
	HARMONIOUSNESS (the elements of the solution fit together in a consistent way)
	SUSTAINABILITY (the solution is environmentally friendly)

(*Continued*)

TABLE 4.4 (Continued)

Criterion of creativity	Indicator
Genesis	FOUNDATIONALITY (the solution suggests a novel basis for further work)
	TRANSFERABILITY (the solution offers ideas for solving apparently unrelated problems)
	GERMINALITY (the solution suggests new ways of looking at existing problems)
	SEMINALITY (the solution draws attention to previously unnoticed problems)
	VISION (the solution suggests new norms for judging other solutions—existing or new)
	PATHFINDING (the solution opens up a new conceptualization of the issues)

Products are evaluated on the CSDS by asking raters to apply each of the 27 indicators using a five-point Likert scale, with values ranging from "not at all" through to "very much." In other words, a given product is evaluated for "performance," for example, by considering the extent to which "the solution does what it is supposed to do" on the five-point Likert scale. These qualitative ratings are converted into numerical scores that then directly quantify the creativity of the product. Although various options exist for exactly how these scores are quantified and interpreted, let us say, for argument's sake, that products are now given a score out of 20 on each factor, for an overall score of 100. A maximum score of 100 then means that the product is highly creative, whereas a score of 0 means that it is not creative at all. The research cited also shows that engineers of varying levels of expertise can use the CSDS to make accurate and consistent judgments of product creativity.

Now, if I were to tell you that your product lacked creativity, it would be possible to explain what this means in much more useful (and quantitative) detail. Imagine two hypothetical products (or design concepts) that we are considering: Product A and Product B. Each scores 70/100 on the CSDS (see Table 4.7).

This not only tells us something about the products and their overall creativity, but also allows us to diagnose what is contributing to that creativity and, consequently, what we might do to improve it.

TABLE 4.5 The Original 30-Item Creative Solution Diagnosis Scale (CSDS) Hierarchy

Product creativity						
	Novelty			Elegance		
Relevance & effectiveness	**Problematization**	**Existing knowledge**	**New knowledge**	**External**	**Internal**	**Genesis**
Correctness	Diagnosis	Replication	Redirection	Recognition	Completeness	Foundationality
Performance	Prescription	Combination	Reconstruction	Convincingness	Gracefulness	Transferability
Appropriateness	Prognosis	Incrementation	Reinitiation	Pleasingness	Harmoniousness	Germinality
Operability			Redefinition		Sustainability	Seminality
Safety			Generation			Vision
Durability						Pathfinding

TABLE 4.6 The Revised 27-Item CSDS Factor Structure

Product creativity				
Relevance & effectiveness	**Problematization**	**Propulsion**	**Elegance**	**Genesis**
Performance	Prescription	Redefinition	Pleasingness	Vision
Appropriateness	Prognosis	Reinitiation	Completeness	Transferability
Correctness	Diagnosis	Generation	Sustainability	Seminality
Operability		Redirection	Gracefulness	Pathfinding
Durability		Reconstruction	Convincingness	Germinality
Safety			Harmoniousness	Foundationality
			Recognition	

TABLE 4.7 Hypothetical Product Creativity Scores

Creativity factor	Product A	Product B
Relevance & Effectiveness	18	15
Problematization	12	15
Propulsion	18	10
Elegance	10	20
Genesis	12	10
Total	**70**	**70**

To do this, we would turn to the indicators, to drill down into each factor score to see where weaknesses, if any, lie. Product A, for example, scored well for effectiveness (18/20). However, there is some room for improvement, and if the indicators are examined, it might transpire that the design was assessed as slightly deficient in *performance*. This knowledge can then inform the design process. Product B, by comparison, scored only 10/20 for *propulsion*. An examination of the indicators might show that it was deficient, for example, in *generation*, in that it did not *offer a fundamentally new perspective on possible solutions*. This also then provides useful information that can be incorporated into the design process in an effort to improve product creativity. Our detailed understanding of product creativity can now feed directly into the engineering design process.

TABLE 4.8 *Product* and the Phases of Problem Solving

		Invention					Exploitation	
	Phase	**Preparation**	**Activation**	**Generation**	**Illumination**	**Verification**	**Communication**	**Validation**
Dimension	**Poles**	Knowledge, problem recognition	Problem definition, refinement	Many candidate solutions	A few promising solutions	A single optimal solution	A working prototype	A successful "product"
Product Phase output	Routine vs. Creative	Routine	Creative	Creative	Creative	Routine	Routine	Routine

INDUSTRIAL DESIGN AND ENGINEERING

Before I close this examination of *Product*, I want to discuss one more related issue. Engineering design, in the sense I have been using it, is not the only discipline that is concerned with technological solutions to problems. A closely related domain is industrial design. You may have heard the phrase "form follows function" (or specifically, "form ever follows function") (Sullivan, 1896); this is another way of describing the top-down approach to design that we discussed in an earlier chapter. What we need to be aware of, as engineers, is that related domains, and in particular industrial design, approach the same design challenges in different but highly complementary ways.

If engineers work, in effect, from the inside out, then industrial designers work from the outside in. Engineers start with a functional description of the system or problem, while industrial designers start by considering the form of the product—its shape and style, for example. The obvious connections between the two disciplines and creativity are *novelty and elegance*. This is not to say that industrial designers do not care about function and *effectiveness*. Rather, it is a question of emphasis. There are several lessons for engineering creativity in this. One is that we should acknowledge and make use of the expertise of industrial designers in rating product creativity, especially with regard to novelty and elegance. Another is that we should use the skills of industrial designers when our products are deficient against the factors of novelty and elegance.

SUMMARY

To close the discussion of the *Product*, I summarize (in Table 4.8) the relationship of the *Phases* of creative/engineering problem solving (from the Extended Phase Model, Table 3.3) and the *Product*. In other words, in the same way that the *character* of each phase oscillates between *convergent* and *divergent*, as engineering problem solving unfolds, the particular nature of the *Product* of each phase changes. Not surprisingly, the Product that results from the *Generation* phase is *creative* (i.e., emphasizing novelty and divergence), whereas the Product that results from the *Verification* phase is routine in the sense that convergence (selecting the best solution) is the goal. Readers will see that this pattern of oscillating emphasis recurs as we examine Process, Person, and Press, and is a key to understanding creativity in engineering.

CHAPTER

5

Process: Generating Creative Ideas

"I never made one of my discoveries through the process of rational thinking." **Albert Einstein, 1879–1955, Theoretical Physicist**

The first thing needed for creativity is the *generation* of novelty. Although this may sometimes occur serendipitously—through luck, intuition, or simply because of letting ideas flow unchecked—it usually results from *processes* that systematically generate novelty. In the context of engineering creativity, this is especially the case. We are not particularly interested in accidental creativity, but in the way that creativity occurs as part of a deliberate problem-solving endeavor. These intentional processes can loosely be labeled as belonging to the family of *divergent thinking*. Such thinking involves *special processes* and *strategies for processing information*. Among the processes and strategies that I will explore in this chapter are linking remote associates, constructing unusual categories, building broad networks, assimilating rather than accommodating, and using creativity-facilitating cognitive styles. In order to produce relevant and effective ideas (and not just large numbers of surprising ones), such thinking has to be steered by metacognitive *heuristics*—rules of thumb—that filter out in advance unpromising lines of attack or help to recognize possible solutions. The good news for engineering creativity is that these processes and heuristics are all things that can be learned and improved.

Although it will seem counterintuitive in a discussion of creativity, we also know that convergent thinking—i.e., analysis—is an integral part of the creative problem-solving process. Because we are interested in understanding not only what can *help* creativity across the phases of the engineering design activity, but also what can *hinder* it, we need to study both divergent thinking *and* convergent thinking. We need to know not only how to engage in these cognitive processes, but also when and why we use each one.

UNSYSTEMATIC CREATIVITY

To set the scene for our exploration of the *deliberate* use of divergent and convergent thinking processes, I will first discuss what we are *not* so interested in—what we could call unsystematic creativity. In fact, we will see that there are some aspects of this that are useful to understanding the systematic processes.

Effortless Creativity

History seems to be littered with examples of apparently effortless creativity—that is, the effortless generation of effective, novel solutions to problems. Charles Goodyear is said to have discovered how to vulcanize rubber—paving the way for hard-wearing and efficient car tires—by accident. He supposedly unintentionally dropped a mixture of rubber and sulfur onto a hot stove. Alexander Fleming was investigating the antibacterial properties of nasal mucus when he noticed that fungus was killing his samples of staphylococci—leading to the discovery of the antibiotic penicillin. Henri Becquerel left an undeveloped photographic plate, a metal object (a Maltese cross), and uranium salts together in a desk drawer and noticed an intense image of the metal cross on the plate when it was subsequently developed—the mystery was explained by naturally occurring radiation in the uranium salts. Alfred Nobel is said to have hit upon a method for combining nitrocellulose and nitroglycerine—to make a more stable, but still powerful, *blasting gel*—after dressing a cut on his finger with the substance collodion. He reasoned that the collodion might be able to bond the two reluctant substances in the same manner that it bonded to his skin to form a protective layer over the cut. Louis Pasteur's discovery of a vaccine for chicken cholera, which led to the concept of attenuated bacteria serving as the basis for vaccinations, resulted from the fact that both he and his assistant went on vacation and unintentionally *ruined* a batch of bacteria that was to be injected in chickens as part of their search for a vaccine.

Although these cases all appear to be the result of luck, chance, or good fortune, you have probably noticed that there is a certain pattern to each. The most striking is that these discoveries were *not* made by random people. These stories *would be* remarkable if, for example, we were told that little Johnny, a five-year-old orphan, had serendipitously hit on the antibiotic properties of mold spores, or if young Mary, a six-year-old schoolgirl, had stumbled on the formula for blasting gel during show-and-tell. Nevertheless, what mechanisms are at work when these cases of *apparently* effortless creativity occur?

Blind Combinations

The prevalence of these examples of apparently effortless creativity has led to a considerable effort among creativity researchers to understand the mechanisms involved. Indeed, some see this as possibly the primary way that effective novelty is achieved—to stumble upon it more or less by accident. Although not overtly advocating blind guessing, Sir Harold Kroto (who won the 1996 Nobel Prize for Chemistry) drew attention to the importance of being *open* to the unexpected: "If it interests you ... explore it, because something unexpected often turns up, just when you least expect it" (Frängsmyr, 1997). Csikszentmihalyi (1996), in his *flow* approach to creativity, also seems to be an advocate of unfettered idea production, while Sternberg and Davidson (1999, p. 68) argue that novel ideas evolve through what they call "haphazard recombinations." According to the *evolutionary* view of creativity (Campbell, 1960), the process of *blind variation* (p. 380) generates novelty, while *selective retention* of effective *hits* yields creativity. Like the examples of lucky discoveries, these models hold clues that suggest the creativity is not as random, or serendipitous, as might be thought.

Simonton (1988a) refined this approach with what he called the *chance configuration* model. He concluded—using the language of the Phases I introduced in Chapter 3—that the generation of effective novelty starts with the acquisition of a large number of *mental elements* (i.e., pieces of information, memories, ideas, or concepts). These are gathered in the phase of *Preparation* and organized to some extent in the phase of *Activation* (see, for example, Figure 3.2). Unfettered associations are then made, more or less randomly or blindly (in the phase of *Generation*), until the chance occurrence of an effective combination that solves the problem in question—a *configuration*. Simonton (1999) equated this *blind-variation-and-survival-of-whatever-proves-effective* approach to creativity with Darwin's position on the origin of species, making the link explicit by titling his 1999 book *Origins of Genius*. Dasgupta (2004, p. 403) pointed out that other writers use the term *evolutionary epistemology* to describe this model of creativity.

This chance approach to creativity has also been explored in nontechnological domains. In art, for example, a *chance-configuration* model might involve splashing paint on canvas in the hope that an acclaimed work would emerge (of course, engineers might argue that this *does* seem to be what happens!). Similarly, in music, the approach implies that merely stringing notes together in the hope of achieving something worth listening to will result in effective novelty. In fact, such forms of music *have* existed for some time. *Aleatoric music* was first identified by Werner Meyer-Eppler at the Darmstadt Summer School in the early 1950s, and has been used by such notable composers as Karlheinz Stockhausen. This form of music either leaves some component of the composition to chance or leaves elements of the performance to the decision of the performer. Once again, although this might seem to be entirely random, it shares an obvious link with the activities of Pasteur, Goodyear, Becquerel, and others. The explanation is the same reason why my 13-year-old son, pounding on a piano keyboard with a baseball bat, is not an acclaimed composer of the Darmstadt School tradition. It is also why improvisation in jazz music, where combinations may appear to be random, is actually not blind variation at all but is highly disciplined activity.

What these examples appear to tell us—a fact that is highly significant to understanding systematic idea generation—is that the randomness of some creativity is, in fact, just a symptom of a deeper and much more structured and nonrandom process. If this were *not* the case, then we probably would never have heard of Edison, Franklin, Newton, Einstein, and hundreds of other acclaimed geniuses.

Luck

Luck has also been put forward as an explanation for the production of effective novelty. A. J. Cropley (1992a) cited several well-known writers, spanning from different cultures and with different scientific backgrounds, who noted a significant role of luck in creativity. Ernst Mach—the German physicist best known for his work on shock waves and for defining the *Mach* number—referred to "the part which accidental circumstances play in the development of inventions and discoveries" (Mach, 1896, p. 163). Etienne Souriau—the French philosopher of aesthetics—concluded that the basis of creativity is chance (Souriau, 1881). Somewhat in contrast, Alexander Bain—the Scottish philosopher and educator—acknowledged the importance of hard work in creativity but saw this work as "energy put forth ... on the *chance* of making lucky hits" (Bain, 1868, p. 196).

Austin (1978) identified four kinds of *luck* leading to creativity:

- *blind chance* (the individual creator plays no role except that of being in the right place at the right time;
- *serendipity* (the creator stumbles upon something novel and effective when not actually looking for it);

- the *luck of the diligent* (a hard-working person finds a solution in an unexpected setting for something that is being sought). Diaz de Chumaceiro (1999, p. 228) called this *pseudo-serendipity* since in genuine serendipity the person would *not be* looking for what was found;
- *self-induced luck* (special qualifications of a person—such as knowledge, close attention to detail, or willingness to work long hours—create the circumstances for a *lucky* breakthrough).

Case studies, not least those examples we have already considered, suggest that genuinely creative people experience a combination of all four kinds of luck. This raises a question I have alluded to in previous sections: is it a matter of luck at all? The *luck of the diligent* and *self-induced luck* clearly involve elements of deliberate, targeted hard work and specialist knowledge, neither of which occurs by chance. Even blind chance and serendipity seem to occur only because the right groundwork has been laid. More importantly, that groundwork is not just convergent preparation and the acquisition of specialist knowledge, but also aspects of personality (which I will discuss in Chapter 6) and the climate (discussed in Chapter 7). Finally, and most importantly for this chapter, that groundwork is systematic divergent thinking processes.

Intuition

A final word on aspects of creative problem solving that appear to be effortless and unsystematic in nature. In the stage model that we discussed in Chapter 3, Wallas (1926) identified a stage of *Incubation* (see Figure 3.1) during which ideas seem to churn and work in a person's head without the person being aware of them, until—apparently out of the blue—an answer pops up. This is the prototypical definition of intuition: a process of fermentation until an idea is suddenly there, even seeming to come from nowhere. The eminent French mathematician, Jacques Hadamard, who is famous for, among other things, the Hadamard matrices, wrote about his own creativity (Hadamard, 1945), arguing that he did not think in words or construct categories (i.e., he argued against a strict cognitive approach to creativity), but depended more on *intuitions* that seemed to pop up at just the right moment. Richards (2010) explains, however, that intuition—"arriving at the solution of a problem without reasoning toward it" (Damasio, 1994, p. 188)—is heavily connected to knowledge, cognitive processes, memory, and the like, even though the manner in which this is utilized may not be hidden from us. In other words, like the other forms of effortless creativity discussed previously, intuition may well *not be* the product of blind or haphazard events at all, but of experience, learning, and knowledge.

SYSTEMATIC PRODUCTION OF NOVELTY

The idea that creativity comes from the effortless, unfettered production of ideas is attractive. It would be nice if it was the case that for engineers "... creative thinking consisted of just letting their minds waft about dreamily, waiting for the muse to strike them" (Kawenski, 1991). Sadly, as I have indicated, even effortless creativity sits on a foundation of expert knowledge, hard work, analytical skills, and an ability to direct these to creative problem solving through the use of appropriate divergent cognitive processes. There is no *free lunch* when it comes to creativity. However, as I indicated at the beginning of this chapter, there is one piece of good news: the necessary skills in divergent thinking are not mysterious and esoteric. We know what cognitive processes are involved in generating novelty, and these can be learned, practised, and improved. The first thing that we need, in order to be proficient at divergent thinking, as a complement to the convergent thinking that we naturally develop as engineers, is an understanding of how novel ideas come into existence.

Generating Variability

Runco (2010) uses the term *ideation* to explain, in more holistic terms, how divergent thinking is part of the larger activity of creative problem solving, and includes decision making, judgment, and evaluation. This ties in closely with arguments I developed in earlier chapters about engineering problem solving. Runco also stresses that a common fallacy is that divergent thinking IS creativity. The cognitive processes that we employ in our problem-solving endeavors—let us loosely call these processes *thinking*—use and manipulate ideas to produce further ideas. As we have already seen, this thinking includes, among other things,

- *selecting* from among the masses of information available at any moment (perception is not simply a passive acceptance of everything that impinges on the senses or is already stored in the mind);
- *relating* elements of information to each other;
- *combining* elements of new and old information;
- *evaluating* these combinations;
- *selectively retaining* combinations (which may then function as new information, returning the process to the phase of relating elements of information);
- *communicating* the results to others.

The phases that I described in Chapter 3 mirror these different kinds of thinking. From the point of view of novelty—that core characteristic of creative products—these processes are most interesting when they lead to *generation of variability*. The cognitive processes—the kinds of thinking—that are likely to give rise to this vital *variability* are

- adapting general concepts to specific circumstances in *unexpected* ways;
- recognizing opportunities to *transfer* specific knowledge to *dissimilar* settings;
- building *broad categories;*
- *crossing boundaries;*
- working with *uncertainty*.

Although there is insufficient space to discuss this in detail here, some of these points could be described in terms of educational theory (for example, Biggs & Tang, 2011) as an extension of *conventional* deep understanding. For example, if the deepest level of understanding—*extended abstract* understanding—is characterized by an ability to generalize knowledge from one domain or situation to another, then an ability to do this between domains that would *not* normally be expected to have an association is unconventional deep understanding that leads to the generation of variability.

Other cognitive processes—*thinking skills*—that generate variability include

- *uniting disparate ideas* by putting them into a common context;
- *imagining almost anything*, at least as a theoretical possibility;
- *fantasizing* and thus enriching one's own thinking;
- *using humor* to enrich to thinking.

All of these cognitive processes are *favorable to generation of variability*, the first step in production of effective novelty. Critically, they are not random or isolated processes, but constituent elements of broader systematic patterns of thinking (thinking *strategies*) that generate variability. What then is special about such thinking?

Divergent Thinking

Divergent thinking, as we have already seen, involves producing *multiple* answers through processes like *shifting perspective* on existing information (seeing it in a new way) or *transforming* it, for instance, through unexpected combinations of elements usually not regarded as belonging together. The answers arrived at via divergent thinking may never have existed before. Sometimes this is true merely in the experience of the particular person or the particular setting, but it may involve what Boden (1995) called "radical originality." The characteristics of the divergent thinking are listed in Table 5.1.

TABLE 5.1 Characteristics of Divergent Thinking

Typical processes	Typical results
• being unconventional • seeing the known in a new light • combining the disparate • producing multiple answers • shifting perspective • transforming the known • seeing new possibilities	• alternative or multiple solutions • deviation from the usual • a surprising answer • new lines of attack or ways of doing things • opening up exciting or risky possibilities

Boden (1994b) contrasted divergent and convergent by concluding that divergent thinking produces ideas that could not have been produced without a leap in thinking. Examples of cognitive processes that produce such leaps are

- retrieving a broader than usual range of facts from existing knowledge;
- building unusual chains of associations;
- synthesizing apparently unrelated elements of information;
- transforming information in unlikely ways;
- shifting perspective so as to see ideas in a new light;
- constructing unexpected analogies.

The opposite processes to these, such as recognizing the familiar, retaining what already exists, or reapplying the tried and trusted, do not generate variability but, on the contrary, orthodoxy—i.e., convergence. As I have stressed previously, this convergence is not a bad thing. It is a vital part of the problem-solving processes. However, the key is that one must apply both divergent and convergent thinking *at the appropriate point in the problem-solving process.*

Other Concepts of Novelty-Generating Thinking

Our focus remains divergent thinking—what it is, how it happens, when it is necessary. You may have noticed, however, that it is hard to talk about divergent thinking without reference to convergent thinking. Although I will deal with some other specific features of convergent thinking later in this chapter, I want to discuss some other *kinds of thinking* that are associated with the generation of variability and novelty. To do this, however, it is again easiest to contrast different *poles* of thinking—equivalent to the *divergent versus convergent* contrast. Examples include *open versus closed* thinking, *primary process* (not bound by strict adherence to reality) *versus secondary process* (conscious, rational, logical, and oriented to reality) thinking, or *reproductive versus productive* thinking.

According to psychoanalytic theory (A. J. Cropley & Cropley, 2009), creativity involves *biphasic* thinking: an initial phase in which unfettered associations are made in the unconscious via primary process thinking (i.e., novelty is generated) followed by a phase in which these associations are admitted into consciousness in the *realistic* form of secondary process thinking (i.e., they are explored). Thus, creativity involves *tertiary process* thinking in which primary and secondary process thinking are combined to yield effective novelty. This is highly reminiscent of the *convergent-divergent-convergent* pattern that I discussed in an earlier chapter, and also emphasizes the importance of understanding not only *how* to think divergently (or convergently) but also *when and why* we do so.

Rothenberg (1983) introduced the idea of *janusian* thinking, naming it after the Roman god Janus, who had two faces and thus could look in two directions at the same time so that he had the ability to deal simultaneously with pieces of information that would not normally be processed together. Rothenburg also introduced the term *homospatial* thinking. This kind of thinking is able to unite apparently conflicting or mutually exclusive ideas by bringing them into the same cognitive space, thus producing novelty. Presumably, homospatial thinking is the opposite of *heterospatial* thinking—which keeps ideas locked away from each other in separate spaces.

Edward de Bono (1993) made an interesting and popular contribution to the discussion of creative thinking. Initially, he emphasized *lateral* thinking. Unlike *conventional* thinking, which is strictly sequential in nature and follows a set of logical steps, lateral thinking involves detours or sidesteps. Marginal characteristics of a concept or object that are not central to its usual definition are emphasized and brought into juxtaposition with similar characteristics of other concepts and objects to yield unexpected associations. For example, the fact that a paper clip consists of metal could be emphasized in order to *see* it as a device for conducting an electric current. A matchbox can be regarded as a nonconductor with movable parts. Seeing these two objects in this way would make it possible to utilize them as the basic materials for the construction of an electric switch.

de Bono extended his model to distinguish between *rock logic* and *water logic*. Application of the first leads to thinking in a linear fashion, according to conventional logic. Decisions on what the next step should be are based on correctness, and this is decided in terms of absolute norms such as *truth*, *justice*, or *beauty*, which change only slowly. Water logic, by contrast, allows ideas to flow together from many directions according to the *natural* pathways in the material in question, just as water flows along cracks and depressions in the ground where there is no resistance, and forms pools and eventually rivers (creative ideas). According to de Bono, the process of flowing together has its own

TABLE 5.2 Examples of Thinking Leading to Production of Variability/Orthodoxy

Thinking that produces variability	Thinking that produces orthodoxy
Divergent thinking	Convergent thinking
Open thinking	Closed thinking
Homospatial thinking	Heterospatial thinking
Primary process thinking	Secondary process thinking
Productive thinking	Reproductive thinking
Lateral thinking	Sequential thinking
Thinking based on *water logic*	Thinking based on *rock logic*

energy: in the case of water in nature, this is gravity; in the case of ideas, it is *creative* or *constructive* psychological energy.

Table 5.2 lists examples of contrasting kinds of thinking favorable on the one hand for production of variability, on the other of orthodoxy. Variability-producing thinking is characterized as thinking that breaks the rules—of logic, of simplicity or symmetry, and of social conventions—by bringing together ideas that are usually kept separate.

THINKING TACTICS THAT GENERATE VARIABILITY

Having explored the underpinning concepts of the generation of variability, and contrasted these to convergent thinking and orthodoxy, I want to discuss particular processes (or *thinking tactics*) that are especially favorable for the generation of variability. Notice that I am avoiding any mention of specific tools in this section. You might expect to see a primer on *how to brainstorm* or the use of other tools and techniques here, but that is something I am avoiding. The reason is twofold. First, consistent with a pedagogical approach that seeks to develop deep understanding, I am attempting to explain the underlying concepts. This is the educational equivalent of "give a person a fish, and you feed him for a day; teach a person to fish, and you feed him for a lifetime." By understanding the underpinning concepts, you are freed from dependence on specific tools and techniques (many of which have limitations).

Constructing *Remote* Associates

The idea that cognition largely involves seeing connections between bits of information has already been mentioned (e.g., recognizing patterns, synthesizing, uniting, relating, combining). This involves *making associations*. Mednick (1962) argued that what is necessary for producing novelty is that such associations go beyond the traditional, conventional, or orthodox, and are *remote*. He described the formation of remote associates and their connection to novelty production in the following way: In the course of experience, people learn a number of possible responses to any given stimulus. Those responses most frequently linked with a particular stimulus whenever it was encountered in the past have a high probability of being selected as appropriate when the stimulus is encountered once again (i.e., they are common). Responses seldom paired with the stimulus in the past have a low probability of being chosen (i.e., they are uncommon or *remote*). When the stimulus recurs in a new situation, most people select a common response (they have often made it to this stimulus in the past). This means that people's reactions to familiar stimuli have the advantage of being consistent but are repetitious: they interpret stimuli in the same way over and over again. In other words, they do not create novelty.

Chicken is a common associate to the stimulus word *Egg*, since these two ideas often occur together. A person with a high preference for common associates might associate *Green* with *Grass*, whereas another who preferred remote associates might create novelty by associating *Green* with *Eggs and Ham*.

In engineering, we see both the benefits and the penalties of different kinds of associations when we examine the functions of common objects. For example, a paper clip's conventional association is with the function *clip paper*. The name of the object itself reinforces this orthodox association. If engineers are asked to devise alternative uses for a paper clip, they must first overcome their *functional fixedness*—that tendency to associate objects with their *customary* function. These traditional associations do have, however, certain advantages to engineers. Standardized electronic components, for example, resistors and capacitors with known values, are extremely useful in speeding up design and manufacturing processes. The penalty is that orthodox associates can become so ingrained that it is difficult to make the transition to remote associates in situations in which novelty is required.

The idea of remote associates applies both in a bottom-up sense, as illustrated by the paper clip example, and in a top-down sense. Whereas the bottom-up remote associate is a matter of finding *other, unconventional functions* for the product (the paper clip), the top-down equivalent would start with a function and seek other, unconventional products that satisfy that function. Thus, the function *clip paper* is normally satisfied by a *paper clip* (the conventional top-down association) but *could be* satisfied by a *staple* (still conventional), a *clothes peg* (somewhat unconventional), or *tree sap* (highly unconventional). Note that we are inherently constrained by the need to find *effective* remote associates, thus ruling out *strand of hair* or *banana* as options, as remote as they might be.

Building Unusual Categories

Bruner's (1964) approach went somewhat further than remote associates, and may remind engineers of concepts found in the world of object-oriented software. Events, concepts, and ideas that repeatedly occur together are recognized as belonging to the same *category*. Properties that are common to a number of individual members of a category are seen as typical for members of that category: for instance, all members of the category of "weapon" are useful for fighting, those of "food" for eating. These generalized, abstract properties are the basis of group, or category, membership.

In the language of psychology, as distinct from the world of object-oriented methods, the process of assigning events, ideas, and concepts to yield categories is referred to as *coding*. A new event (I will use *event* to mean ideas and concepts as well) is seen to have distinctive and recognizable properties. When these are judged to match the definitive properties of a particular existing category (i.e., *pattern recognition* occurs), the new event is encoded into that category. The new event is then treated as though it had all characteristics of the category, even characteristics that have not been directly observed. Thus, coding is a special form of "going beyond the information given" (Bruner, 1964). Furthermore, once an object has been coded into a category, it is difficult to see it as anything other than a member of this category. A simple example is the difficulty people have in seeing a hammer as anything other than a device for driving in nails. It could also be a weight, a hook, or a can opener, but encoding it as "tool" shuts out most other interpretations.

This may seem no different from the discussion of remote associates; however, it differs in one key regard. The addition of categories is an overlay on the relationship between object (e.g., paper clip or hammer) and function (e.g., clip paper or insert nail). The recognition of a category, e.g., that the hammer is an example of a *tool*, can act to *reinforce* the conventional association (insert nail) and make it even harder to find remote associates.

Coding is very useful in everyday life. Without it, every situation would have to be dealt with anew as though the person concerned had had no prior experience with the external world. It makes it possible to deal with the familiar swiftly and efficiently by activating the category to which a well-known stimulus belongs and thus knowing what it is and what to do about it. It also makes it possible to deal with the unfamiliar: once something new has been assigned to a familiar category, it has a meaning and the person knows how to deal with it. This gives life consistency and predictability and engenders a high level of confidence in one's own behavior. Consistency and predictability are, however, characteristics of orthodoxy, not variability. When new stimuli are coded into existing categories and the categories remain intact in their existing form, no novelty is generated. Under these circumstances, going beyond the information given simply protects the *status quo*.

To produce novelty, coding needs to go beyond the obvious and dominant properties of a stimulus so that its membership of categories other than the most obvious can be recognized. Coding also needs to be flexible so that a stimulus can be seen to be capable of belonging to more than one category, or of being recoded as the situation demands. The way that a stimulus is coded not only depends on the properties of the stimulus itself, but is also strongly affected by contextual factors. For instance, in a library, rectangular paper objects are likely to be coded as books, while a *hungry* person is likely to code a spherical object about the size of a tennis ball, yellowish red in color, and with indentations like those on a golf ball on its surface as an orange. Contextual factors act like a key that increases a category's accessibility by unlocking and leaving the door to it ajar. Codings usually reflect past experience in a particular context and are thus usually commonplace and lacking in novelty. They may, of course, be sensible and socially acceptable as well as readily available, so that coding into these categories trades off production of novelty for ease of processing and avoidance of "cognitive strain" (Bruner, 1964).

As just described, the context predisposes people to code new events into certain categories. This induces a *set*, i.e., a tendency to see the world in fixed ways. The conventional coding of a wristwatch would be to class it as a device for telling the time. However, it is possible to break sets and code stimuli into unexpected categories such as coding the watch as an object with weight instead of as a timepiece. This recoding of the watch draws attention to previously ignored properties that it possesses, and the person is then in a position to use the watch to solve problems in which a weight is required but time of day is irrelevant, for instance, by using it as a sinker on a fishing rod. Indeed, being consciously aware of categories opens up the possibility of recoding.

Building Broad Networks

Another general approach to thinking that is associated with the generation of variability is the idea of building networks of concepts. Miller (1992), in a study of the creative thinking of French mathematician Henri Poincaré and German-born physicist Albert Einstein, concluded that the essence of novelty generation is "network thinking." Building broad networks involves combining apparently disparate concepts. Miller defined the mechanism of combination as "proper choice of mental image or metaphor." In the same way that coding represents an

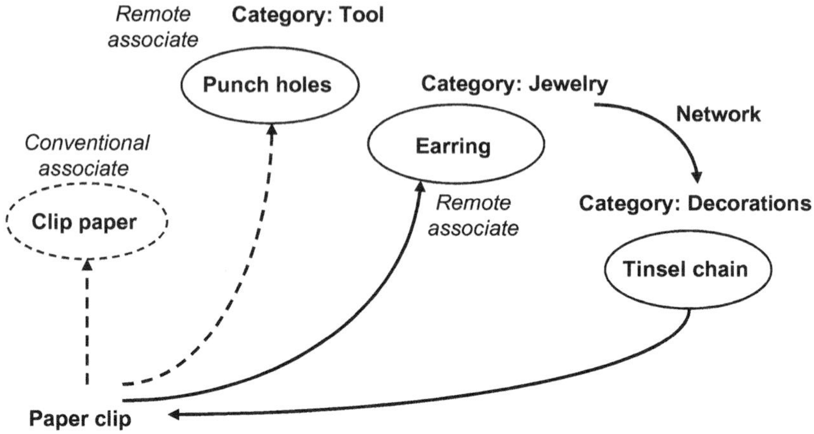

FIGURE 5.1 Remote associates, categories, and networks.

extension of remote associates, the building of networks is an extension of the process of coding. Individual ideas are generated through a process of forming remote associates. These unusual ideas belong to categories, and this can stimulate us to find other members of any given category. *Building broad networks* recognizes that *different categories* may share properties, and that new categories can be developed that yield new solutions (Figure 5.1). A paper clip can be used to clip paper (conventional associate) or to punch holes (remote associate) or as an earring (remote associate). The earring can be coded as jewelry. The category jewelry shares some properties with another possible category: *decorations*. Forming that new category and attempting to populate it leads to a new idea for using a paper clip: making a chain of clips and using that as a form of tinsel to decorate a Christmas tree! The two categories—jewelry and decoration—overlap and can thus be combined to form a system or *network*. In this case, the network might be characterized as *aesthetic uses of a paper clip*. For a relevant discussion of networks in thinking, see Anderson (1976).

The concept of networks of interlocking categories was stated somewhat differently by Köstler (1964), who saw knowledge as existing in *matrices*. Information processing usually involves linking elements from *within* the same matrix and thus produces no novelty. By contrast, when two matrices are linked via "bisociation," or three are "trisociated," variability is produced.

Accommodating Rather than Assimilating

The last of the *thinking tactics* that I will consider is derived from Martinsen (1995). He identified two *strategic dispositions* in thinking: some people consistently seek to deal with the new by reapplying existing knowledge and tried-and-trusted solution strategies (think of the *routine solutions* and *stasis* that I discussed in Chapter 2), whereas others try to construct new approaches when dealing with new situations. There are thus two contrasting ways of reacting to change, uncertainty, or ambiguity: The one seeks to absorb the change into what is already known (i.e., to make it familiar and preserve what already exists), the other to alter the already known (i.e., to produce novelty). This disposition can be characterized by borrowing from Swiss developmental psychologist Jean Piaget and emphasizing *assimilating* and *accommodating*. In the present context, assimilating involves fitting change in with existing mental structures and thus preserving the *status quo*. We saw, in Figure 2.5, pathway 4, how this leads to *stagnation*. Conversely, *accommodating* is based on recognizing that current structures are not adequate and revising them. Intuitively, assimilating (making change fit the known) is related to production of orthodoxy, whereas accommodating (altering the known in response to change) is related to production of novelty.

However, Piaget himself pointed out that it is the interaction between assimilating and accommodating that produces a creative product. The external world must first be assimilated, for instance, by recognizing that familiar things may appear in a number of diverse situations, or take diverse forms in the different situations. This involves recognizing the familiar, seeing connections, etc., and requires knowledge of the field (i.e., in Piaget's terms, *assimilated experience*). The second step is accommodation: adapting the already known to solve a new problem.

USING CREATIVITY-FACILITATING COGNITIVE STYLES

Cognitive styles give us another general approach to the generation of effective novelty. Cognitive styles are consistent and stable differences between people in the way they obtain information from the world around them, sort, organize and recall information, and cope with demanding situations. They are often stated in the form of bipolar dimensions such as *leveling* versus *sharpening*, *focusing* versus *scanning*, *field-dependence* versus *field-independence*, *preference for wide* versus *narrow categories*, or *seeking cognitive complexity* versus *seeking simplicity*. Some of these styles (e.g., field-dependence, wide categories, or preference for complexity) are favorable to production of novelty. Some well-known bipolar styles are summarized in Table 5.3. It is important to notice that no particular style of thinking is *universally* favorable for

TABLE 5.3 Examples of *Bipolar* Cognitive Styles

	Cognitive style	Result
Leveling	Paying little attention to subtle or minor differences among the elements of a configuration	Perceiving globally Focusing on similarities Homospatial thinking
Sharpening	Exaggerating minor differences and overlooking anything the elements may have in common	Getting bogged down in details Difficulty in seeing "the big picture" Heterospatial thinking
Scanning	Looking at the entire situation Seeing the forest but hardly noticing the trees	Broad grasp of the big picture Broad, ill-defined categories Associating elements with only low similarity Homospatial thinking
Focusing	Homing in on precise details Seeing the individual trees, not the forest	Precise knowledge of details Ignoring similarities Heterospatial thinking
Field-dependence	Interpreting stimuli in terms of the context in which they occur	Omnisociation Homospatial thinking Broad coding, but discourages seeing things in a new way Difficulty in breaking out of the corset
Field-independence	Interpreting stimuli in terms of their individual properties, regardless of the context Failing to take account of the context	Awareness of contrasting or unexpected properties of stimuli Heterospatial thinking
Simplicity seeking	Reducing a complex situation to simple, harmonious, closed terms	Undifferentiated understanding of the "big picture" A harmonious (even if simplified) view of the way things are Ignoring fine differences Broad categories Low tolerance for uncertainty
Complexity seeking	Tolerance, even preference for "messy" situations	Attention to fine detail Sharp differentiation of stimuli High tolerance for uncertainty

production of effective novelty. A style may be favorable in some respects, unfavorable in others. Thus, to take a single example, field dependence favors broad coding (good for production of effective novelty), but makes it difficult to see things separately from their context and thus encourages stereotyped perception (bad for production of effective novelty).

META-COGNITION

In my earlier discussion of *unsystematic* creativity, I pointed out that some authors have argued that creativity results from random processes. I suggested, however, that the apparent effortlessness is merely a symptom of underlying processes. In other words, cognitive processes leading to effective novelty do not proceed by "brute force" (Simon, 1989) through processes of perceiving, blindly associating, and occasionally recognizing, perhaps by luck, that a new combination happens to offer the required solution. Apart from anything else, this would lead to a *combinatorial explosion* involving huge numbers of empty trials and causing *cognitive strain* on the problem solver. Thus, thinking processes must be guided by *procedural and conditional knowledge about how to acquire, organize, or apply knowledge*: heuristics, strategies, hunches, or "rules of thumb" (Rickards, 1999, p. 219), or what is sometimes called "meta-cognition" (Flavell, 1976, p. 232).

Heuristics, in fact, play an important role in engineering, although they often come only with experience. Rechtin and Maier (2000) cataloged a large number, many of which will be familiar to engineering students (e.g., *if it ain't broke, don't fix it*). Other meta-cognitive processes that help to keep the creative problem-solving activity manageable include the ability quickly to discern promising lines of attack, or blind alleys.

Meta-cognition involves the *executive processes* in thinking that allow people to organize and keep track of their own cognitions. In the case of creativity, these include

- redefining plans where necessary;
- monitoring one's own progress;
- changing existing line of attack if necessary;
- being aware of alternative routes;
- recognizing conditions that make a change of approach necessary;
- possessing insight into the costs and benefits associated with the various possible changes; and
- recognizing opportunities.

The ability of people to articulate their own meta-cognitions (i.e., to state in words the results of the processes described here) is also of considerable importance. Articulation permits conscious self-reflection, identification of precise differences between approaches, discussion of progress with other people, and the like.

A striking description of the way in which creative thinking is systematically guided by meta-cognition is found in the writings of Henri Poincaré (2003 [1908]). He regarded production of novelty as a *process of selection* in which knowledge elements from widely separated domains are combined. Poincaré himself possessed vast knowledge of mathematics, and could in principle have produced an endless string of combinations, but somehow homed in on the effective ones. Apparently, nearly all of the theoretically possible combinations are *filtered out*, leaving only the ones that meta-cognition suggests offer prospects of a solution.

Avoiding the *Wrong Approach* Barrier

Olken (1964) argued strongly that creativity is partly a trial-and-error process, thus apparently adopting a blind variation position. According to him, initially possible solutions are imagined as mental images and are matched with the desired end result. If this matching process shows that the solution envisaged in the mental image does not match the desired result, it is abandoned. (Of course, if trial solution and desired end result match, the trial solution is retained.) More complex problems are broken down into small steps or subproblems, and these are solved one at a time using the trial-and-error method mentioned in the previous sentence. However, Olken emphasized that the mental images are not produced randomly, as would happen in a blind combination approach, but are selected from among the large number of possible images on the basis of "hunches," or as Olken points out, what are now called "heuristics" (meta-cognitions). The process is iterative: solving one subproblem provides a new jumping-off point for the next, and the process restarts.

Olken extended his discussion by focusing attention on the problem of avoiding dead-ends. In an examination of the development of innovations such as the triode vacuum tube, the Astron machine (a reactor for producing power by controlled nuclear fusion), or a device for the simultaneous production of multiple photocopies, he identified the "wrong approach barrier." He pointed out how dangerous it is to start along a wrong approach because the difficulty of getting out of the *cul de sac* increases as the researcher invests more and more resources (money, time, ego-commitment, difficulty of disengaging without loss of face, etc.) in the dead-end. In effect, a barrier builds up that blocks finding a productive approach.[1] Thus, the first step in generating effective novelty is to produce variability, to be sure, but to avoid becoming trapped in dead-ends. Olken identified three ways of breaking the barrier: getting a lucky break (see earlier discussion), "letting the work lie fallow" (incubating), or continuing along the fruitless line of attack but gradually "veering around" to find a new direction (what Sternberg called "redirection"—see p. 86). To this, I would add finding the courage and the openness of mind to recognize when one has entered a *cul de sac*.

MEASURING DIVERGENT THINKING

The dominant approach to testing creativity has involved tests of divergent thinking. Such tests emphasize the underlying cognitive processes thought to be involved in production of novelty. The tests are interesting because they represent cognitive traits thought to be connected with creativity in concrete tasks. This means that they offer what is in effect an operational definition of creative thinking, and give hints on how to recognize latent creativity. Research on the reliability (stability of scores) and validity (link between test scores and actual creativity) of the tests provides a degree of empirical support for the view that their representation of creative thought is useful.

Many of the tests are based on Guilford's original concept of divergent thinking that I first introduced in Chapter 1. Tests of divergent thinking typically ask respondents to generate multiple answers to open-ended questions. There are no correct answers; there may be many equally good answers to the same test question, and the answers are sometimes unknown until the people being tested give them. Examples of items from divergent thinking tests are as follows: "Write down as many interesting and unusual uses as you can think of for a tin can." "What would the consequences be if it started raining and never stopped?" "What problems can you imagine in connection with birds nesting in a tree in your garden?" There are also nonverbal tests asking for unusual titles for pictures, for completion of partially completed drawings, or for interpretation of schematic drawings, to give a few examples. The people taking the tests are encouraged to give as many answers as possible (the *fluency* that I discussed in Chapter 3);

[1]This problem is probably most acute for experts, who may have invested a lifetime into a particular approach, so that recognizing this as a dead-end means acknowledging that a lifetime's work has all been in vain. Senior professors and managers are probably most at risk of becoming unable to recognize the wrong-way barriers in their own work.

different types of answers (*flexibility*); to try to produce unusual or unexpected answers (*originality*) and to develop these (*elaboration*).

The best known and most widely used of the tests based on Guilford's concept of divergent thinking are the *Torrance Tests of Creative Thinking* (TTCT). These tests were originally published in 1966 and have since been revised several times (Torrance, 1998). They are described in detail by Kim (2006), and subjected there to an extensive evaluation of their reliability and validity. Considerable effort has also been expended in analyzing test results over the period that the TTCT has been used, with interesting conclusions, for example, about a pattern of declining creativity scores in children (Kim, 2011). The TTCT is most frequently used with children, despite the belief that it is suitable for a wide age range, probably because the test materials are more interesting for children. The test consists of a verbal section "Thinking Creatively with Words," and a nonverbal or figural section, "Thinking Creatively with Pictures," both of them having two forms, A and B. There are seven verbal activities (Asking, Guessing Causes, Guessing Consequences, Product Improvement, Unusual Uses, Unusual Questions, and Just Suppose) and three figural activities (Picture Construction, Picture Completion, and Lines/Circles). The verbal activities yield scores on three dimensions (referred to by Torrance as *mental characteristics*) and now familiar to us: Fluency, Flexibility, and Originality. The nonverbal activities yield scores for five mental characteristics: Fluency, Originality, Elaboration, Abstractness of Titles, and Resistance to Premature Closure. In addition, the figural tests can be scored for 13 *creative strengths* (e.g., Storytelling Articulateness, Synthesis of Incomplete Figures, and Fantasy).

The TTCT test manual reports a median inter-rater reliability derived from a number of studies of the verbal activities of the TTCT of as high as .97, and other research (Sweetland & Keyser, 1991) indicates that the figure is commonly greater than .90 for both parts. According to Treffinger (1985) test–retest reliabilities of the various subdimensions commonly lie between .60 and .70. Mumford et al. (1998) asked judges to use a 5-point rating scale ranging from "low" to "high" to rate answers on a version of Guilford's *Consequences* test (similar to the *Guessing Consequences* subtest of the TTCT) on, among other things, "quality" (in essence, effectiveness), "originality," "complexity," and "realism." After a practice run and a meeting to discuss the basis of ratings, the judges achieved inter-rater reliabilities of .90 for quality, .86 for complexity, and .84 for originality. The figure for realism was somewhat lower at .65. Although the Consequences test is not identical with the TTCT, data from this test are highly suggestive of what might be expected of the TTCT.

Turning to validity, Plucker (1999) analyzed 20-year longitudinal data originally collected by Torrance, and reported long-term predictive validity of the TTCT of about .7. The test scores differentiated well between students who subsequently went on to achieve public acclaim as creative and those who did not. In the study mentioned in the previous paragraph, Mumford et al. (1998) showed that the originality scores of Consequences correlated .49 with the originality of over 1,800 U.S. Army officers of both sexes on simulated problem-solving tasks. These correlations remained almost unchanged when IQ and expertise were partialed out. The test scores for Originality also correlated .36 with real-life leadership achievement as estimated by the officers themselves and .47 with job success as indicated by the level of seniority reached in the organization in question (the army).

Scoring Divergent-Thinking Tests

The most widely applied approach to scoring the tests focuses on three aspects of divergent thinking: *fluency* (quantity of answers), *flexibility* (variability of idea categories in the answers) and *originality* (uncommonness of answers). Fluency requires mere counting of the number of different answers given by a particular individual. Flexibility and originality, however, focus on the quality or style of answers. Flexibility involves the number of separate categories defined by the person's answers, whereas originality assesses the uncommonness of individual answers. Some tests extend scoring by including dimensions such as *elaboration* (complexity and completeness of answers) or *effectiveness* (link to the constraints of the real world). It is worth noting that relationship between fluency, flexibility, and elaboration can be thought of as broadly similar to the relationship between *remote associates*, *categories*, and *broad networks* depicted in Figure 5.1.

To illustrate these dimensions, consider the following example. As a response to the test item *Write down as many uses as you can think of for a tin can*, the four answers *saucepan*, *milk jug*, *kettle*, and *suit of armor for a mouse* would each score one point for *fluency*, yielding a total of *four*. However, there are two basic idea categories in the four answers: *container*, on the one hand (saucepan, milk jug, and kettle); *protective covering* on the other (suit of armor). Thus, the four answers would score two points for *flexibility*. With regard to *originality*, the answers *saucepan*, *milk jug*, and *kettle* are commonplace and would score nothing, whereas *suit of armor for a mouse* is uncommon and would score several points, exactly how many depending on the particular scoring method used (e.g., two points using Torrance's approach [1998], four points according to A. J. Cropley [1967]). The response *suit of armor* is, in addition, more *elaborate* than *kettle* or *saucepan* and would attract additional points if this dimension were applied.

Scoring for *originality* can be somewhat labor-intensive. Torrance (1998) gives lists of common answers for his tests. Common answers receive no points. He also gives lists of fairly uncommon answers that receive some recognition (one point). All other answers that do not appear in the manual are regarded as original and worth two points. However, although this approach makes scoring for originality less tedious, it limits the usefulness of the tests in cultures other than the one where the word lists were compiled. For example, a child in Australia might suggest using a tin can as *the stumps in a game of cricket*.[2] Unsurprisingly, this answer will *not* be found in the American test manual's list of common answers and will therefore receive two points—i.e., it seems to be highly original. However, in Australia this response, especially for a *large* tin can, would likely be quite common.

The whole issue of cultural bias is a very interesting one, not only for creativity testing, but also for other psychometric variables—and not least, intelligence. An approach to dealing with this *cultural relevance* of answers is to incorporate it more formally into the scoring procedures (A. J. Cropley, 1967). This can be done, for example, by calculating the relative frequency of each answer to a given item in a specific group; *saucepan*, for instance, might appear on 30% of the protocols of a particular group of 200 people being tested on Tin Can Uses; *armor for a mouse*, by contrast, on perhaps 0.5%. Answers given by only a few members of the group or only a single person (such as *armor*) are regarded as original as a result of their statistical rarity. Scoring can be adjusted to reflect this in a more differentiated way by assigning different values to answers according to their *relative* frequency/infrequency. For instance, *zero* might be assigned for answers occurring on 15% or more of tests, *one point* for answers on from 7% to 14% of tests, *two points* for 3%–6%, *three points* for 1%–2%, and *four points* for less than 1%. These values correspond approximately to the proportions lying beyond half-a-standard-deviation intervals along the X-axis of a normally distributed trait.

This approach defines originality in the specific context of a particular group via a statistical procedure. The percentages for a given answer *may be quite different* in different groups: in an Australian sample, for instance, the *cricket* answer for the use of a tin can might be given by 20% of children and receive zero points, whereas in an American group, it might occur only rarely and receive four points. The originality score of a particular answer given by a specific child *depends on the answers of the rest of the group*, which makes intuitive sense if the social

[2]Very roughly speaking, *cricket* is to *baseball* as *stumps* are to *home plate*!

definition of creativity is borne in mind (something I will discuss later in the context of *Press*), but is unsatisfactory if we are looking for a universal concept of novelty.

There are a number of limitations and disadvantages in the scoring methods described. They can be laborious, culturally biased (at least across significantly different cultures), context dependent, contingent on the makeup of the group being tested, and based on an assumption that the capacity to produce novelty is normally distributed. They also do not differentiate between answers with regard to effectiveness, and might therefore seem to encourage pseudo-/quasi-creativity (see the discussion in Chapter 4).

A Creativity Quotient?

Although scores for the other dimensions (e.g., flexibility, originality) tend to correlate substantially with fluency, it was demonstrated quite early (A. J. Cropley, 1972) that the various dimensions are *not* the same. Some people taking the tests produce only a small total number of answers, of which many are highly original (low fluency but high originality). Others produce large numbers of answers, of which few or none are original (high fluency, low originality). Still others produce a consistently high proportion of original answers throughout a large number of answers (high on both fluency and originality) or a consistently low level of originality in a small number of answers (low on both). From the point of view of differential diagnosis, this kind of information is very useful, although the problem remains of how to summarize and compare the different possible combinations of high and low fluency, flexibility, and originality.

Snyder and colleagues (2004) suggested dealing with this situation by calculating what they called (p. 415) a "creativity quotient (CQ)," calculated using a procedure derived from information theory.[3] Focusing on fluency and flexibility, they analyzed the real example of two people, one of whom gave 23 uses for an object, the other 19. However, the first person's 23 answers involved seven separate categories (flexibility = 7), whereas the second person's 19 answers were clustered in only two categories (flexibility = 2). The first person obtained a rounded CQ score of 14, and the second a coefficient of 6 (see Table 5.4).

How would a third, hypothetical, example compare to these? Imagine that this person gave only *eight uses* but spread these equally *over four separate categories*. When compared with the scores of the

[3]Many readers will be familiar with Claude Shannon's contributions in this field, and its importance to modern digital communications. Shannon also played an important role at the Bell Labs for many years.

TABLE 5.4 *Creativity Quotient* Scores

Person	Fluency	Flexibility	CQ
A	23	7	**14**
B	19	2	**6**
C	8	4	**6**
D	8	4	**5.5**

second person above (CQ = 6), the higher flexibility (four categories as against two) might be expected to balance out the lower frequency (only 8 uses suggested as against 19). Indeed, the Snyder et al. procedure yields a rounded CQ of 6 for this constellation too (i.e., the same CQ score as the second person—see Table 5.4), a result of greater variability in the smaller number of suggestions. A fourth person who also gave 8 responses in four categories, but with 5 exemplars of one category and only one in each of the remaining three would obtain a CQ of approximately 5.5, the massing of uses in one category (as against spreading them evenly over categories) reducing the CQ.

The Creativity Quotient approach opens up the possibility of comparing people with different *combinations* of fluency, flexibility, and originality. In Table 5.4, Person B, for example, did not lack in the capacity to generate ideas (a trait favorable for creativity) but found it difficult to branch out to a broader set of categories (unfavorable). To increase overall creativity, one might counsel Person B to try to look for a greater range of unusual categories. Using our concepts of top-down and bottom-up design (see Chapters 2 and 3), Person B could achieve this by expressing the dominant category of ideas as a verb–noun pair (e.g., hold liquid) and then looking for *other categories* in terms of verb–noun pairs, rather than focusing on other *solutions*. In the same example, Person C showed a greater ability to generate a diverse set of ideas (a trait favorable to creativity) but suffered because of the small number of answers (unfavorable). Person C, in contrast to Person B, might be counseled to practice generating *more* answers, perhaps by consciously suspending a tendency to prejudge and therefore discard ideas (convergent thinking).

Other Tests of Creative Thinking

A frequently cited *creativity* test of the foundation period in the 1960s was the *Remote Associates Test* (RAT) developed by Mednick (1962). This test is now out of print, but because of its seminal influence on creativity testing, it will be discussed here. One important technical characteristic of the test is that it involves correct or incorrect answers.[4] This means that it can be administered to a single person and is quick and straightforward to score. It is based on the fact that some people are more successful than others at finding remote associates to stimulus words: these people are rated more creative. Each of the 30 items, for which 40 minutes are allowed, consists of three apparently unrelated words (e.g., *moon, cheese,* and *grass*), and the task is to find a fourth word that links these words—the more remote, the better (in the case of the example just given, *blue* would be appropriate). The score is the number of correct solutions.

Mednick reported internal consistency coefficients of .91 and .92, respectively, when the test was administered to samples of male and female undergraduates. The correlation with instructors' ratings on a university-level design course was .70, and the scale distinguished significantly between psychology students rated as creative researchers and those rated as low on creativity. Scores on the RAT also distinguished between students with liberal social attitudes and those with conservative attitudes, as well as between those with artistic and those with mechanical-agricultural vocational interests. However, as Kasof (1997) summarized, the RAT has not shown more than moderate correlations with creative behavior in nontest situations.

Guilford himself developed a series of tests derived from his complex model of intellectual ability known as the SI (structure of intellect) model (Guilford, 1976). The *Structure of the Intellect Learning Abilities Test: Evaluation, Leadership, and Creative Thinking* (SOI: ELCT) (Meeker, 1985) measures eight cognitive activities—all involving divergent thinking. These cognitive activities are *divergent symbolic relations, divergent symbolic units, divergent figural units, divergent semantic units, divergent semantic relations, divergent semantic transformations, divergent figural relations,* and *divergent figural transformations.* Factor-analytic studies support the construct validity of the test (i.e., it seems to measure what it claims to measure), and inter-rater reliabilities are often as high as .99 (different people scoring test responses reach identical conclusions).

Another influential creativity test to appear in the early period was that of Wallach and Kogan (1965), whose major contribution was perhaps their emphasis on a game-like atmosphere and the absence of time limits in the testing procedure. This test contains three verbal subtests (*Instances, Alternate Uses,* and *Similarities*) and two subtests consisting of ambiguous figural stimuli (*Pattern Meanings, Line Meanings*). Probably the most widely applied subtest is Alternate Uses (also featured in the TTCT), which, as the name suggests, asks respondents to give as many

[4]The inherently convergent nature of this process is noted!

unusual uses as they can for various common items (e.g., newspaper, knife, car tire, button, shoe, key). Originally, the test was scored by counting the number of responses (fluency) and by identifying responses that were unique to a specific person within the group being tested (uniqueness). Nowadays, some users also score the test for flexibility, originality (in the sense of *statistical uncommonness*), and usefulness (practicality and relevance to reality). Whereas fluency and flexibility require, as we have already seen, merely counting numbers of responses, originality and usefulness involve rating answers on a 7-point scales (ranging across *not original–very original* and *not useful–very useful*). Kogan (1983) listed many studies supporting the validity and reliability of this test. More recently, Vosburg (1998) reported inter-rater reliabilities of .92 for originality ratings and .83 for usefulness. An overall alpha (internal consistency) reliability of .86 was reported by the same author (i.e., people taking the test a second time obtain scores similar to those they achieved the first time). It will be clear to the reader that there are a number of broadly similar tests that have demonstrated they are both reliable and appear to bear a clear connection to actual creativity in those whom they test.

The Test of Creative Thinking—Drawing Production

Perhaps the most convenient test of creative thinking ability is the *Test of Creative Thinking—Drawing Production* (TCT-DP) (Urban & Jellen, 1996). The name of this test suggests that it is a divergent thinking test, and its acronym (TCT-DP) makes it easy to confuse with Torrance's TTCT. However, it is based on an approach that differs substantially from divergent thinking (Torrance) or divergent production (Guilford)—a Gestalt[5]-psychology theory of creativity. The test has two forms, A and B, on each of which respondents are presented with a single sheet of paper containing a set incomplete figures (for example, a semicircle, a wavy line, and a dotted line), supposedly part of an unfinished drawing. The respondents' task is to complete the drawing containing the fragments in any way they wish. Scoring assesses what the authors call *image production*: respondents' drawings (productions) are rated not according to statistical infrequency of occurrence but according to dimensions such as *Boundary Breaking*, *New Elements*, and *Humor and Affectivity*. These are properties of a particular test answer sheet, and do not depend on other people's drawings for their points tally. The test can thus be given to an individual person if desired.

[5]Gestalt psychology looks at the mind and behavior in holistic terms. In contrast, a behaviorist approach attempts to understand cognitive processes in a reductionist fashion. Think top-down versus bottom-up!

Studies in a number of different countries indicate that the inter-rater reliability of the test is above .90, while test–retest reliability is about .70–.75. The test manual reports correlations up to .82 with teacher ratings of creativity, and correlations with real-life criteria show that TCT-DP scores distinguish between people who follow acknowledged creative pursuits and those who do not. My practical experience (D. H. Cropley & Cropley, 2000b) confirms that TCT-DP is also suitable for administration to university students (in fact, engineering students) and is readily accepted by them, is easy to administer and score, and can be used for counseling purposes. The scores can be used either at the highly and perhaps artificially differentiated level of the 13 dimensions suggested in the handbook or by combining subtest scores to form the 3 more complex dimensions *Productivity*, *Novelty*, and *Unconventionality* that have been demonstrated factor-analytically:

- *Productivity* involves continuation of what already exists in an existing direction: It produces novelty, but only in the sense of *more of the same* (recall our use of this idea in the context of *routine solutions*). The difference between two individuals' responses lies in the *amount* of additional material, and the difference between highly and less *creative* responses is a question of *amount*.
- *Novelty* involves addition of *new elements* to what already exists, but along existing lines. The new material may differ from person to person, but is a logical and predictable extension of the source material (reminiscent of Sternberg's *forward incrementation*—in the Propulsion Model). In Chapter 2, I characterized this as *new solutions satisfying old needs* or *better, faster, cheaper* (see Pathway 2 in Figure 2.5).
- *Unconventionality* occurs when new, surprising elements are added. They may differ sharply from person to person, and the main difference between people is the degree of surprisingness of responses (reminiscent of Sternberg's *redirection*—extension in a new direction). I described this in terms of *new solutions satisfying new needs* (or *technology push*) in Figure 2.5.

Thus, the TCT-DP makes it possible to differentiate novelty even further, and in a manner that is consistent with the development and progression of engineering solutions, by distinguishing between novelty through *extension of what exists* and novelty via *generation of previously unknown elements*.

Tests Based on Problem Solving

Runco, Plucker, and Lim (2001) argued that ideas are, in effect, *products* yielded by creative thinking. They went on to suggest that it should be possible to specify observable, relatively objective behaviors that indicate the extent to which a person gets ideas, and infer from these the presence of creative thinking. In other words, they aimed at measuring the *effects*, or output, of creative thinking, as opposed to the thinking itself. They argued that such a test would provide a criterion against which to validate tests based on assessing thinking (in other words, their test would make it possible to show that the cognitive tests discussed in the preceding sections are, in fact, measuring creativity). The result was the *Runco Ideational Behavior Scale* (RIBS). This has 23 items such as "I often have trouble sleeping at night, because so many ideas keep popping into my head" (i.e., large number of ideas), "I am good at combining ideas in ways that others have not" (i.e., unusual combinations of idea), or "My ideas are often considered *impractical* or even *wild*" (i.e., unexpected ideas). The reliability of the scale was .91 and .92 in two college student samples. Its factor structure suggested that it measures one or two dimensions. Test scores correlated scarcely at all with GPA, indicating that production of ideas is not related to academic achievement.

In examining creative thinking, Mumford and colleagues (1996; 1997) focused on problem solving. They developed tests of Problem Construction, Information Encoding, Category Selection, and Category Combination and Reorganization. The category combination test, for instance, involves problems consisting of sets of four exemplars of each of three categories. Consider this example in the style of Mumford et al. (1997)—a problem could consist of the following three sets of exemplars:

- table, chair, lamp, bed;
- banana, pineapple, orange, peach;
- telephone book, search warrant, marriage certificate, map.

These are given without naming the categories defined by the exemplars. The respondents' task is then as follows:

- to *identify the categories* defined by the exemplars;
- to combine these categories to *create a new, super-ordinate category*;
- to provide a *label for the new category* and write a brief, one-sentence description of it;
- to list as many *additional exemplars of the super-category* as possible; and
- to list *additional features* linking the exemplars combined in the new category.

A respondent might identify the three subordinate categories in the preceding example as furniture, fruit, and printed documents. She might then combine these to form the super-category of tree products, supporting this with the explanatory sentence, *All the furniture could be made of wood, all the documents of paper (which is made from wood), and fruit and wood come from trees.* The sequence is reminiscent of the progression from remote associates, through categories to broad networks that I illustrated in Figure 5.1.

In the Mumford et al. (1997) study, five judges rated the respondents' products on a 5-point scale for quality and originality of solutions. After the judges had a brief discussion to iron out discrepancies, inter-rater reliabilities of .84 and .81 were achieved for quality and originality, respectively. When Category Combination scores were compared with a criterion consisting of originality of solutions to simulated management and advertising problems, correlations of .32 and .40 were achieved. Similar coefficients were obtained for Problem Construction, Information Encoding, and Category Selection with the same criteria. When Problem Construction, Information Encoding, Category Selection, and Category Combination scores were combined in a regression approach, the multiple correlations with originality of the solutions to the advertising task was .45, with originality on the management task .61.

A second test based on problem solving, but one that adopts a novel approach, is the Creative Reasoning Test (CRT) (Doolittle, 1990). This test has two levels: Level A for children in grades 3–6 and Level B for secondary and college-level students. There are two forms of each level (Form 1 and Form 2), each with 20 items. A novel aspect of this test is that the problems to be solved are presented in the form of riddles. At Level A, for instance, these riddles take the form of four-line rhymes, in which some animal or object gives clues to its identity, and respondents must work out what the animal or object is. An example in the style of this test would be

I grow in the park,
where I stand tall and green.
For birds, I am home.
When the wind blows, I lean.

Respondents are required to find the correct answer, and a scoring key is provided that lists these answers. According to Doolittle, the test, which is in some ways reminiscent of the RAT (see earlier), requires associative, inductive, and divergent thinking. Since answers are specified in the scoring key, inter-rater reliability is not an issue. The author reported split-half reliabilities of .63–.99 for Form A and .90 for Form B, and validity (correlations with scores on the RAT) of .70, the latter scarcely surprising in view of the similarity of contents.

Divergent thinking, and ideation, is a defining characteristic of creativity. However, we have seen in our earlier discussions of engineering problem solving that this divergence must be complemented, at the right time, with analysis. This analytical thinking is therefore as important to the process of creativity as divergent synthesis, and we will now consider some of the key points of convergent thinking before we leave the discussion of cognitive processes.

CONVERGENT THINKING: THE PREPARED MIND

At the start of this chapter, I used examples of *apparently effortless* creativity to develop a picture of some of the key elements of ideation. A key concept that emerged in that discussion was the fact that luck was anything but random. In each case—Goodyear, Pasteur, Fleming—there was a logical explanation for the apparent luck—it could be described as *self-induced*. In this section, I will briefly consider the role that knowledge plays in preparing creative problem solvers for their task and how this knowledge facilitates convergent thinking and analysis.

Intuition and Convergent Thinking

I mentioned earlier in this chapter that *intuition*, although sometimes seen as one of the serendipitous pathways of creativity, is, in fact, heavily dependent on knowledge. The particular knowledge in question is acquired via *implicit learning* (Seger, 1994). In the course of everyday life, people acquire knowledge without becoming aware they possess it; chances are, you could *make* your own morning coffee at Starbucks if you had to, because you have unconsciously acquired that knowledge over months or years as a customer. As a result, although they have not consciously prepared themselves to solve a problem, people sometimes already have in their head a rough outline of the solution they are seeking, even though they may be unaware of this. This is referred to as *tacit knowledge*.

The basis of a mechanism like intuition—which appears at first glance to be the epitome of creativity coming from nowhere—is *knowledge*. In recent years there has been increasing recognition that actual creative production does not derive solely from divergent thinking, but also requires convergent thinking (e.g., Brophy, 1998; A. J. Cropley, 2006; Rickards, 1993). The *joint contribution* of the two kinds of thinking is critical to creativity, not least in the context of engineering problem solving.

The *Prepared* Mind

Charles Goodyear spent five years searching for a solution to the problem of stabilizing rubber so that it could be used more effectively in the manufacture of shoes. He did not simply take any two substances at random and throw them on a stove. The fact that he was *heating* the rubber (as against cooling it) was also the result of his earlier findings. In other words, his discovery of the stabilizing effect of sulfur on the rubber only came about because of a combination of the application of particular knowledge to a particular problem.

TABLE 5.5 Characteristics of Convergent Thinking

Typical processes	Typical results
• being logical • recognizing the familiar • combining what "belongs together" • homing in on the single best answer • reapplying set techniques • preserving the already known • being accurate and correct	• greater familiarity with what already exists • better grasp of the facts • a quick, "correct" answer • improvement of existing skills • closure on an issue

The factors that made it possible for Henri Becquerel to capitalize on the opportunity that chance presented to him when he left the photographic plate, Maltese cross, and uranium salts together in a desk drawer included

- *general knowledge* that permitted him to realize that the fogging of the photographic plate was unusual and important;
- *specific knowledge* that told him that some kind of *emitted energy* had caused the phenomenon; and
- *research skills* that enabled him to further investigate and clarify the whole observed phenomenon.

Had Becquerel *not already* been engaged in relevant research, the uranium salts and the photographic plate would not have found themselves in the drawer together in the first place, along with a personal object (the Maltese cross). He benefited from his knowledge *after* the *chance* event occurred—in being able to exploit the opportunity—but also *beforehand,* in that his knowledge *created the lucky combination* of materials. This was *not* an example of effortless creativity at all, but one of a *prepared mind* combining convergent knowledge with divergent thinking.

It becomes clear that none of the examples given was accidental or effortless creativity. As Louis Pasteur put it in a frequently cited aphorism he uttered in a lecture in 1854 (Peterson, 1954): "Chance favors only the prepared mind" (p. 473). Creativity does not come from nowhere, but (a) rests on a foundation of knowledge and (b) requires effort. To be a creative engineer, you first need to be a capable, *technical* engineer!

The characteristics of convergent thinking that are vital in supporting our overall process of creative problem solving are summarized in Table 5.5.

The Problem of *Too Much* Knowledge

The preceding discussion reminds us of the importance of technical knowledge in engineering problem solving. However, too much of a good thing can be dangerous! It goes without saying that a *lack* of knowledge, *incorrect* information, and *mis*understanding of ideas or principles can cripple creative problem solving: if you don't know what you are doing, how can you solve any problem creatively? However, a *high level of knowledge* can also lead to problems in generating, exploring, and introducing novelty. Some researchers have gone as far as to suggest that learning facts and applying strict logic, accuracy, and the like conflict with or *block creativity* (Getzels & Jackson, 1962). Others suggested that convergent knowledge is sometimes bad, or at best *a necessary evil* that is greatly exaggerated in education and business (A. J. Cropley, 1967). However, from the point of view of creative problem solving in domains such as engineering, convergent thinking can be both a facilitator *and an inhibitor*.

It is easy to see why this dual role of knowledge occurs. Working successfully in an area over a long period of time—e.g., engineering—and becoming an expert can result in a substantial knowledge base that is then available to be manipulated to yield effective novelty; i.e., it can benefit divergent thinking. How much easier must it be to generate remote associates, diverse categories, and broad networks, for example, if you have so much experience and expertise to draw on? However, as Gardner (1993) pointed out, there may be "tension between creativity and expertise" (p. 52): The pre-existing knowledge of an expert can also act as a corset that blocks novel ideas, so that thinking leads only to production of tried and trusted, correct answers.

In addition to making it difficult to recognize effective novelty when it occurs, extensive knowledge can also channel information processing into a narrow range of approaches—possibly even unconsciously—and limit the variability of what is produced (via divergent thinking), or even block generation of novelty altogether. Research has looked at this interesting problem: although working successfully in an area over a long period of time (i.e., becoming an expert) can provide a knowledge base both of the subject matter and also of the organization, it can also produce a kind of tunnel vision that narrows thinking and restricts it to the conventional (Ericsson & Smith, 1991; Root-Bernstein, Bernstein, & Garnier, 1993). Thus, despite possessing precisely what seems to be required for generating effective novelty, very knowledgeable people can inhibit creativity. Even when they are assessing the creativity of products, it can be harder for experts to reach consensus than it is for quasi-experts who possess an intermediate level of domain knowledge (Kaufman et al., 2013). To be creative, experts must not only know the facts, but also be capable of seeing them in a fresh light. Root-Bernstein (1989) spoke of the "novice effect" in this regard. The novice is not inhibited by the negative effects of prior knowledge.

The processes that can lead to a negative correlation between creativity and expertise are sometimes cognitive in nature (e.g., sets, functional fixity, and confirmation bias)—hence, our focus in this chapter. However, they can also be noncognitive—encompassing Person and Press, for example. For instance, the well-prepared expert may have a vested interest in maintaining the status quo. Radical new solutions to old, apparently already-solved problems may threaten the self-image of experts who invested enormous effort in a particular way of looking at a problem. By exposing a weakness in a given line of attack, a new way of doing things may render an expert's lifetime of work irrelevant and make the person look outdated and incompetent, with dramatic effects on social status. The result may be that experts resist the introduction of novelty, either consciously or unconsciously. For an experienced engineer to return to novice status, or even to admit to lack of expertise in a team setting, requires a high level of self-confidence.

In an empirical study of almost 1,000 employees and managers, van der Heijden (2000) showed that *expertise* has five dimensions, including special knowledge and specific skills, but also two further important dimensions: *meta-cognitive* knowledge and *growth and flexibility*. The former involves self-insights of the type discussed earlier in this chapter. The latter concerns the combination of fields (referred to earlier in this chapter as *forming associates*, or *coding*). The fields combined may be adjacent, with the result that combining them involves forming commonplace associates or narrow coding (orthodoxy) or, more interesting for present purposes, the fields may be remote, in which case combining them involves forming remote associates or coding broadly, which leads to generation of novelty (see Figure 5.1). In other words, deep knowledge is favorable for creativity only when it is accompanied by insight, flexibility, and similar characteristics. For a more detailed discussion of the role of personal characteristics in production of effective novelty, see Chapter 6.

The *Un*prepared Mind

Notwithstanding the factors I have just discussed, even diligent and effective problem solvers, equipped with many, or all, of the right cognitive abilities and personal characteristics, and working in the right place at the right time, are not always able make use of the chances that come their way.

Eugen Semmer (1870) *accidentally* cured two dying horses in a novel way but did not recognize their return to good health as an important discovery at all. He noted that horses, which had been admitted to the Institute of Veterinary Science in Riga (Latvia) suffering from infections, had been exposed to spores of the fungus *penicillium notatum*. However, this interfered with his intended approach to the problem and prevented him from studying the horses once they had died. To him, the *cure* was a problem that made it impossible for him to investigate why the horses had died!

Apparently, both Semmer and the distinguished readers of the journal failed to recognize that he had discovered a novel (and, as we now know, extremely effective) curative agent (i.e., antibiotics), and medicine had to wait another 60 years for Fleming to discover penicillin. We can speculate that Semmer's mind was *unprepared*, not because he possessed *insufficient knowledge*, but the *wrong kind* of knowledge. He was thorough and skillful enough to discover the presence of mold spores in his lab, and to see this appeared to be linked to the horses, and was thus well on the way to discovering penicillin. However, unlike Becquerel or Fleming, Semmer failed to appreciate the *significance* of his observations. He did not recognize the presence of a *solution* because he was thinking about a *different problem*.

KNOWLEDGE AND CREATIVITY

As I have described, knowledge—as a basic ingredient of convergent thinking—is a necessary part of creativity. It can either help or hinder *the process*. Too little, and our creativity is directionless and irrelevant; too much, and we risk becoming blinkered and stale. Some writers (Hausman, 1984) have argued that true creativity is always so novel that it is *unprecedented*, and thus has no connection to anything that went before—therefore, knowledge is irrelevant. Others, such as Bailin (1988), have argued that creative products are *always* conceived by both the creative person and external observers in terms of existing knowledge. In other words, this line of argument reasons that many novel ideas are based on what already exists, even if existing knowledge is transferred to a field quite different from the one in which it is already known.

Lubart (2001) expressed the link between knowledge and creativity another way: He suggested that there may well be no difference between the *processes* of divergent and convergent thinking, but that differences in *outcome* may depend instead on " . . . the quality of the material (e.g., knowledge)" (p. 301). Lubart extended this thought with the concrete metaphor: "The engine is the same, but some people use better grade fuel" (p. 301). Those who have only limited or narrow knowledge (the poorer grade of fuel) are not able to combine ideas, make unexpected associations between pieces of knowledge, or synthesize apparently unrelated facts, since they do not possess the ideas, knowledge, or facts upon which to operate.

Scott (1999) listed a number of creativity researchers who all give a prominent place to knowledge in creativity. Ericsson and Lehmann (1999) summarized the link between knowledge and creativity by concluding that "... the empirical evidence on creative achievement shows that individuals have not been able to make generally recognized creative contributions to a domain unless they had mastered the relevant knowledge and skills in the course of a long preparatory period" (p. 706). Ericsson and Lehmann (1999) also repeated (p. 700) the idea of a 10-year rule. This idea is frequently attributed to Gardner (1993) in connection with creativity, but was already discussed in other domains more than 100 years ago by Bryan and Harter (1899), and more recently by Simon and Chase (1973). It suggests that an *apprenticeship* of at least 10 years is necessary for acquiring the fund of knowledge and skills necessary for creativity.

Knowledge, as a component of convergent thinking, seems to have three particular effects on creativity. First, it represents the pool of ideas that the problem solver brings to the process of generating novelty. Second, it helps us to recognize creativity. Third, it defines the boundaries and constraints that enable us to generate novel, effective, elegant, and generic solutions.

Knowledge as the Source of Ideas

Even before the beginning of the modern, post-Sputnik era, the idea that creativity draws from the wellspring of conventional knowledge was well established: Rossman's (1931) study of inventors, for instance, concluded that they "manipulate the symbols of ... past experience" (p. 82). He also showed that they combined "known movements" (p. 77). Feldhusen (1995, p. 255) and other writers have made an important point by emphasizing the *knowledge base* of creativity. As Bailin (1988) put it, novelty "always arises out of what already exists" (p. 5). The position of knowledge as the basis of creativity has been put in more formal terms by Boden (1994a), using the language of artificial intelligence. What we call *knowledge*, she calls "cognitive maps" of a "conceptual space" (p. 8). The greater the number of *structural features* of a conceptual space such as engineering that are represented in a person's mind (i.e., the more the person *knows* about engineering), the more creative the person can be. Boden gave the example of Mozart and concluded that his creativity arose from his vast musical knowledge. Of course, an emphasis on the *importance of knowledge* raises the specter of the *wrong* knowledge, or *too much* knowledge, as inhibitors, as I discussed earlier.

Knowledge Defines *What* is Creative

As Sternberg and Lubart (1999) put it, a creative product must be adapted to "task constraints" (p. 3). Boden (1994b) made this point very strongly by arguing that it is dealing with the task constraints that makes a product or idea creative instead of merely original (occurring for the first time). Without task constraints, ideas could not cause surprise, since there would be no expectations from which they would deviate. Thus, paradoxically, novelty is determined by existing knowledge and not just by the product itself—it is *relative*, in other words.

Csikszentmihalyi (1999) extended the idea of existing knowledge as defining creativity when he described creativity as a novel variation in a domain of practice that *experts in the domain* recognize as novel and effective, and regard as worth incorporating into it. The experts judge according to their knowledge of their domain, which they have acquired through convergent thinking. This line of argument touches on a relatively recent idea in creativity research—the 4Cs model (Kaufman & Beghetto, 2009). This defines creativity relative to the creator. Thus, a child discovering that a screwdriver can be used as a lever to open a can of paint might find this extremely novel, even though, in the wider world of mechanics and physics, it is not novel at all.

Although it goes beyond the limits of this chapter, it is interesting to note that, since knowledge in a domain changes with the passage of time (usually by increasing), whether or not novelty is judged as effective—and thus creative—may change with time. Indeed, once incorporated into existing knowledge, novelty, by definition, ceases to be novel! In other words, one era's creativity is the next era's orthodoxy.

Creativity and knowledge are therefore related as follows:

- creativity derives from what is already known;
- novelty is judged as effective (or not) in terms of the already known;
- novelty passes into the body of knowledge if it is judged to be effective, and thereupon becomes orthodoxy; and
- creativity, having lost its status as effective novelty over time, influences the assessment of later novelty.

D. H. Cropley, Kaufman, and Cropley (2008) characterized this change from novelty to orthodoxy as the *decay* of novelty and pointed out that this decay begins from the moment the novelty is first introduced: as knowledge of the creative product increases, in other words, its novelty begins to evaporate. Once everybody has seen a given product—i.e., once it is widely known—it is, by definition, no longer original and surprising. Thus, to some extent, creativity is a race against time!

Knowledge can also exert the opposite effect. Rather than an accumulation of knowledge destroying novelty, new knowledge can reveal the effectiveness of earlier novelty that was initially not regarded as effective (or possibly not even novel). Although this phenomenon can result from factors such as changes in conventions, social values, or taste (thus touching on elegance and style), it can also result simply from increased factual knowledge. The idea of a product *ahead of its time* is an example of a mismatch between existing knowledge and the product.

Knowledge Guides and Shapes Creativity

Existing knowledge indicates which kinds of attack on a problem are likely to be fruitful (or are already known to be fruitless); defines the pathways, methods, and tools through which progress can be made; and specifies the nature of acceptable solutions (recall our discussion of *constraints* and design in Chapter 3). Although this can be largely a question of technical knowledge and the product, in fact, knowledge in this sense shapes and guides creativity across all four Ps: Person, Product, Process, and Press. In each case, however, knowledge may serve to help or hinder creativity.

THE INTERACTION BETWEEN DIVERGENT AND CONVERGENT THINKING

To close out our discussion of cognition processes, I will return to the theme of the *joint role* of convergent and divergent thinking in generating effective novelty. How do the two combine to produce creativity?

Generating and Exploring Variability

Finke, Ward, and Smith (1992) distinguished between two broad processes in the production of effective novelty: on the one hand, *generating* novelty, and on the other, *exploring* this novelty once it has been generated. The first kind of process produces novelty, but on its own, it can easily lead to quasi-creativity or pseudo-creativity. Suppose that a civil engineer noticed that both steel reinforcing rods and spaghetti are long, tubular, and flexible. Spaghetti has *some similarities* to steel rods. This involves a changed perception of spaghetti (i.e., *generation* of novelty). There *really are* similarities between steel reinforcing rods and spaghetti, and it is imaginable that settings exist where this variation from the usual perception of steel and spaghetti really *could* lead to *effective* novelty (for example, there may be circumstances in which it is preferable to build a bridge from spaghetti instead of steel). However, most civil engineers would probably reject the actual use of spaghetti instead of steel, and would predict a catastrophe if spaghetti were used as reinforcing rods (i.e., they would *explore* the novelty and would decide against it). This rejection of the novelty would be based on the engineers' knowledge of basic principles of civil engineering, such as strength and properties of materials. The message is simple: converting mere novelty into *effective novelty* (i.e., creativity) requires both *generation* (via divergent thinking) *and exploration* (via convergent thinking). I described the same basic concept earlier in the book when I used the terms *synthesis* and *analysis*.

Lonergan, Scott, and Mumford (2004) summarized recent thinking in this area, and concluded that the idea of a two-step process of production of effective novelty is now widely accepted. In our terms, this would involve novelty generation followed by (or accompanied by) exploration of the novelty from the point of view of workability, acceptability, or similar criteria, in order to determine if it is effective. Only then would we speak of creativity. It is tempting to think of *exploration* as essentially a process of evaluation, and Runco (2003) supported this view. He argued that creativity requires a combination of divergent and convergent thinking, and argued further that the convergent thinking involves "critical processes" (p. 432)—*critical* meaning not merely that the processes are necessary for creativity, but also that they involve *criticism* of the results of the divergent thinking, i.e., what I have just called *evaluation*.

Continuing with the example of making a link between spaghetti and steel reinforcing rods via divergent thinking, Table 5.6 gives examples of processes of divergent and convergent thinking in both *generation* and *exploration* phases of idea production. It also suggests what the results might be if divergent thinking were not tempered by convergent thinking. The point here is that both forms of thinking *could be* utilized in both phases. However, there is an ideal combination. Divergent thinking should be the basis of the generation phase, while convergent thinking should be the basis of the exploration phase. That approach will yield the best result. When we deviate from this, we risk failure. In other words, we think divergently when it is appropriate to think divergently, and we think convergently when it is appropriate to do so (indicated by the shaded boxes in Table 5.6)!

Divergent thinking is critical to the production of effective novelty. However, although necessary, it is not sufficient on its own. Convergent thinking is necessary too, because it makes it possible to explore, evaluate, or criticize variability and identify its effective aspects. In the enthusiasm for divergent thinking, it is important not to forget the contribution of convergent thinking. Equally, it is also important not to overemphasize the importance of convergent thinking, as may sometimes be done in schools and universities, and in management settings of various kinds.

TABLE 5.6 Processes of Divergent and Convergent Thinking in Generating Novelty

	Generation of variability		Generation of orthodoxy	
	Divergent thinking processes	**Result**	**Convergent thinking processes**	**Result**
Generation Phase	Linking Transforming Reinterpreting Branching out	The engineer *sees* a similarity between steel rods and spaghetti.	Recognizing the familiar Reapplying the known Sticking to the rules	The engineer *does not see* a similarity between steel and spaghetti.
Exploration Phase	Shifting contexts Exceeding limits Crossing boundaries Creating surprise	The engineer concludes that steel rods *can be replaced* with spaghetti. **Novelty**, but in this case, a disaster!	Avoiding risk Being certain Staying within limits Seeking simplicity Assessing technical and financial feasibility	The engineer sticks to steel rods to reinforce concrete. **No creativity**, but the building does not fall down!

MODELS OF CONVERGENT AND DIVERGENT INTERACTION

In Chapter 3, I described the interaction of convergent and divergent thinking, in the context of engineering problem solving, by working from first principles. How do people go about solving problems? What do they actually have to do? In this chapter, I have added further detail to this, drawing on concepts from the literature of psychology and creativity. What other formal models of this interaction between convergent and divergent thinking exist, and how do these add to our understanding of the role of each in successful engineering problem solving?

Prerequisite Models

The simplest explanation of the joint roles of divergent and convergent thinking in production of effective novelty is based on the idea that convergent thinking is a *prerequisite* for effective divergent thinking.

The simplest prerequisite approach is the summation model: divergent thinking and convergent thinking seem to add something to each other, or even to compensate for defects in each other. As an engineer, you need both, and you can only hope to be creative through a combination of both. This fits the arguments I have made so far, but seems simplistic. How much of each is needed? Is there an upper limit as I suggested with the case of too much knowledge inhibiting creativity?

More dynamic is the threshold model: below some lower limit of convergent thinking, effective divergent thinking is thought to be impossible. Once this lower limit has been exceeded, increasing convergent thinking is associated with increasing divergent thinking. Above some upper limit, convergent thinking has no further effect on divergent thinking. In other words, there is a Goldilocks zone within which convergent thinking facilitates divergent thinking. Once an engineer possesses sufficient technical knowledge, we see a relationship to divergent thinking. However, beyond a certain point, enough is enough (or may start to harm creativity). This model is a better explanation of what I have suggested seems to be the way that the two forms of thinking actually interact.

A further elaboration of the threshold approach is the channel model, according to which convergent thinking provides the channel or pathway through which information reaches the systems responsible for divergent thinking, and thus determines how much and what kind of information is processed. A variant of this is the capacity model, which argues that convergent thinking determines the amount of information that reaches cognitive systems, divergent thinking then being applied to whatever information becomes available. As in the sense of Lubart (see previous discussion), convergent thinking would thus influence the level of performance by providing high or low octane fuel (channel model) or sufficient or insufficient fuel (capacity model), but divergent thinking (or absence thereof) would influence the kind of performance.

The Super-Ordinate Ability Approach

Ward, Saunders, and Dodds (1999) identified two approaches to conceptualizing the relationship between *intelligence* and creativity that differ somewhat from the supplementation models just outlined. Here, we are treating intelligence as broadly synonymous with convergent thinking. According to Renzulli's "three ring" approach (1986), creativity and intelligence—together with motivation—are *separate subcomponents* of a superordinate ability we usually call *giftedness*. According to the *overlapping skills* model, (convergent) cognitive skills such as problem definition, selective encoding, shifting context, or transcending limitations are *common* to both intelligence and creativity (Finke et al., 1992; Sternberg & Lubart, 1995). Hassenstein (1988) went so far as to argue that a new term is needed to refer to the superordinate intellectual ability that combines both divergent and convergent thinking—he suggested *Klugheit* (cleverness).

Style Models

An alternative way of conceptualizing the interaction between production of variability (divergent thinking) and production of orthodoxy (convergent thinking) is the *style* approach. According to this, convergent and divergent thinking act *jointly* on the product, but do not directly influence each other. Both involve application of a superordinate ability to acquire, process, and store information; form abstract, general networks; develop knowledge matrices; form systems; and the like. Whether convergent or divergent products result depends on the *style* in which this thinking power is applied. A *convergent style of application* would use ability to generate orthodoxy (more of the same), whereas a divergent style would use it to generate novelty. An early example of this approach is seen in the work of Hudson (1968). A. J. Cropley (1999) discussed the interaction of creativity and intelligence in some detail, especially from the point of view of using *ability* to generate orthodoxy versus using it to generate novelty.

This conceptualization of the interaction between convergent and divergent thinking treats differences between them as *qualitative* rather than quantitative: regardless of level, mental ability can be applied in a convergent or a divergent style. Hudson (1968) raised the possibility of people who are good at both and can switch from one style to another. In terms of facilitating the process of creativity, this may be important, especially as we move from a discussion of process to a discussion of *the Person* engaging in the cognitive process.

Facaoaru (1985) related such all-round cognitive ability to creativity by showing that *creative* engineers could move freely between divergent and convergent thinking. In other words, the most creative engineers in her study were those who were *technically* good, but also creative. In a more recent report, Heller (2007) showed that Japanese members of a cross-national sample of engineering students predominated in a cluster of people who were above the total sample's average in *both* convergent and divergent thinking. By contrast, American students tended to be below the full group's average on both, while German students were below the group's average on convergent thinking although above it on divergent thinking. Cultural issues therefore may play an important role on top of process (and I will explore this issue further in Chapter 7).

TABLE 5.7 *Process* and the Phases of Problem Solving

		Invention					Exploitation	
	Phase	**Preparation**	**Activation**	**Generation**	**Illumination**	**Verification**	**Communication**	**Validation**
Dimension	**Poles**	Knowledge, problem recognition	Problem definition, refinement	Many candidate solutions	A few promising solutions	A single optimal solution	A working prototype	A successful "product"
Process Thinking Style	Convergent vs. Divergent	Convergent	Divergent	Divergent	Convergent	Convergent	Mixed	Convergent

Heller linked this cognitive *all-rounder* status to achievement by suggesting that Japanese engineers' balanced ability in convergent *and* divergent thinking may well be the cause of that nation's "astonishing technological successes" (p. 223).

SUMMARY

In a similar fashion to the end of Chapter 4 (Product), I close this chapter by summarizing the relationship between the *Phase* of the problem-solving process and changing nature of the thinking *Process* at each stage (Table 5.7). This helps to reinforce the idea that creativity is not exclusively divergent thinking, but requires phases of convergent thinking at key stages of the process.

CHAPTER

6

Person: The *Who* of Creativity

"If you're not prepared to be wrong, you'll never come up with anything original." ***Sir Ken Robinson, 1950–, Author***

As engineers, we are used to dealing with the deterministic world of machines. Cars do not need to be encouraged in order to work. Cell phones do not respond to threats (even though we may sometimes shout at them). Laptop computers do not get tired. For this reason, we are not trained to think in terms of internal, psychological factors. Machines do not have personalities, and anyway, that is the domain of psychologists! Yet, much of what we do as engineers is closely tied to people. Whether we are attempting to extract user needs from a customer and translate them into the language of the engineer, or working in a team, we are dealing with consequences that trace at least some of their origins back to the *Person*.

When we consider the task of solving engineering problems, is it simply the cognitive processes involved in thinking (Chapter 5) that enable people to generate effective novelty, or are other aspects of the *Person* also involved? Could it be that *personality*—the "unique and relatively enduring set of behaviors, feelings, thoughts, and motives that characterize an individual" (Feist, 2010, p. 114)—plays an important role in determining who is, and can be, creative?

Although it is *not clear* that there is a single, standard set of personal characteristics that are *always* found in all creative people, and not in the less creative, there is still a high level of agreement among researchers that certain personality traits help people to become creative and that particular motivational states are also helpful. Creativity is also linked with certain emotions and feelings. At the same time, the link between personal characteristics and creativity is paradoxical: creativity seems to require simultaneous possession of apparently contradictory traits (e.g., openness for fantasy associated with a strong sense of reality). As we will see, like the cognitive processes of divergent and convergent thinking, different aspects of personality are important at different stages in the engineering problem-solving process.

In the sections that follow, I will introduce the core personal characteristics that research has shown are associated with helping and hindering creativity.

THE SEARCH FOR THE CREATIVE PERSONALITY

The first step I will take in examining the question of creativity and personality is to ask whether there is a creative personality at all. The idea of a unique personality type that can be labeled *creative* has been questioned strongly by some psychological researchers: Helson (1996) concluded that there is no single, unitary, differentiated personality profile that is typical of all highly creative people and also distinguishes them as a group from the less creative. However, this does not negate the idea that certain traits are related to production of effective novelty, either in a positive way (i.e., they facilitate its appearance) or in a negative way (i.e., they inhibit it). The point here is not to discover a personality constellation that makes people creative (indeed, I have just suggested that this may not exist), but to look at aspects of personality that make it easy (or difficult) for people to become creative.

It is generally agreed that discussions of creativity and personality can be broken down into three or four distinct areas. Feist (2010), for example, argues that the "personality traits most consistently connected to creativity are clustered into..." (p. 120) the following groups:

- *Cognitive Personality Traits*—Those traits (distinguishing features) that address how people process information, for example, *openness to experience*;
- *Social Personality Traits*—Behaviors and attitudes of a social nature, for example, *introversion, arrogance*;
- *Motivational-Affective Personality Traits*—Those traits that address, for example, *persistence, ambition,* and the *feelings* associated with creativity;
- *Clinical Personality Traits*—Those aspects of *mental health*, for example, psychoticism, and their association with creativity.

The latter category, though not as directly relevant to engineering problem solving, is nevertheless important to developing an understanding of the interaction of personality with creativity. If nothing else, a discussion of clinical personality traits will help to demystify some of the terminology for an engineering audience and dispel some preconceptions. For these reasons, I will begin with a discussion of creativity and mental illness.

CREATIVITY AND MENTAL ILLNESS

One prevalent opinion is that creativity is strongly related to mental illness (Kottler, 2005; Nettle, 2002). Indeed, this topic is one of the oldest issues in psychology, and was already a subject of empirical investigation more than 100 years ago. The stereotypes of the mad genius and the tormented artist, whether justified or not, do nothing in popular culture to help the image of creativity. The theme has continued to be the subject of serious psychological investigation right up to the present day. Kaufman and Baer (2002) summarized much of the more recent research in their study of mental illness and female poets, while Silvia and Kaufman (2010) looked closely at the evidence both for and against a relationship between creativity and mental illness. Weeks and Ward (1988) demonstrated the relevance of the discussion not just to literary/artistic domains but also to practical fields such as engineering. They claimed, for example, a relationship between eccentricity and technological patents. The latter case notwithstanding, the popular stereotype seems to be associated primarily with the fine arts.

Contemporary research has adopted two approaches to investigating the question of creativity and mental illness. Studies typically either focus on acknowledged creative people to see if they are more frequently mentally disturbed than chance would predict, or they focus on people already regarded as mentally ill, to some defined degree, in order to see if they show more creativity than the general population. Across these two approaches, we can examine at least three categories of what I will loosely term *mental illness* (I hasten to add that I am *not* a clinical psychologist and that the field is both complex and highly nuanced. Mental illness can be regarded as any *impairment* of a person's *normal* thoughts, feelings, and behaviors. The important point for this discussion is that mental illness is not an all-or-nothing phenomenon, and I am not talking about exclusively *crazy* behavior. A person can be slightly mentally unwell, in the same way that he can be slightly physically unwell. Equally, mental illness, like physical illness, can be severe). The three categories of mental illness that I will consider in relation to creativity are

- *Thought disorders*—e.g., the schizophrenia spectrum of disorders, which may include, for example, delusions and hallucinations;
- *Mood disorders*—e.g., depression, anxiety, and bipolar disorders;
- *Substance abuse*—e.g., alcoholism, illicit drug use.

Thought Disorders

Research has shown that there are some similarities between schizophrenic (schizotype) thinking and creative thinking: for instance, schizophrenics make more remote associations (see Chapter 5) and think more divergently than uncreative people without schizophrenia (Schuldberg, 2001). A particularly interesting finding is that Nobel Prize winners, schizophrenic patients, and creative college students all show patterns of schizotype thinking that differ from those of less creative students (Rothenberg, 1983).

However, although creative people resemble schizophrenic patients in some ways, they also differ sharply from them in others. In a study comparing architects and members of other creative professions with schizophrenic patients, A. J. Cropley and Sikand (1973) showed that both groups displayed schizotype thinking (e.g., remote associations, unexpected combinations of ideas) by comparison with members of a control group. However, this divergent thinking did not favor production of effective novelty in the schizophrenic patients, because they were frightened by their own unusual ideas. The creative people, on the other hand, were inspired by them. This suggests that, although atypical cognitions may be part of creative thinking, on its own cognitive (thought) disturbance does not produce creativity unless supported by other qualities. Earlier findings supported this view: Barron (1969), for instance, showed that creative writers and architects scored in the upper 15% on all psychopathology scales of the MMPI (a psychological test assessing clinical conditions). However, because of their high ego-strength (their objectivity and insight), they were able to harness the unusual associations and elevated mood to generate and explore variability (i.e., to generate effective novelty). What would be pathological (i.e., problematic, in a mental health sense) in conjunction with low ego-strength actually enriched the thinking of people with high ego-strength, and led to production of novelty.

The picture that emerges is that there *does not* seem to be a *straightforward* cause-and-effect relationship between cognitive (thought) disturbance and creativity. One theory, supported by empirical evidence (Kinney et al., 2001), suggests that the relationship between thought disturbances and creativity has the form of an inverted U. This stipulates that *extremes* of thinking (e.g., both very narrow and very broad), *extremes* of categorizing, *extremes* of making associations, and so forth are associated with *low levels of creativity*. At one extreme, the thinking generates no novelty, while at the other extreme, the novelty generated lacks effectiveness. Both cases cannot achieve the desired combination of effectiveness and novelty that we know if required for creativity. As with many things in life, moderation is optimal! A modest amount of deviant thinking, however, allows production of novelty without abandoning effectiveness.

Mood Disorders

Research has also suggested that mood disturbances such as depression are much more common among acknowledged creative people than in the general public (Andreasen, 1987). Jamison (1993) reported the results of a study carried out with famous British artists and authors in which she found that manic-depressive disturbances (mood fluctuations ranging from depression to heightened excitement) were six times more common in this group than in the general public. This link appears to be unequivocally established: indeed, Jamison concluded that mental states such as elation are vital for creativity.

The question that now arises is whether mood disturbance *makes* people creative. This might occur, for instance, by driving a person to a state of despair in which he sees the problems differently (and in a more novel fashion) compared to other people. Alternatively, it might take the form of a depressed mood encouraging a person to retreat from his pain into fantasies. Lastly, it could take the form of an elevated mood that leads a person to regress to infantile kinds of thinking that are free from the rigid rules of everyday adult thinking. It is also imaginable that creativity is related to emotional lability (uncontrolled emotional displays, e.g., crying) and greater sensitivity to external stimuli or internal mood fluctuations: the mood swings associated with emotional disturbance may provide a rich source of feelings, motives, and unusual ideas for those who are also creative, but give only pain to others. It is also possible that the apparent link is an artifact: for instance, mood states such as manic disorders could reduce creative people's fear of embarrassing themselves or promote self-confidence. Reduced fear or increased self-confidence may then encourage people to behave in atypical ways, thus creating an erroneous impression that the manic disorder causes the creativity.

Generally, the position of clinically oriented researchers in creativity is that it requires a *high* level of mental health (Maslow, 1973; May, 1976; Rogers, 1961). Helson (1999) demonstrated empirically that sound mental health was necessary for the realization of creative potential. Women who as students had shown creative potential but had problems in areas such as sense of identity did not fulfill in their adult lives the promise they had shown 30 years earlier. It can even be argued (A. J. Cropley, 1990; Kaufman, 2009) that creativity promotes mental health. Studies of highly creative people indicate that creativity is connected with psychological properties such as flexibility, openness, autonomy, humor, willingness to try things, or realistic self-assessment (see below). These are prerequisites for the emergence of creativity. However, research on normal personality development also emphasizes similar properties as core elements of the healthy personality. Adopting a psychoanalytic position, Anthony (1987) argued that creativity is related to ego-autonomy, ego-autonomy to mental health, with the consequence that creativity promotes mental health. Krystal (1988) showed that uncreative people had difficulty in "self-caring" and lacked "self-coherence." Fostering creativity in such people assists them in these areas and thus promotes self-realization.

The Line between Normal and Abnormal

The reasoning that concluded the previous section is compatible with Schuldberg's (2001) conclusion that creativity is linked with six *subclinical* patterns of personal adjustment (i.e., patterns of characteristics that in *pronounced* degrees are pathological—that is, problematic—but at modest levels are not). In other words, the line between what is *normal* (and possibly beneficial) and what is *abnormal* (and frequently hindering) in terms of thought, mood, and behavior is a fine one. However, three of the patterns of personal adjustment identified by Schuldberg are *positively related* to creativity when present *in moderation*. These are

- *positive schizotype cognitive symptoms* (similar to *divergent thinking* in Chapter 5);
- *hypomania* (elevated mood); and
- *impulsivity* (behavior based on sudden desires, whims).

The first of these is associated with the thinking aspects of creativity, the latter two with personal properties such as courage (even if it is false courage), risk taking, self-confidence, lack of concern about social norms, and the like. The other three that Schuldberg (2001) discussed inhibit creativity:

- *negative schizotype cognitive symptoms* (i.e., very rigid, inflexible thinking);
- *negative schizotype affective symptoms* (flat affect, or emotion); and
- *depression*.

This approach accords well with earlier remarks about the beneficial effects of properties associated with cognitive (thought) disturbance (e.g., broad categorizing, wide associating, and the like) combined with others related to mood disturbance (reduced inhibitions, positive mood), within nonclinical limits. In other words, once again, we are talking about disturbances within the context of a normal range. Why is this important? It will lead directly on to a discussion of those normal, subclinical, nonpathological personal traits that can either help or hinder creativity. By first discussing these factors in the context of mental illness, we establish something of the boundaries and limits of thought patterns, moods, and behaviors. We also develop a better understanding of what is broadly normal and how this can support creativity or work against it.

Ludwig (1998) provided a related perspective on creativity and personality. He categorized creative fields (domains) according to four bipolar (in this case, in the sense of having two *poles*) criteria: impersonal–emotive; objective–subjective; structured–unstructured; formal–informal.

His findings suggest that the more impersonal, objective, structured, and formal a field (e.g., science, in comparison to performing arts), the lower the incidence of psychopathology (mental disorders) in practitioners in the field. This may have two causes: on the one hand, those who show the emotivity, lack of structure, subjectivity, and informality associated with psychopathology will have difficulty making creative achievements in these fields, so that successfully creative individuals in these fields will show low levels of cognitive and mood disturbance. On the other hand, in some fields creativity is aided by subjectivity, emotion, informality, and lack of structure, all of which are encouraged by cognitive and affective disturbance, so that psychopathology will encourage creativity in such areas. A second possibility is that subjective, unstructured, emotive, informal fields are attractive to people who have difficulty with structure, formality, impersonality, and objectivity, thus causing a higher incidence of people displaying psychopathology in such areas. This does not mean that everybody in these fields is mentally ill or that mental illness is a necessary prerequisite for success in the fields. This is, of course, important to our discussions with regard to engineering creativity: although some aspects of mental disorders may be associated with factors that favor creativity, in general the benefits seem to arise most where the thinking, mood, and behavior remain in the normal range. Furthermore, the analytical abilities required to make productive use of any novelty remain more firmly associated with normal patterns and levels of thinking, mood, and behavior. All of this seems to support the argument that normal, subclinical personality traits are most favorable to engineering creativity, on the whole, and that within a general band of normal traits, we can teach engineers to accentuate the positive and minimize the negative in order to boost creativity. I will now turn to personality traits within that normal range, and we will see what the traits are, how they are studied, and how they are measured.

STUDYING PERSONALITY AND CREATIVITY: METHODS

The logical way of investigating relationships between creativity and personality is to focus on people who have already produced highly acclaimed products (i.e., to concentrate on those who have successfully introduced effective novelty). As Runco and Charles (1997) pointed out, focusing on real-life examples of functional creativity is "safer," since it is certain the people involved really are producers of effective novelty. Although a focus on *acclaimed* creativity has the advantage of concentrating on people whose generation of effective novelty is self-evident, this approach has the disadvantage that it risks setting the bar too high. This should *not* create the impression that I am concerned only, or even primarily, with *sublime* creativity—namely, the creativity of geniuses. While it is interesting and important to understand how people like Edison, Pasteur, Fleming, and Goodyear achieved their success, my interest is in opening up creativity to professional engineers. In fact, I already mentioned, in the previous chapter, that creativity can occur at different levels for the individual (the 4Cs model). In the context of the present discussion, my focus is really on *Pro-C* creativity (Kaufman & Beghetto, 2009)—that is, the creativity of the professional engineer. That said, we learn a great deal about creativity and personal factors by examining the creativity of Big-C creators.

Case Studies

Case studies are an excellent device for obtaining information about, for instance, distinguished creators. A common approach is the intrinsic case study, which focuses on a single person because of the overpowering interest of the case in question. Gardner (1993), for instance, carried out case studies of Freud, Einstein, Picasso, Stravinsky, Eliot, Graham, and Gandhi. Such people offer examples of overpowering cases. The data obtained from a case study may be in the form of a narrative provided by the person in question (e.g., an interview) or notes, audiotapes, videos, and the like yielded by observation of the case. In some studies, the data have been second-hand in the form of diaries and letters, autobiographies or biographies, or similar documents.

The data can also be works of the person being studied (e.g., the paintings of Picasso), which could be analyzed, for instance, for the presence of certain themes such as violence, joy, despair, and the like, or for other kinds of content such as original scientific ideas or prophetic statements (e.g., the writings and technical drawings left by Leonardo da Vinci).

The case study approach is beset by a number of problems. The data may be highly idiosyncratic, i.e., specific or peculiar to a single respondent or even to the interaction between a particular respondent and a particular researcher. A case study of the development of Picasso's creativity would obviously require only a single case (Picasso himself), and would obviously be extremely interesting and informative. However, such a study demonstrates two problems centering on the representativeness of case studies: idiosyncrasy and generalizability. Nonetheless, case studies provide a rich source of hypotheses about creativity and the person. It is well known, for instance, that women experienced severe difficulties in intense relationships with Picasso, who seems to have been uncaring, selfish, cruel, and exploitative of them. This raises questions such as whether these are typical characteristics of men hailed as creative geniuses.

More generally, without prejudging the nature of the relationships, it could be asked whether there is a systematic connection between interpersonal relations and creativity. To the best of my knowledge, there is no such systematic investigation of eminent *engineers* with regard to creativity and personality, although some work has been done in relation to *science* more generally (Gruber & Barrett, 1974; Simonton, 2004).

Occupational Creativity

One way of investigating systematic relationships between interpersonal relations and creativity is to focus, more broadly, on people who pursue occupations regarded as inherently creative. This approach studies, for example, writers, musicians, or actors, treating them as creative simply by virtue of the area in which they work, regardless of level of achievement (i.e., occupational creativity). Examples of this method include the work of Barron (1972), Cattell and Drevdahl (1955), Drevdahl and Cattell (1958), Eiduson (1958), Götz and Götz (1979), or MacKinnon (1983). This method could also be applied to engineering. One finding of these studies is that such people possess special personality characteristics that set them apart from people in less creative occupations. Although far less common, a technologically focused example of this method was early research on the psychology of *patent holders*, such as that of Prindle (1906) and Rossman (1931). This research also indicated a link between personality and creativity, in a domain that is more closely aligned to our interests as engineers.

A variant of this approach is to study people in occupations not regarded (at least, by the unenlightened) as inherently creative, but offering opportunities for creativity (e.g., architecture, research science), or even occupations thought (rightly or wrongly) to offer few opportunities for creativity, such as business, the armed forces, or engineering. Although we might grind our teeth in frustration at this characterization of engineering, it does not matter too much. If the assumption is invalid, the results will soon show this. Creative members of these professions are then compared with less creative colleagues. Participants are often identified as creative by means of ratings of their creativity in their job by colleagues or other qualified persons.

MacKinnon (1983) showed that there were personality differences between creative and less creative architects, while Helson (1983) reported similar differences for creative mathematicians, and Barron (1969) studied, among others, Air Force officers. Facaoaru (1985) investigated engineering students and engineers, contrasting those rated creative by their peers with others rated less creative, and showed that there were differences in thinking, personality, and motivation.

Some researchers, e.g., (Cattell & Butcher, 1968; Roe, 1953) looked at possible differences in personality between people who had achieved creative eminence in different fields (e.g., creative chemists versus creative psychologists, creative social scientists versus creative physical scientists, creative scientists versus creative artists). This involves comparing creative people with other equally creative individuals, the difference being the field of eminence. Certain traits seem to differentiate between people who are creative in aesthetic fields such as art or literature and those who are creative in science. Examples include radicalism and rejection of external constraints. Art and literature people tend to be radical and to reject social constraints, whereas engineers (even those who generate variability) tend to be more conformist and more restricted by external factors. This finding emphasizes once again differences between *functional* and *aesthetic* creativity, along lines I discussed in earlier chapters.

Unacclaimed Behavior

Our focus is rightly on the Pro-C level of creativity in the context of engineering problem solving. However, in the same way that we can learn a great deal about creativity and personality by studying the Big-C creators, we can also learn about this relationship by studying the more humble activities that nonetheless produce effective novelty A. J. Cropley, 1990; Kaufman & Beghetto, 2009). We can label this little-c or mini-c creativity.

Finally, it is possible to study people—especially children—who have not yet displayed creative behavior but seem likely, for instance, on the basis of test scores (especially tests of creative thinking—Chapter 5), to become creative if they receive appropriate encouragement (i.e., potential or latent creativity). Whatever approach is adopted, what do we know about creativity and personality in this normal range?

STUDYING CREATIVITY AND PERSONALITY: RESULTS

I have already indicated that there is no single profile of the creative person. However, our discussions so far make it clear that certain personal characteristics, in some way, do help people to generate effective novelty or to learn to generate it. Several comprehensive reviews of research on creativity and personality traits have appeared over the years. These summaries confirm that a reasonably stable set of findings has emerged, although different authors name traits somewhat differently or give differing weight to particular traits, according to their own areas of interest. The next sections outline some of the more important findings. Of course, new research is constantly emerging, much of which adds weight to the importance of the traits that we already know to be important. Some, however, draw in new aspects of the existing traits that add new insights into the relationship between personality and creativity. An example of this is the growing interest in *self-efficacy* and creativity describe by Feist (2010, p. 121) and others.

PERSONALITY-FACILITATING TRAITS

A number of comprehensive reviews of research on creativity and personality traits have appeared over the years, including Dellas and Gaier (1970); Farisha (1978); Barron and Harrington (1981); Motamedi (1982); Treffinger, Isaksen, and Firestien (1983); Dacey (1989); Albert and Runco (1988); Eysenck (1997); and Feist (2010). Falling into three main categories—cognitive, social, and motivational-affective—is a relatively stable set of personal characteristics that seem to be particularly helpful for creativity. These *positive*, or enabling, *traits* are

- nonconformity (both in attitudes as well as in social behavior);
- autonomy/inner directedness;
- intuitiveness;
- ego strength;
- tolerance of ambiguity/preference for complexity;
- flexibility;
- openness to stimulation/breadth of interests;
- risk taking;
- androgyny (possession of both stereotypically male and female characteristics);
- acceptance of being different (i.e., self-acceptance);
- a positive attitude to work.

However, there are also some creative traits that are less positive: for instance, a lack of concern for social norms, and antisocial attitudes.

Other, more specific, studies have given particular weight to a smaller number of traits that are thought to be of central importance. For example, Barron and Harrington (1981) identified

- preference for complexity;
- autonomy;
- self-confidence;
- the ability to tolerate contradictory aspects of one's own self; and
- high evaluation of aesthetic qualities.

Parloff et al. (1968) emphasized autonomy and complexity, whereas Albert and Runco (1988) focused on independence. Barron (1969) showed the importance of ego-strength, including acceptance of conflicting aspects of one's own self.

Interestingly, Dellas and Gaier (1970) concluded that the personalities of young creative people are similar to those of creative adults, and this is supported by studies focusing on children. Heinelt (1974) studied schoolchildren identified on the basis of test scores as highly creative and came to the conclusion that they were significantly more introverted, more self-willed, intellectually more active, more flexible, and possessed of greater wit and a stronger sense of humor than less creative youngsters. According to Neff (1975), they are flexible, tolerant, and responsible, as well as sociable and success-oriented. They are also characterized by being less satisfied and less controlled than children who display lower levels of creativity. In social situations, they are less willing to conform and less interested in making a good impression.

Openness

Basadur and Hausdorf (1996) drew attention to the fundamental importance for creativity of placing a high value on new ideas; this is true both for individuals and for societies, although I will concentrate here on individual people (for a discussion of society and creativity, see Chapter 7). One of the basic personal characteristics associated with creativity thus seems to be a general willingness to work with or an ability to tolerate novelty itself—openness. A. J. Cropley (1992a), for instance, particularly emphasized "openness to the spark of inspiration." Openness was defined by McCrae (1987) as interest in novelty for its own sake: the open person likes to go beyond the conventional and enjoys the unexpected, even without any observable payoff. There is mounting evidence to support this relationship. Feist (2010) reported on the particular relationship between *openness to experience* and creativity demonstrated by numerous empirical studies. Linked with openness are traits such as tolerance for ambiguity and self-confidence.

Gough (1979) described the opposite personality configuration from the one just described, calling it "cautious." In fact, *openness* versus *caution* seems to be a fundamental dimension of personality. In fact, differences in openness from person to person are already visible in early childhood: many people, even as children, reject out of hand anything that departs from the familiar by more than a small amount, preferring novelty to arise, if at all, out of small, barely perceptible changes to the status quo (i.e., a *cautious* personality). Others welcome variability or even seek to generate it (i.e., an *open* personality). The size of the maximum tolerable change in the status quo may correlate with creativity. In view of the well-known importance of *collative variables* in the environment (such as incongruity, unexpectedness, unpredictability) for maintenance of optimal cognitive and emotional functioning (Berlyne, 1962) and the negative effects of monotony, as seen in, for instance, the hospitalization syndrome in small children, the drive in many people for maximum sameness and stability seems to be unhealthy. In studying the effects of stimulus deprivation on mental and emotional functioning, Heron (1957) described the "pathology of boredom," while Burkhardt (1985) referred to society's *Gleichheitswahn* [sameness psychosis]. For engineers, this may manifest itself not so much as a personal trait hurdle to overcome, but as a characteristic of others (individuals and organizations) that can act as a limiting factor on creativity.

Play and Humor

Linked with openness are play and humor. Since early studies (Getzels & Jackson, 1962), these have been emphasized as personality characteristics associated with creativity. More recently, Graham, Sawyers, and DeBord (1989) demonstrated the relationship between playfulness and creativity in schoolchildren. Isen, Daubman, and Nowicki (1987) showed that children did better on creativity tests after they had seen a comedy film. According to Bruner (1975), a playful approach fosters creativity because play is not chained to the strict rules of reality and is freed from social pressures. Play is also less risky than real life because situations imagined in play can be canceled out if they prove too problematic, and everything returned to what it was before. This means that in play novel situations can be tried out without risk. Picasso's well-known observation that he *played with ideas* is often cited as support for the importance of play even for acknowledged creators. Torrance and Safter (1999) saw play as a central element in what they called "making the creative leap."

Motivation

Studies of famous creative people from the past have confirmed that motivation plays an important role in their achievements. For instance, Cox and Terman (1926) showed that geniuses such as Newton, Copernicus, Galileo, Keppler, and Darwin were marked by tenacity and perseverance, in addition to high intelligence. Goertzel, Goertzel, and Goertzel (1978) also showed the importance of motivation in their case studies of historical figures, while Hassenstein (1988) too commented on the obsessive nature of the work of gifted individuals.

Biermann (1985) concluded that fascination with the subject matter and consequent extreme motivation were among the most important characteristics of creative mathematicians of the 17th to 19th centuries. Facaoaru (1985) showed that creative engineers were characterized not only by special intellectual characteristics but also by motivational factors such as determination. Among more recent acknowledged creative people, Sir Harold Kroto, winner of the Nobel Prize for Chemistry in 1996 for the discovery of fullerenes (*Bucky Balls*, stable carbon spheres consisting of as many as 60 individual atoms), greatly emphasized intrinsic motivation in discussing his own work. William Phillips, winner of the 1997 Nobel Prize for Physics for the development of techniques for cooling and trapping atoms with laser light, was repeatedly described as possessing "insatiable curiosity." Both Kroto and Phillips were characterized by a high level of ability to work in teams, and both emphasized how important this was to their innovative work.

According to Perkins (1981), creativity is the result of six elements, of which four are closely related to motivation:

- the drive to create order out of chaos;
- willingness to take risks;
- willingness to ask unexpected questions; and
- the feeling of being challenged by an area.

Henle (1974) gave a Gestalt psychology (holistic) perspective to the drive to create order out of chaos by emphasizing that perception of "dynamic gaps" (inadequacies, inconsistencies) in existing knowledge leads in creative people to a drive to build a good gestalt (i.e., a good whole) by reorganizing knowledge. Einstein's (Miller, 1992) description of how his recognition that existing theories of thermodynamics were inadequate motivated him to develop the special theory of relativity is an example of this phenomenon. Einstein continued to be dissatisfied with his own theory, and worked on it for much of the rest of his life. Mumford and Moertl (2003) described two case studies of innovation in social systems (management practice and student selection for admission to university), and concluded that both innovations were driven by "intense dissatisfaction" (p. 262) with the status quo.

Linked with *dissatisfaction with the status quo* as motivation to produce something new and better is belief in one's own ability to do better. People who are dissatisfied with gaps in what exists but do not believe that they can do anything about it are hardly likely to be motivated to generate effective novelty, especially if this requires long years of toil or even hardship. Thus, self-confidence, or what Ajzen (1991) called "creative self-efficacy," is necessary. This involves the personal perception of a task as lying within the ability of the individual to solve. In the present context of creativity, this is a person's image of herself as capable of generating the necessary effective novelty. Creative self-efficacy intuitively seems to be related to openness, tolerance for ambiguity, risk taking, and the like. It is one of the most important personal properties that managers and faculty can foster.

Park and Jang (2005) investigated motivation for scientific creativity by interviewing both theoretical and applied physicists. They concluded that, in addition to more affective (i.e., emotional) motives such as interest or curiosity (states more related to affective conditions within the mind of the person in question), these scientists were also affected by what they called "cognitive" motives—essentially deriving from their knowledge about phenomena in the external world. In particular, they identified (a) recognition of gaps in existing knowledge ("incompleteness"); (b) a drive to round out recently emerging novelty ("development"); and (c) identification of contradictions in accepted knowledge ("conflict/discrepancy") as cognitive motives for creativity. They gave examples from statements by Albert Einstein that indicate that he experienced all three of these motivating forces at various times. It is apparent that the cognitive motives identified by Park and Jang have a great deal to do with discovering problems.

Hennessey (2010) provided a recent comprehensive overview of the connection between creativity and motivation. Many of the motivation-related factors that I have discussed arise in the external (to the individual creator) world—i.e., the motivation is *extrinsic*. What is particularly important in this discussion of personality is the intrinsic motivation that is associated with creativity.

Intrinsic Motivation

The factors such as personal dissatisfaction with the status quo or intolerance of incompleteness and the like that are discussed in the preceding sections suggest that the motivation for creativity may arise not only in the external world but also within the individual. A widely accepted position is that creativity is based on *intrinsic* motivation (Amabile, 1996): the wish to carry out an activity for the sake of the activity itself, *regardless* of hope of external reward. This can be contrasted with working for external rewards such as praise, awards, pay raises, promotion, even avoidance of punishment (*extrinsic* motivation). In the case of extrinsic reward, it is argued that people become active only in order to gain the reward, and shape their behavior in order to conform to whatever is necessary to receive that reward, usually generation of orthodoxy. Even where variability is generated, it is done only in accordance with external directives—i.e., generation of variability as a form of conformity (or generation of variability *to shut people up*).

This is a vital issue in discussions of fostering creativity in engineering education and fostering it in engineering organizations. Can it be encouraged by faculty/managers through a system of external rewards? If external encouragement actually blocks creativity, there does not seem to be any chance of actively fostering it, especially in individuals who do not produce novelty more or less spontaneously. The only possibility of fostering creativity would then seem to be avoiding blocking it. This apparent paradox may not be as limiting as feared. Research has demonstrated (Eisenberger & Armeli, 1997) the possibility of fostering creativity by the application of external rewards. These authors showed that extrinsic reward led to enduring improvements even in a creative area such as music, provided that

- teachers (in this case) knew precisely what they wanted to foster;
- students knew what was required of them;
- students were rewarded for specified creative behaviors such as incorporating unexpected elements into a problem solution or producing alternative possibilities; and
- students were NOT rewarded for uncreative behaviors.

Further research on creativity and rewards has been described by (Eisenberger & Byron, 2011; Eisenberger & Rhoades, 2001).

Research has, however, also yielded contradictory reports on the effects of external rewards, some studies indicating that they can increase creativity, some that they block it. This is attributable to the fact that external supervision of activities supposed to generate variability can take two forms: it can focus on (a) controlling behavior or on (b) providing information. In the latter case, feedback, rewards, and the like do not inhibit production of variability and are perceived by the people receiving them as facilitating. This finding provides an important hint on how teachers and managers in an engineering setting should go about facilitating creativity. It also reinforces the importance of a clearly defined concept of creativity that specifies

- what behaviors are necessary in order to be creative;
- where and how each individual's behavior needs to be changed; and
- what aspects of personality, attitudes, and motivation are facilitating as well as blocking of such behaviors.

D. H. Cropley and Cropley (2000b) demonstrated that engineering students who received a concrete definition of what was meant by creativity and who were also counseled individually on the basis of a personal profile of their own specific strengths and weaknesses (determined from the TCT-DP; see Chapter 5) were more original on creativity tests and built more creative models in a laboratory exercise, despite the fact that they were working for grades (extrinsic motivation). This study will be discussed again greater in Chapter 10 in the context of education and creativity.

From the point of view of motivation, creative people are the ones who recognize and explicitly identify the defects in what exists and are driven to try to do something about them. Why do they do this? There are many possibilities: to gain fame, to get paid (extrinsic motivation), because they just assume that it is their job, because they are annoyed, because they are curious (intrinsic motivation). Aspects of motivation that therefore serve to *enable* creativity in the individual include

- a dissatisfaction with defects, problems;
- a desire to eliminate those problems;
- confidence in their own judgment and a willingness to act on it;
- the courage and other personal properties needed to act;
- the ability to generate variability (knowledge and skill, plus risk taking and the like);
- the ability to explore the variability (knowledge and skill, plus self-evaluation, strong reality orientation);
- the ability to communicate the new solution and to interact with other people (both supporters and detractors); and
- the ability to deal with negative, as well as positive, feedback.

Preference for Complexity

Early research in aesthetics (Eysenck, 1940) showed that two fundamental dimensions of visual preference are involved in judging the pleasingness of works of art: on the one hand, good taste; on the other, preference for simplicity versus *preference for complexity*. The latter is of particular interest here. Research (Götz, 1985) has shown that this dimension is stable and can be measured reliably. Gestalt psychologists also emphasized preference for complexity and developed instruments for measuring it (Welsh, 1975). Shaughnessy and Manz (1991) reported a substantial number of studies that showed that preference for high complexity and asymmetry is an indicator of creativity. In creative people, complexity and asymmetry energize behavior aimed at creating good gestalts (wholes) that nonetheless contain original and unexpected combinations. Nardi and Martindale (1981) showed that the preference for asymmetry goes beyond visual perception. They found that creative people preferred dissonant tones when passages of music were played to them.

The material just reviewed demonstrates the existence of two fundamental dimensions of individual difference in motivation that are central to the present discussion: openness versus closedness (i.e., willingness to accept the different) and preference for complexity versus preference for simplicity. Taken together, these seem to define two basic approaches to life. The one involves welcoming the new and different and being positively motivated by incompleteness, disharmony, and uncertainty; the other, rejection of novelty and a drive to maintain neatness, harmony, and closure. Not surprisingly, the first combination is favorable for production of novelty, while the second favors orthodoxy. I am not suggesting that this distinction provides an exhaustive definition of the motivational aspects of creativity, but it appears to be well founded both in terms of research and also of practical observation of people confronted with novelty.

It is also interesting to speculate on these dimensions in relation to engineering. Engineering is increasingly confronted by complexity. While we are familiar with aphorisms like *keep it simple stupid* (KISS), in many cases the goal is not to reduce complexity to simplicity, but to find effective ways to deal with the complexity that is inherent in modern systems. Thus, while engineers may not prefer complexity and may not actively seek it out, it is important that they do not avoid it. For this reason we can hypothesize that a disposition *toward* preference for complexity (or at least, *away from* preference for simplicity) is also an enabler for engineering problem solving. In other words, a *tolerance* for complexity may be important in engineering.

Figure 6.1 shows how people could be rated on a grid defined by closedness–openness on one axis and level of tolerance for complexity on the other. Person 1 on the grid represents an idealized combination of high tolerance for complexity paired with great openness—a state of affairs highly favorable for creativity in general. Person 2 represents the negative stereotype of a person who cannot tolerate complexity and is closed. All other combinations are theoretically possible, but only some seem likely in practice: for instance, Person 3 (high tolerance for complexity paired with extreme closedness) and Person 4 (low tolerance for complexity paired with high openness) seem intuitively to be unlikely.

Feelings and Emotions in Creativity

Shaw (1989) expressed concern that creativity research has not paid sufficient attention to a further noncognitive aspect of novelty production: the feelings and emotions people experience when they generate effective novelty. Basadur and Hausdorf (1996) emphasized a related aspect of the personal correlates of creativity: attitudes favorable to creativity (e.g., placing a high value on new ideas, believing that generating variability is an appropriate thing to do, admiring creative people in one's immediate environment and not just in the abstract). In a study of acknowledged creative engineers, Shaw showed that at various points in the process of production and insertion of effective novelty,

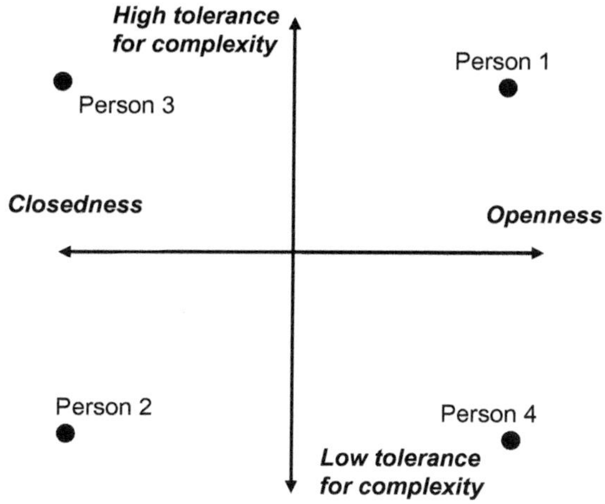

FIGURE 6.1 The openness–complexity grid.

they experienced feelings such as fascination, self-confidence or self-doubt, frustration, relief, excitement, and satisfaction. These various feelings can be regarded as part of the *joy* of creating. Shaw's respondents seldom mentioned negative feelings and emotions such as aggressiveness or triumph at having beaten someone else, perhaps because they knew that these are socially undesirable but perhaps because it really is more a matter of joy in creating.

THE DYNAMICS OF PERSONALITY AND CREATIVITY

I have outlined some of the key findings regarding personality traits and creativity. However, up to this point, the discussion has been confined largely to *associations*. Intrinsic motivation, for example, seems to be *linked to* creativity. In some cases, I have also suggested that certain traits are *enablers of creativity;* this points to something more than a correlational relationship. The next logical question in our discussion of creativity and personality is *how* personal factors affect production of effective novelty. Is one the *cause*, the other an *effect*? Do they merely co-occur? The general thrust of my discussion so far—and something usually assumed by researchers—is that *personality influences creativity*, although it is theoretically conceivable that the reverse is true, i.e., that creativity influences personality. For instance, the experience of producing novelty and having this accepted by other people seems likely to increase self-confidence, willingness to deviate from the commonplace, openness for new ideas, and similar traits. However, in the first comprehensive review of research on personality and creativity in the modern era, Dellas and Gaier (1970) concluded that personality traits affect creative behavior, rather than the reverse. The question of the effects of personality on the production of effective novelty is of central interest in this chapter. As interesting as this may be, we are concerned *not* with creativity's capacity to make you happy or fulfilled, but with fostering *functional* creativity. Therefore, our interest is, quite naturally, on the personality aspects of how we improve our capacity to solve problems in a creative manner; in other words, we want to know more about if and how personality *causes* creativity.

A Cause-and-Effect Relationship?

There are two ways, logically, that a causal relationship between personality and creativity might work. The first is as a *threshold effect*, and the second is as a *linear relationship*. According to the threshold model, there is a minimum level (a threshold) of certain key personality traits and that these must be reached, in an individual, for creativity to occur. In other words, if you have the right personality, you are creative; if you do not, bad luck! According to the linear model, certain special characteristics of personality increase the likelihood of creativity in a proportional manner. The more strongly a person possesses them, the more likely it is that she will be creative. The linear approach is more interesting for our purposes, since it offers the possibility of strengthening existing positive traits (or weakening negative ones) as a means for enhancing creativity. The threshold model is less favorable, as it raises the possibility that it may be impossible for some people to reach the required threshold, no matter what they do. Of course, it is not a matter of *choosing* an option. Different aspects of personality will have existing relationships to creativity, whether we like it or not. Our task, in understanding personality and creativity, is to understand what the actual relationships are for the traits that we know are important so that we can do our best to enhance these in the individual.

In the following sections, I will discuss five possibilities for the causal relationships between personality traits and creativity. The first two correspond to our discussion so far, and the latter three extend these somewhat. The relationships I will examine are

- a *direct causal relationship* (i.e., certain personality traits *actively trigger* creativity);
- a *threshold relationship* (i.e., certain personality traits are *necessary* for creativity);
- a *facilitatory relationship* (i.e., certain personality traits make creativity *easier*);

- a *common source relationship* (i.e., personality and creativity both derive from the *same fundamental roots*); and
- an *interaction relationship* (i.e., personality and novelty production *mutually affect* each other).

Personality as a Compelling Cause

It is not hard to imagine that certain personality traits may *directly trigger* generation of variability and the like (i.e., creativity). The most obvious examples of such characteristics would be negative properties such as *lack of impulse-control* or *rejection of social norms*. These would lead to surprising, unusual behavior as antisocial impulses were expressed; i.e., they would cause generation of variability. Positive characteristics could also more or less *force* people to generate variability. For instance, a strong sense of justice could impel a person to turn to literature, the theater, or music in the hope that these would provide a pathway to righting what the person regarded as social wrongs. An example might be the creativity embodied in a book or piece of music written to draw attention to a particular cause. Other positive traits too, such as determination and strength of character, could energize a person who had experienced misfortune and hardship to turn to literature or art to communicate to others the sorrow and disappointment these experiences had caused. At the more scientific end of the spectrum of domains, a sadness, anger, or disappointment resulting from a setback, such as failing to win a research grant or failing to secure a patent, could drive a person to achieve a major scientific breakthrough against all the odds.

Of course, creativity is not always a reaction to negative factors either in the individual's personality or in the environment, as might be inferred from the examples just given. Positive personality characteristics such as the drive for self-realization or generative motives (the desire to build something up) may lead to the production of effectively novel products even without deprivation, injustice, or the like. Indeed, some acknowledged creators are born into environments of wealth and privilege, or at least of acclaim and success, and seem to become creative either because the production of effective novelty is simply a natural part of the ethos of their environment, or because their social privileges give them the time and facilities to pursue their interests.

It is important to notice, however, that effective novelty can occur only if the cognitive elements such as knowledge, skills, and divergent thinking are also present. In other words, whether a personality acts in a linear fashion, or as a threshold, in relation to creativity, it is not enough simply to have the right personality. This mirrors the argument

I made in the previous chapter—simply banging on the piano keys with a baseball bat *still won't lead to creativity*, even if the situation is now supplemented with a positive openness to new ideas or a burning desire to be creative. You still need the knowledge, etc. You will notice a theme that is building as we move through a discussion of the 4Ps: *Process* and *Person* interact and are mutually supportive in the production of creativity.

Personality as a Facilitator/Blocker

Personality may function not so much as a direct cause (whether linear or threshold) but rather as an *assister*. This raises the possibility that personality traits might be necessary, or at least helpful, for creativity without actually *causing* it. In the case of some negative traits, this assister model means that it may be their *absence* that is necessary or helpful to creativity. This approach involves the idea of personality as a *necessary but not sufficient* cause. Relevant personality characteristics may include courage, interest in the novel, self-confidence, a generative or growth orientation, and similar factors.

This way of conceptualizing the relationship between personality and creativity can be grasped easily by examining negative personality characteristics that seem to inhibit production of effective novelty. For instance, a person could conceivably be cognitively equipped to produce effective novelty and even highly motivated to do this, but be inhibited by personality characteristics such as fear of looking foolish, excessive need for certainty, or exaggerated social conformity. In these cases, the personality traits in question can be thought of as *blocks* to creativity. Facilitating creativity is then a case of *weakening* the effects of such characteristics to remove their blocking effect.

The Common Source Explanation

Although she was studying the possible relationship between creativity and psychosis, Jamison's (1993) conclusion that mental states such as elation are vital for creativity—but do not *cause* it—is of considerable interest here. Personality traits such as excitability, nonconformity, or risk taking may well be an expression of more fundamental characteristics such as emotional lability, greater attentiveness to internal mood fluctuations, or greater sensitivity to small changes in external stimuli. These also underpin production of novelty. The result is an *apparent* direct relationship between excitability, nonconformity, and risk taking and creativity. In a more concrete example, elevated mood could lead to flamboyant (nonconforming) behavior in people by reducing fear of embarrassing themselves (i.e., it could remove a blocker). Simultaneously, it could promote making remote associations and thus lead to production of novelty. It could then seem logical to conclude that the nonconformity had led to the production of novelty, whereas in reality both might be the result of euphoric mood.

Although he was referring specifically to creativity and psychopathology, Schuldberg (2001) also emphasized that an apparent link between personality and creativity could result not from a direct cause-and-effect relationship but from a "diathesis" factor—an earlier risk factor that leads to later behaviors of which some are viewed as pathological symptoms, others as creativity. This kind of approach introduces the interesting idea that some later deviations from the norm resulting from a single risk factor may be labeled pathological (problematic) by observers, but other deviations, as creative. In other words, what is creative may be decided by the surrounding society (see Chapter 7 for a more detailed discussion). This model makes the phenomenon of pseudo-creativity easier to conceptualize. A fundamental state—for example, impulsivity—might lead some people to behave in ways that the majority regards as rude, wild, or antisocial. At the same time, this impulsivity might promote generation of variability. If the antisocial behavior were repeatedly seen co-occurring with generation of variability, it might come to be regarded as essential for creativity or even be regarded as a cause of it. Ultimately, it might even be mistaken for creativity itself, thus reducing creativity to nonconformity (pseudo-creativity). The interaction relationship is seen as a feature across all of the other relationships.

THE PARADOXICAL PERSONALITY

You may feel, from the arguments presented so far in this chapter, that the relationship between personality and creativity is neither simple nor straightforward. While I agree that this is the case, it *does not* mean that it is incomprehensible. I have built up a picture of a number of personality traits that, by various means, lead to, or can prevent, creativity. We know that these fit into broad categories—motivation, for example—and we know a great deal about the specifics within these categories. Some of the lack of simplicity must stem from a theme that we have already encountered in earlier chapters—something tied to the phases of creative problem solving. To understand this better, we can use the findings of McMullan (1978), who showed that the link between personality and creativity is characterized by seven "polarities":

- openness combined with a drive to close incomplete gestalts;
- acceptance of fantasy combined with maintenance of a strong sense of reality;

- critical and destructive attitudes together with constructive problem solving;
- cool neutrality combined with passionate engagement;
- self-centeredness coexisting with altruism;
- self-criticism and self-doubt together with self-confidence; and
- tension and concentration side by side with relaxedness.

These polarities appear to be *mutually incompatible* and yet, despite this, they seem to occur together in creative people. This means that such people are marked by what McMullan called a "paradoxical" personality. More recently, Csikszentmihalyi (1996) made a similar point when he emphasized the importance in creativity of a "complex" personality, combining, among others, sensitivity with toughness or high intelligence with naiveté.

The phase model that I developed and introduced in Chapter 3 is now vital to making sense out of the paradoxes of personality and creativity. In fact, we can take that a step further and incorporate *Process* as well. When the *cognitive processes* (Chapter 5) of production of novelty and the *personal characteristics* and motives associated with creativity are mapped on to the expanded phase model, specific processes, motives, personal traits, and feelings can be associated with specific phases. This highlights the fact that apparently contradictory ones are necessary in other phases, thus creating an *apparent paradox*. Focusing on personality again, in the stage of *Preparation*, extrinsic motivation is of paramount importance, whereas in the stage of *Illumination*, intrinsic motivation predominates (see Table 6.1).

A DYNAMIC SYSTEM

An important overlay on the role of the phase model in creativity—not only for personality—is the fact that the stages are not always followed in a linear fashion. An understanding of the dynamic relationship between production of novelty and personality is of great importance in the deliberate fostering of novelty production. As Eisenberger and Armeli (1997) showed, even young children can be taught to generate variability through the application of external rewards. Crucial, however, is that professors, supervisors, and managers know precisely what it is they wish to promote, and that students and/or engineers know what it is that they are supposed to do differently and how they are to do it. A global, undifferentiated model of novelty production would be restricted to general statements such as "Be daring in your thinking!" and would not be able to reconcile apparently contradictory principles of novelty production. For this reason, a phase approach, such as the one outlined in Chapter 3

TABLE 6.1 Creativity-Enabling Personality Traits

Phase	Motivation	Personal properties	Feelings
Preparation	hope of gain willingness to work hard	optimism self-discipline openness	interest curiosity
Activation	preference for complexity problem-solving drive (intrinsic) dissatisfaction with the status quo	critical attitude willingness to judge and select self-confidence	dissatisfaction excitement hopefulness
Generation	freedom from constraints tolerance for ambiguity willingness to take risks	relaxedness acceptance of fantasy nonconformity adventurousness	determination fascination
Illumination	trust in intuitions willingness to explore ideas resistance to premature closure	sensitivity openness flexibility	excitement
Verification	desire for closure desire to achieve quality	hardnosed sense of reality self-criticism	satisfaction pride in oneself
Communication	desire for recognition (intrinsic) desire for acclaim or reward (extrinsic)	self-confidence autonomy courage of one's convictions	anticipation hope fear
Validation	desire for acclaim mastery drive	toughness flexibility	elation

and applied to each of the 4Ps, is central to understanding how creativity comes into existence and how to foster its production.

DIAGNOSING THE CREATIVITY OF PEOPLE

In Chapter 5, I looked at measurement in the context of cognitive processes. The rationale was that we need to *measure* in order to understand, develop, and change. Chapter 5 focused on tests of divergent thinking, and while the focus was *Process*, it was hard to avoid the role that these tests play in telling us something about the creative *Person*. The purpose of any testing of creativity, however, is not simply to label or classify; it is as a means for *diagnosis* leading to focused and differentiated efforts to foster creativity. If we want engineers to be more creative, then we need to know something about the way they think (Chapter 5) and their personal characteristics (this chapter) so that we can bring about positive changes. Genco, Hölttä-Otto, and Seepersad (2012) discussed many of the issues at the intersection of creative potential and creative production, for the particular case of engineering students. Faculty need to be able to recognize creativity in their students, in order to foster it, just as managers need to be able to identify creativity in their engineers. Just as there are reliable and valid instruments (i.e., tests and scales) for assessing the creativity of products and cognitive processes like divergent thinking, there are also tests for assessing the relevant personality traits in adults.

PSYCHOLOGICAL DIMENSIONS OF CREATIVE POTENTIAL

The difficulty in tackling measurement and assessment in relation to Person is deciding what to include. The theme of this chapter has largely been the *personality* of the creative person—i.e., the constellation of traits spanning the cognitive, the social, the motivational/affective, and mental health. We can therefore focus on how these aspects of personality are measured, but this seems to risk overlooking other aspects of the Person that are relevant. For example, what about past creative behavior? If I want to understand a person's *potential* to be creative so that I can develop and foster it, wouldn't it be useful to know not just *how motivated* he is, or *how positive* he feels about creativity, but also if he has actually been highly creative in the past?

Plucker and Makel (2010) give a very useful overview of approaches to the assessment of creativity in the Person (and also in relation to Product, Process, and Press). They divide the assessment of the Person into three categories:

- *Personality Scales*—i.e., personality factors like openness, introversion, self-confidence, impulsivity;
- *Activity Checklists*—i.e., biographical inventories, past creative achievement;
- *Attitudes*—i.e., beliefs about the importance of creativity, self-efficacy.

Helson (1999) was similarly aware of the difficulty of where to place the boundary when considering assessment of the person. She resolved this by differentiating between creative productivity and creative potential. In general, psychological tests, especially personality tests, measure only the latter. Consequently, a number of authors (e.g., Helson, 1999; Kitto, Lok, & Rudowicz, 1994) have suggested that creativity tests are best thought of as tests of creative potential, not of creativity. In recent years, this view has been presented with considerable force by Proctor and Burnett (2004). Productivity, in contrast, is perhaps more the domain of Process (Chapter 5) through things like divergent thinking tests, as well as activity checklists. Nevertheless, they all relate to the Person in some fashion, and I am not dismissing any of these categories, regardless of the general focus I have taken.

Another perspective on this comes from Proctor and Burnett (2004), who brought out clearly that there is widespread (although not universal) agreement that measuring creativity requires *more than* simply testing thinking. Among other things, they quoted Sternberg's (1985) conclusion that thinking tests (especially tests of divergent thinking) run the risk of measuring only "trivial forms of creativity" (p. 126), and emphasized the need to take account of *other aspects of the person*, in addition to cognitive processes. Although I looked at knowledge in Chapter 5, in the context of Process, knowledge is as much a characteristic, or property, of the person. Table 6.2 attempts to clarify the situation. In the case of assessment, both Person and Process give us tools to assess both potential and productivity. *Process*, furthermore, is something executed or utilized by the *Person*.

In the sections that follow, I will attempt to give an overview of a number of additional means of assessment of the Person. However one or two of these stray across the boundary of Process, as you will see.

Personality and Potential

I have already discussed the importance of *openness* when considering the Person. This is now borne out by many studies of creativity—for example, McCrae (1987), King, Walker, and Broyles (1996), Feist

TABLE 6.2 Possible Combinations of Psychological Prerequisites for Creativity

	Person			Process	
	Personality scales	**Activity checklists**	**Attitudes**	**Thinking skills**	**Knowledge**
Creative Potential	✓		✓	✓	✓
Creative Productivity		✓		✓	✓

(1998) and Dollinger, Urban, and James (2004)—so much so that the inclusion of a measure of the Big Five factors of personality (Costa Jr. & McCrae, 1992)—openness, extraversion, agreeableness, conscientiousness, and emotional stability—is a useful check on the quality of data in any study of creativity. The strong relationship between openness and creativity has also been found in engineering students (Charyton & Snelbecker, 2007).

Another procedure based on properties of the individual is the *Myers-Briggs Type Indicator*, or *MBTI* (Myers-Briggs & McCaulley, 1992), a test that has achieved considerable popularity in business circles in recent years. This procedure measures four bipolar personality types:

- Extraversion (E) versus Introversion (I);
- Sensing (S) versus Intuiting (N);
- Thinking (T) versus Feeling (F);
- Judging (J) versus Perceiving (P).

A dimension, for example, Sensing versus Intuiting, is a bipolar scale on which some individual people are rated as falling at one pole (Sensing) while some are at the other (Intuiting). In the context of our discussions of creativity and personality so far, *Extraversion* corresponds to being more attentive to external stimuli, while *Introversion* corresponds to attending more to internal information. *Sensing* involves focusing on information delivered by the senses, whereas *Intuiting* involves internal hunches and the like. *Thinking* focuses more on thinking about evidence, whereas *Feeling* gives greater weight to things feeling right. *Judging* involves weighing up and evaluating, whereas *Perception* leads to proceeding on the basis of the way things look. We recognize in these many familiar concepts.

Individuals are rated on each dimension of the MBTI, according to the pole they are closer to. The four bipolar dimensions produce 16 possible combinations or *types*. These are represented both by their letter description—e.g., I-N-T-J—and also given descriptive archetypes such as *pedagogue, field marshal, inventor*, or *administrator*. Of the various possible combinations, there are some that are of particular interest to creativity. The profile I-N-F-P, for example, involves looking into oneself and not constantly checking what others think or are doing, playing hunches and the like, being open to what feels right regardless of logic, and taking in all available information without censoring some out. This profile is referred to as the *questor* archetype and is thought to be most favorable for production of variability. Conversely, the profile I-S-T-J (the *trustee*) involves being dominated by the way things are always done, looking to others for information and feedback, concentrating on hard information and knowledge, and puzzling things over and intellectualizing; it favors the production of orthodoxy or singularity.

A number of experts have suggested that creativity is particularly related to the Sensing (S)–Intuiting (N) dimension, with creative people very frequently being intuiters (N). However, it has also been shown (Walk, 1996) that creative graduate students showed a strong tendency towards the Intuiting (N)–Perceiving (P) combination (i.e., open for uncensored information *and* inclined to interpret it in terms of intuitions) as against the Sensing (S)–Judging (J) combination (i.e., inclined to focus on concrete information and process it on the basis of strict logic, correctness, and the like).

Activity and Productivity

This category of assessment of the Person is based on a widely held concept—that "the best predictor of future creative behavior may be past creative behavior" (Colangelo et al., 1992, p. 158). There are a number of options for this category. Csikszentmihalyi, Rathunde, and Whalen (1993), for example, showed in a 5-year longitudinal study of adolescents that early absorption and fascination with an area successfully predicted later adult creativity. Milgram and Hong (1999) conducted 15-year and 18-year longitudinal studies of the potency of predictors of later creativity, and showed that teenage leisure activities predicted adult creativity much better than IQ or school achievement, although the latter were good predictors of undergraduate grades. Numbers of similar studies exist (see A. J. Cropley, 2001). Indeed, this approach traces its origins back to the earliest days of the modern creativity era (Buel, 1960; Buel & Bachner, 1961; Buel, 1965; Buel, Albright, & Glennon, 1966). On the basis of this connection between life circumstances, interests, hobbies, etc. and adult creativity, a number of procedures have been developed for assessing such factors.

Michael and Colson (1979) developed the *Life Experience Inventory* (LEI) for assessing potential creativity on the basis of early life experiences. The 100-item inventory concentrates on factual information (e.g., number of changes of address in childhood, composition of family, education, hobbies, and recreation). As the authors pointed out, this approach enhances reliability. In an initial study of 100 electrical engineers who had also been classified as creative or noncreative on the basis of whether or not they held patents, 49 items differentiated between creative and noncreative participants. An intuitive grouping of these items by the authors indicated that they cover four areas:

- *self-striving or self-improvement* (e.g., enjoying competition, displaying curiosity, being committed to an area of interest);
- *parental striving* (parental emphasis on getting ahead, perceived need to do well in order to satisfy parents);

- *social participation and social experience* (membership of organizations, helping other students with their schoolwork);
- *independence training* (being allowed as children to choose their own friends, being allowed to set their own standards in judging their own accomplishments).

In a cross-validation study, again based on the real-life achievements of 98 engineers, a validity coefficient of .62 was obtained (criterion = possession—or not—of patents). No less than 83% of the engineers above the cutoff point on the inventory were indeed creative according to the criterion (i.e., correctly identified), although 29% of those not identified were, according to the criterion, actually creative (false negatives).

A different approach to the study of the creative person involves identifying not thinking (hence my comments about the boundary between Process and Person in relation to assessment) but personal characteristics whose presence is thought to increase the likelihood of creativity or even to be essential for its appearance. The *Creativity Checklist*, or CCL (Johnson, 1979), can be used for rating people at all age levels, including adults in work settings. On a 5-point scale ranging from "never" to "consistently," observers rate the behavior of the people being assessed on eight dimensions: In addition to the by now familiar cognitive dimensions Fluency, Flexibility, and Constructional Skills, personal properties such as *Ingenuity, Resourcefulness, Independence, Positive Self-Referencing*, and *Preference for Complexity* are assessed. Inter-rater reliabilities ranged from .70 to .80, and the test correlated between .51 (with the Remote Associates Test) and .56 (with the Torrance Tests of Creative Thinking).

Colangelo et al. (1992) developed the *Iowa Inventiveness Inventory*, initially by studying inventors who held industrial or agricultural patents. The final instrument consists of 61 statements (e.g., "Whenever I look at a machine, I can see how to change it") with which respondents indicate level of agreement on a 5-point scale. The inventory distinguished significantly between acknowledged creative individuals and other people, for instance, sorting into the expected order acknowledged inventors, "young inventors" rated as inventive by teachers, and noninventive academically talented adolescents. The test–retest reliability of the inventiveness score reported by Colangelo et al. was .66 and internal consistency was .70.

The *Creatrix Inventory*, or *C&RT* (Byrd, 1986), is of considerable interest because it integrates both cognitive (thinking) and noncognitive (motivation) dimensions of creativity. It is based on the concept of "idea production," the ability to produce unconventional ideas, creativity being regarded as the result of an interaction between creative thinking and the motivational dimension of risk taking. The test consists of two blocks of 28 self-rating or attitude statements, one block measuring creative thinking, the other risk taking. These are answered using a 9-point scale ranging from complete disagreement to complete agreement (e.g., "I often see the humorous side when others do not," "Daydreaming is a useful activity"). Scores on the items of each dimension are summed and the total score for the dimension rated as high, medium, or low. Each person's scores are then plotted on a two-dimensional matrix (creativity versus risk taking) and the person assigned to one of eight styles: *Reproducer, Modifier, Challenger, Practicalizer, Innovator, Synthesizer, Dreamer*, and *Planner*. The Innovator is high on both creative thinking and risk taking, the Reproducer low on both, the Challenger high on risk taking but not creativity, the Dreamer high on creativity but not risk taking, and so on. Byrd reported a one-week test–retest reliability of .72 for this scale. He argued that the scale possesses face validity but provided no data on other forms.

Most recently, two approaches to the assessment of creative production have emerged, both of which attempt to tackle a wide range of domains, including engineering. The Creative Achievement Questionnaire, or CAQ (Carson, Peterson, & Higgins, 2005), measures creativity in 10 domains, including *science* and *invention*, with a series of self-rating items concerning past achievements. The instrument has demonstrated good reliability and evidence that it is able to differentiate between respondents who would be expected to be different (i.e., concurrent validity). In their own research, Carson et al. found also that the CAQ was related (statistically) to other measures of creative personality and showed a good relationship—predictive validity—to divergent thinking tests. The Runco Ideational Behavior Scale, or RIBS (Runco, Plucker, & Lin, 2001), has also shown good reliability and validity, and also now includes subscales appropriate to engineering. Furthermore, studies have also established a relationship between the RIBS and other personality traits, such as openness (Batey, Chamorro-Premuzic, & Furnham, 2010).

Attitude and Potential

Basadur and Hausdorf (1996) emphasized a somewhat different aspect of the personal correlates of creativity: *attitudes* favorable to creativity (e.g., placing a high value on new ideas, believing that creative thinking is not bizarre). The 24-item *Basadur Preference Scale* consists of statements with which respondents express their degree of agreement/disagreement on a 5-point scale ranging from strong agreement to strong disagreement. Items include "Creative people generally seem to have scrambled minds," "New ideas seldom work out," and "Ideas are only important if they impact on major projects." Factor analysis yielded three dimensions when the scale was administered to university students and young adults working in business settings: *Valuing New Ideas, Creative Individual Stereotypes,* and *Too Busy for New Ideas*. Test–retest reliabilities of the three dimensions ranged from .58 to .63, while alpha coefficients ranged from .58 to .76. Basadur and Hausdorf reported validity coefficients involving correlations with other creativity tests of about .25.

Kirton's (1989) *Adaptation–Innovation Inventory* (KAI) does not mention creativity in its title but is frequently cited in creativity research and is becoming particularly well known in organizational settings. This test distinguishes between people who seek to solve problems by making use of what they already know and can do (*adaptors*) and people who try to reorganize and restructure the problem (*innovators*). Kirton's view is that both adapting and innovating are involved in generating novelty, but the innovative style (which is accompanied by greater motivation to be creative, higher levels of risk taking, and greater self-confidence) leads to higher productivity.

The scale consists of 32 items (e.g., "Will always think of something when stuck," "Is methodical and systematic," "Often risks doing things differently") on which respondents rate themselves, indicating how difficult it would be for them to be like this on a 5-point scale ("very easy" to "very hard"). The procedure yields an overall score and scores on three subscales: Originality, Conformity, and Efficiency. Kirton himself reported KR20 reliabilities of from .76 to .82 for the subscales and .88 for the total score, and test–retest reliability over seven months of .82 for the total score. Puccio, Treffinger, and Talbot (1995) reported alpha reliabilities for the total score of .86 to .88, and from .61 to .83 for the subscales. The same authors reported correlations ranging from about .25 to .47 for the subscale Originality with the rated originality of products.

A more recent development in the area of attitude and creative potential is the concept of *creative self-efficacy*. This addresses people's beliefs about their own capacity for creativity. Studies have shown that the construct can be measured with acceptable reliability and validity, using items such as (Beghetto, 2006)

- *I am good at coming up with new ideas;*
- *I have a lot of good ideas;*
- *I have a good imagination.*

More recently, researchers such as Tierney and Farmer (2011) have been able to show that increases in creative self-efficacy correspond to increases in creative work performance in a professional work setting. Equally importantly, they were able to show that creative expectations from supervisors contributed to this effect.

Multifaceted Approaches to Diagnosing Creativity

Many instruments available for diagnosing creativity have been criticized for their poor predictive validity—that is, the extent to which the instrument is found to predict a future outcome. However, Milgram and Hong (1999) and Plucker (1999) concluded that creativity test scores (see following text) are better predictors of creative life achievements than IQs or school grades. Plucker (1999) used sophisticated statistical procedures to reanalyze 20-year longitudinal data on predictive validity originally collected by Torrance. He concluded that composite verbal (but not figural) creativity scores on the TTCT (obtained by averaging scores on three testings) accounted for about 50% of the variance of scores on the criterion of publicly recognized creative achievements and participation in creative activities obtained several years later, and predicted about three times as much of the criterion variance as IQs. This corresponds to a predictive validity coefficient of about .7—in other words, pretty good! More recently, Plucker and Makel (2010), in their analysis of the various categories of assessment and specific instruments, reported many examples of at least satisfactory validity of various types.

Helson's (1996, 1999) studies are also informative here. Her findings are particularly important because

- they are *longitudinal*, stretching over more than 30 years;
- they use a criterion of creativity derived from real-life behavior, indeed behavior related to earning a living, rather than another creativity test or self- or observer ratings.

Helson showed that almost all creativity scores obtained from female college students aged 21 at the time of testing correlated with ratings of the degree of creativity of their occupations at age 52. These ratings differentiated between (a) "conventional" and "realistic" occupations (lowest level of creativity—1 point); (b) "social" occupations (an intermediate level—2 points); and (c) "artistic" and "investigative" occupations (highest level—3 points). People in an artistic or investigative occupation who had achieved substantial recognition as creative (socio-cultural validation of acclaimed creativity) were placed in a higher category, receiving 4 or 5 points according to the level of acclaim. Examples include writers, artists, dancers, and musicians. She reported correlations of .38–.48 with the occupational ratings for measures of personality (e.g., originality) on the one hand and self-ratings of interests on the other, obtained no less than 30 years earlier.

One of the difficulties with predicting creativity is that actual creative *achievement*, as we have seen, requires more than simply the cognitive potential. The very fact that creativity in a blend—a system—of *Process* (cognitive factors), *Person* (characteristics and traits), and *Press* factors (that I will discuss in the next chapter) means that any attempt to examine predictive validity in a piecemeal fashion (i.e., by looking at only a single tested dimension and future outcomes) is likely to miss the complexities of what actually leads to creative performance. Of course, there are statistical techniques that help to unravel these complexities; however, both the range of contributing factors and their subtle interactions may still make the task of predicting creativity inherently difficult.

More systems-level research is needed to understand better the *interactions* not only between Process, Press, and Person, but also *within* these categories. We know that a range of factors play a major role when it comes to real-life achievements in creativity. Some of them are psychological (mental health and ego strength, diligence, technical skill, or knowledge of a field, presumably acquired via convergent thinking). Some are as mundane as luck or opportunity, or even something as apparently simple as good timing. It is also clear that a major psychological moderator of real-life creative achievement is noncognitive factors such as personality. Helson (1999) showed that youthful openness and unconventionality (typical characteristics emphasized in creativity tests) predict adult creative achievement *only when* they are associated with depth, commitment, and self-discipline. When accompanied by unresolved identity problems, lack of persistence, and self-defeating behavior, *they do not*. This finding brings out once again the need for psychological approaches to creativity to be multidimensional and differentiated in nature—in other words, the "systems" approach introduced in Chapter 1.

Using Assessment of Personal Creativity

A theme of this chapter—and indeed, the whole book—is using our understanding of creativity to improve engineering problem solving. Our interest in personality and creativity should therefore be focused on how we use this knowledge not merely to *identify*, but to drive change. Both faculty in engineering schools and managers of engineers in organizations must use this knowledge to develop strategies and action, that boost the positive, creativity-enhancing aspects of personality, while minimizing those aspects that inhibit creativity. An example of how creativity assessments can be used in an educational setting to

TABLE 6.3 Test-Defined Characteristics that Are Favorable for Creativity

Motivation	Personality
Goal-directedness	Active imagination
Fascination for a task or area	Flexibility
Resistance to premature closure	Curiosity
Risk taking	Independence
Preference for asymmetry	Acceptance of own differentness
Preference for complexity	Tolerance for ambiguity
Willingness to ask many (unusual) questions	Trust in own senses
Willingness to display results	Openness to subconscious material
Willingness to consult other people (but not simply to carry out orders)	Ability to work on several ideas simultaneously
Desire to go beyond the conventional	Ability to restructure problems
	Ability to abstract from the concrete

provide *creativity counseling* is given in D. H. Cropley and Cropley (2000b). This particular study first discussed with students in a university engineering class the 4Ps of creativity and how they influence the production of effective novelty (i.e., material similar to that in Chapters 4, 5, 6, and 7). Scores on the TCT-DP were then used to construct a personal profile for each student. The students in this class were then individually counseled about their strengths and weaknesses in areas thought to be of relevance to creativity. The focus of the counseling was on identifying for the students what they could do to improve their creativity—e.g., *you gave lots of ideas here* (high fluency), *but they were all generally the same* (low flexibility). *When you solve a problem, try to think up lots of different kinds of ideas, and not just variations on the same theme.* As part of their course, students undertook a creative design task, presenting an opportunity to put the counseling advice into practice. Furthermore, as part of their assessment, they were also asked to comment on social (*Press*) factors in the groups in which they worked, as well as describe the project outputs in terms of the characteristics of creative products.

Creativity counseling, of course, presupposes the ability to differentiate personal properties relevant to creativity in different people. The creativity tests I have described in this chapter provide the means for doing this, thus making it possible to construct individualized programs for fostering creativity. As we have seen, a number of important personal characteristics and traits can be strengthened in different individuals, and can be identified with tests. These characteristics are summarized in Table 6.3.

TABLE 6.4 *Person* and the Phases of Problem Solving

		Invention					Exploitation	
Dimension	**Phase** **Poles**	**Preparation** Knowledge, problem recognition	**Activation** Problem definition, refinement	**Generation** Many candidate solutions	**Illumination** A few promising solutions	**Verification** A single optimal solution	**Communication** A working prototype	**Validation** A successful "product"
Motivation	Reactive vs. Proactive	Mixed	Proactive	Proactive	Proactive	Mixed	Reactive	Reactive
Personal Properties	Adaptive vs. Innovative	Adaptive	Innovative	Innovative	Innovative	Adaptive	Adaptive	Adaptive
Feelings	Conserving vs. Generative	Conserving	Generative	Generative	Generative	Conserving	Conserving	Conserving

SUMMARY

The different aspects of the *Person* (properties, feelings, and motivation) are now mapped onto the *Phases* of problem solving (see Table 6.4). As was the case with *Product* and *Process*, we see a pattern of oscillation between two poles. The personal properties that favor creativity during *Preparation*, for example, are different from those that favor creativity during *Generation*.

CHAPTER

7

Press: Creativity and the Role of the Environment

"Human resources are like natural resources; they're often buried deep. You have to go looking for them, they're not just lying around on the surface. You have to create the circumstances where they show themselves." ***Sir Ken Robinson, 1950–, Author***

Everything that I have described so far—engineering problems, end users, creativity, innovation, products, processes, and the people engaged in these activities—occurs as part of a *system*. A system (Blanchard & Fabrycky, 2006) consists of a set of interacting parts organized to achieve one or more stated purposes. The parts interact not only with each other—exchanging energy in various forms—but also with the *environment* (Figure 7.1).

Whether the system is a complex artifact, or the system *that produces the creativity*, the role of the environment is critical. So far, we have considered a system in the context of the Product. Now I want to turn your attention to a system in the context of *Press*—the environment in which the creativity takes place. There are two different, but related, contexts that I will use to discuss this *environment*. One is the *social environment*. You can think of this as the broader form of environment in which creativity takes place—the influence of various aspects of society on creativity. The other context is the *institutional environment*. Think of this as the narrower form of environment—the day-to-day workplace environment, or the college environment, that has a more immediate impact on creativity (Figure 7.2). Each of these environments has a *climate*—in other words, the particular conditions in the environment. In our terms, the question is not "is the climate hot, cold, wet, or dry?," but "is it favorable to creativity or unfavorable to creativity?" Both of these

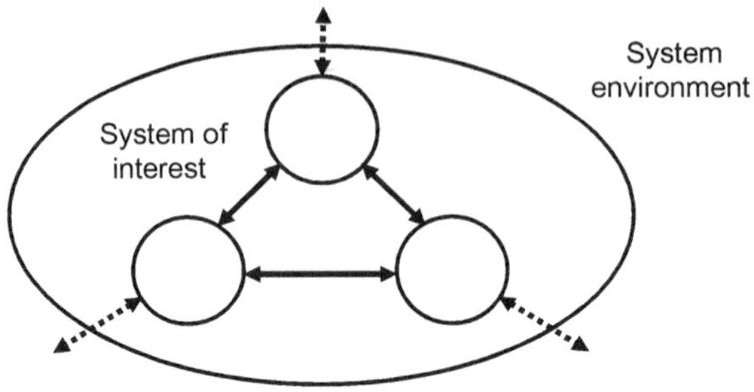

FIGURE 7.1 The system and its environment.

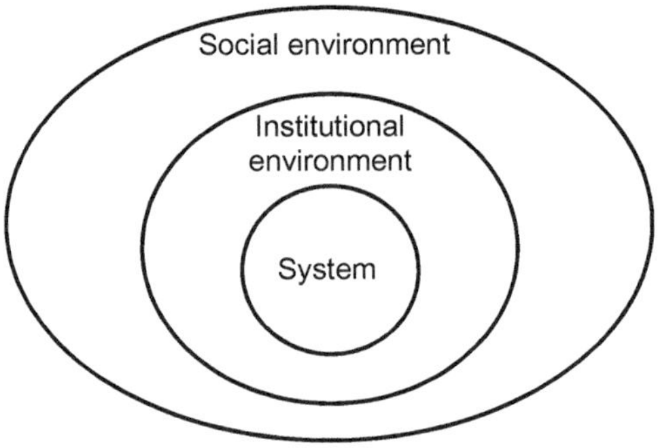

FIGURE 7.2 Social and organizational environments.

environments define the *Press*, and we need to understand how they impact on our attempts to develop creative solutions to complex technological problems.

Puccio and Cabra (2010) give a detailed review of Press and describe a more nuanced model of levels of the environment. For our purposes, it is sufficient to distinguish between the environment outside organizations and the environment inside organizations. I will start by discussing the social environment and then follow this with a discussion of the institutional (or organizational) environment.

THE SOCIAL ENVIRONMENT

Creativity is defined by the society in which it occurs, and creativity is, at least to a degree, socially motivated. It can also be facilitated (or blocked) by the surrounding social environment. This facilitation or blocking occurs because social settings differ in the degree to which they will allow deviation from the usual. Some societies accept more variability (i.e., novelty, surprise, unusualness) than others or are more willing to accept change. The *openness* of a society for variability also depends on the person generating the variability: is the person respected or admired, or does she have license to be different? The willingness of society to accept variability is also contingent on the effects of the variability on *other* people, and the area of the society's way of life into which variability is to be introduced. In addition, the environment is not simply passive, either supporting or blocking whatever innovators choose to offer, but itself influences the amount and kind of novelty that are generated in the first place. In fact, many of the blocking or facilitating characteristics of a society are a societal version of the factors we have already encountered in our investigation of the creative *Person*. This is hardly surprising given that societies are a reflection of the people who comprise them, and vice versa.

A Social Approach to Creativity

We begin by looking back briefly to the *Person*. Creativity has frequently been treated as a form of self-expression or a way of understanding or coping with life that is intimately connected with personal dignity, expression of one's inner being, self-actualization, and the like (Maslow, 1973; May, 1976; Rogers, 1961). Moustakis (1977) summarized the individualistic approach to creativity by seeing it as the pathway to living your own life your own way. Barron (1969) even concluded that creativity requires *resistance* to socialization, and Burkhardt (1985) took the theme of the individual against society further by arguing that the creative individual must fight against society's pathological desire for sameness. Sternberg and Lubart (1995) called this fight "defying the crowd," and labeled the tendency of certain creative individuals to resist society's pressure to conform "contrarianism" (p. 41). Although it may not have been the intention of the writers just mentioned, or others who took a similar view, creativity theory has thus sometimes involved "the glorification of individuals" (Boden, 1994a, p. 4). Although this is a valid way of looking at the interaction of society and the individual, this selfish[1] approach is not really the one that we are interested in for engineering problem solving. Our interest in *Press* is not a matter of fighting against society's attempts to squash our creativity, but simply a question of how we maximize creativity in situations in which the environment can sometimes act to hinder it.

This optimistic view of creativity and Press has a long tradition. Even as far back as the time of the Chinese Emperor Han Wu-di, who

[1]I use the term *selfish* not in a pejorative sense (as in *mean*), but to denote, literally, *focused on the self*.

reigned until 87BCE, there has been interest in creativity as a socially useful phenomenon. The Emperor was intensely interested in finding innovative thinkers and giving them high rank in the civil service, and reformed the method of selection of mandarins to achieve this. Both Francis Bacon (1909) and René Descartes (1991 [1644]), two of the founders of modern science, saw scientific creativity as involving harnessing of the forces of nature *for the betterment of the human condition.*[2] Nowadays we would recognize this as the *human capital approach*, and this view of creative people has become well known (Walberg & Stariha, 1992). Indeed, the original modern-day burst of popular interest following the Sputnik Shock emphasized the possible consequences *for society* of lack of creativity.

However, psychologists and educators in the post-Guilford phase of the modern era tended to emphasize themes deriving from the psychology of the individual such as cognitive aspects of creativity (i.e., Chapter 5) and creativity and personality (Chapter 6), and this may have encouraged an individualistic approach to creativity. Nonetheless, in recent thinking, creativity is increasingly seen as a force for developing society in desirable ways, not least as "the lever of riches" (Mokyr, 1990) as I indicated earlier in the book. This approach, entirely in keeping with our focus on engineering problem solving, gives greater weight to the social aspects of creativity. Creativity is the *servant* of society.

Analysis of the social aspects of creativity may be considered as having three dimensions:

- understanding creativity as a social force with social responsibility;
- defining what is creative in social rather than individual terms; and
- attributing the driving force for creativity (i.e., motivation) to social rather than intra-individual factors.

I will address these three dimensions in the sections that follow, before turning to the organization.

Ethical Aspects of Creativity

A good starting point, with a strong link to engineering, is the question of ethics and creativity. Sternberg (2003b) argued that creativity (along with intelligence) must be balanced or tempered by *wisdom*, and

[2]Interestingly, and not insignificantly, this statement is very close to modern definitions of the discipline of engineering. For example, the U.S. Accreditation Board for Engineering and Technology (ABET, 2011) defines engineering as "... the profession in which a knowledge of the mathematical and natural sciences gained by study, experience, and practice is applied with judgement to develop ways to utilize economically, the materials and forces of nature for the benefit of mankind."

assumed that creative people's wisdom will ensure that their creativity serves the common good. Several authors have proposed a *moral creativity* (Gruber, 1993; Runco, 1993; Runco & Nemiro, 2003; Schwebel, 1993). However, even well-intentioned creativity does not always produce unmitigated benefits for society. Such unintended negative effects of the introduction of novelty are not uncommon. Even the highly acclaimed discoveries of Edward Jenner and Louis Pasteur about the transmission of disease, to take one example, laid the foundations for germ warfare! A more concrete example for engineers consists of a raft of *negative, unintended consequences* that has resulted from the development of automobiles. Tens of thousands of people around the world are killed each year in car accidents. Creativity has a "dark side" (McLaren, 1993) that needs to be explored and understood.

Creativity's dark side includes not only *accidental harm*, but also creative activities that are carried out to satisfy personal vanity or pride, or that benefit narrow, short-term interests. Even more disturbing, it is also possible for effective novelty to be introduced in the full knowledge that it *will* damage others. This can occur without the damage being the primary object of the exercise, as for instance in business, when a new product is introduced in order to make a profit in the full knowledge that it will inevitably harm a rival product. It is also often seen in criminal behavior. Fortunately, as Eisenman (1999) showed, prisoners rated by guards and other inmates as creative typically generated little or no effective novelty, but rather showed lack of inhibitions and low levels of social conformity, i.e., *pseudo-creativity*. This suggests that *unsuccessful* criminals (i.e., those who have been imprisoned) are not particularly novel or innovative. As a result, anti-crime measures are reasonably successful, even without high levels of novelty, elegance, and generalizability. These issues have been explored in depth in D. H. Cropley and Cropley (2013) for both crime and terrorism. In the broader context of ethics, this has been explored further by a range of contributions in Moran, Cropley, and Kaufman (2014), while D. H. Cropley (2014a) explores issues of ethics, creativity, and engineering in more detail.

Unfortunately, it is also possible for the negative consequences of creativity not only to be fully intended by the person or group introducing the effective novelty, but also to be their central purpose—harm to others as the main goal of creativity. Obvious examples of such creativity are seen in war, while D. H. Cropley, Kaufman, and Cropley (2008) have argued that terrorists, such as the perpetrators of 9/11, are capable of generating highly effective novelty and successfully inserting it into a functioning system. As unsavory as it seems, they are creative, and indeed innovative, despite their evil intentions. D. H. Cropley (2005) coined the term *malevolent* as one way to describe such creativity. These examples confront us with particularly difficult issues. Suppose, for example, that an employee found effective, novel ways of bullying a colleague, through harassment or mockery, simply out of malice. Imagine a hacker applying his skills in software to a novel and effective way to distribute a virus that caused economic loss to millions of Internet users, simply for the thrill of damaging others. Along with particularly deadly weapons of war—the atomic bomb even—and effective, novel acts of terrorism, both of the situations just mentioned might well involve generation of novelty that was highly effective in achieving the goals of a particular individual. Are they, however, really examples of *creativity*?

A strictly individualistic approach might indeed conclude that *they are*. In the same way, effective and novel techniques of a mass murderer might be regarded as, *in principle*, having the same virtues as the innovative work of a creative engineer, since both reflect the workings of a mental ability—creativity—to generate effective novelty. However, such a conclusion is also unsatisfactory for the overwhelming number of teachers and managers: few of them will be interested in fostering the creativity of, let us say, an ax murderer!

It is apparent that the generation of novelty really does require deviating from norms, so that in a sense it requires social deviation! The answer to the question in the previous paragraphs—e.g., were the 9/11 terrorists creative?—can only be answered by going beyond a purely individualistic approach to creativity. An understanding of the *social aspects of creativity* helps to clarify where creativity comes from, what factors facilitate its appearance, how it can be fostered or applied in groups (whether, for example, a university classroom or an engineering firm), and so on. Emphasis on the social aspects of creativity does not deny the importance of the cognitive and personal aspects discussed in Chapters 4, 5, and 6, but adds an additional dimension to these.

THE SOCIAL DEFINITION OF *WHAT* IS CREATIVE

The essence of any kind of creativity is production of effective novelty. In earlier chapters, I defined novelty and effectiveness from a *psychological* point of view. The decisive property of novelty is that it causes "surprise" in beholders (Bruner, 1962); i.e., it is people's surprise that defines novelty, not the product itself. Surprise occurs when something is unexpectedly different from the usual; that is, it deviates from what things have been like up to that point. It is the *contrast* with what already exists that yields the surprise. In other words, the production of novelty does not occur in a vacuum, but in a *social context*. Thus, we might say that it is not the product or the process itself that determines novelty, but the particular setting (the contrast of the novelty with the existing state of the art or the constraints of the external world). Without existing external norms, there would be no such thing as novelty, only variability (i.e., differences).

As was emphasized in Chapter 1, and later in Chapter 4 (the *Product*), the term *creativity* is not simply applied to *anything* that surprises people: what is crucial for converting novelty into creativity is *effectiveness*. This too is determined by the surrounding environment. Indeed, Csikszentmihalyi (1999) described creativity as requiring "acceptance by a particular field of judges" (p. 316), thus arguing that creativity is essentially a positive category of judgment in the minds of observers, a term they use to praise products that they find exceptionally good. When a number of observers agree that a product is creative, then it *is*. Csikszentmihalyi called this social definition of effectiveness "socio-cultural validation."

Although social recognition or acclaim defines effectiveness, and is thus necessary for creativity, the *judges* need not be experts. For example, it is not necessary to be a civil engineer to be capable of recognizing the effectiveness of a bridge. In other words, the everyday users of many products may well be in the best position to determine their effectiveness. As Wernher von Braun is reported to have said, "The eye is a fine architect, believe it!" (Rechtin & Maier, 2000). In effect, good products are *recognizable* as good products. I discussed this in more detail in Chapter 4 in the examination of the measurement of product creativity. It appears, therefore, that socio-cultural validation can be carried out by a wide range of *judges*—novices, quasi-experts, and experts.

The Problem of Changing Standards

A practical problem with socio-cultural validation as the means for establishing effectiveness—the *criterion* of effectiveness, in other words—is that what is regarded as creative in one era, or society, can be uncreative in another. There are many examples across many domains of creative activity. In music, for example, the composer Johannes Brahms was unable to obtain the post of director of the philharmonic orchestra in his native Hamburg because his music was initially judged too conservative. He had to go to Vienna to find acclaim. In Georgian England, Shakespeare's plays were regarded as indecent, and had to be edited to make them respectable; in 1818 Dr. Thomas Bowdler published the *Family Shakespeare*, in which he removed expressions that could not, with propriety, be read aloud in a family setting (he *bowdlerized* Shakespeare's work). Engineering is also no stranger to this phenomenon—cases in which we might say that a product was ahead of its time or that the consumer was *not ready* for a particular product. These can be seen as examples of *society* making a judgment about the effectiveness of a product, and rejecting it, regardless of any *technical effectiveness*.

The Social Definition of *Who* is Creative

In addition to deciding *what* is creative, the social environment also identifies certain people who generate novelty as creative, others as *strange*, *mentally ill*, or even *criminal*. One mechanism through which society determines who is creative can be demonstrated by returning to Schuldberg's (2001) discussion of psychopathology and creativity, especially his concept of "diathesis." Some cause in an individual's development leads to psychological states that encourage behavior that differs from the average or normal, such as linking ideas usually kept separate, coming up with unexpected suggestions, freely expressing excitement and elation, and so on (recall the material I discussed in Chapter 6 about the *Person*). Quite apart from the individual definition is society's reaction. If society *applauds* the resulting behavior, the person is regarded as creative. If society it frowns upon the behavior, the person may be regarded as crazy or even criminal. This occurs despite the fact that the underlying *behavior* may be the same! A salient example is Andres Serrano's infamous *Piss Christ* photograph in 1987 (D. H. Cropley & Cropley, 2013) that aroused considerable controversy—so much so that some people made death threats against the artist, while some organizations presented him with prizes. Thus, it would appear that it is not so much the *actual deviation* from the usual that determines creativity, but how the social environment *reacts* to the deviation.

The precise social group in which the creator displays the deviation may be decisive. For example, someone who is active in a setting where *unbridled* expression of impulses and ignoring the conventions may be regarded as odd or incompetent (e.g., engineering!) would be treated differently for the same behaviors from someone in a setting where such behavior was admired (e.g., *avant-garde* theater or modern dance). The latter person might be fortunate enough to have the behavior accepted as not only surprising, but also effective, and thus creative. Of course, this does not mean that we should abandon attempts to be creative in engineering for fear that the social environment might reject it. Rather, it further reinforces the importance of understanding, and being able to articulate to others, what engineering creativity is. If we, as engineers, fully appreciate the importance of balancing novelty with effectiveness, and can demonstrate that balance, then we have a mechanism to overcome any possible adverse social judgments of creativity in engineering. This is also why we need to understand creativity in the context of different domains—the domain extends to the *social domain*, as well as the domain of activity.

Social Determination of *Amount* and *Kind* of Creativity

Another filter through which to study the relationship between the social environment and creativity is the *qualitative* versus *quantitative* points of view. For example, the very large departures from norms that might be labeled mental illness or criminality, compared to more moderate departures looked upon as creativity, can be seen as a question of the *amount* of deviation being decisive. This seems to be the case with the *diathesis* example described earlier. It is also possible, however, that it is not the *amount* of departure from norms, but the *kind* of departure that is decisive in determining whether deviation is condemned or acclaimed. This seems to be more obvious in the case of the engineers versus dancers example.

It seems to be the case that *society* can tolerate only a certain amount of variability, suggesting that the quantitative point of view is a better model of the relationship. This is also true of narrower social settings such as the family, everyday social life with peers, recreational settings (e.g., clubs), educational institutions, work settings, and so on. Indeed, we blur into *Press* as *institutional environment* if we drill down a little. However, *within* any setting, there may be considerable differences in the *kind* of variability (in addition to the amount) that is tolerated or applauded: one family may tolerate wild or undisciplined behavior, whereas another will not. Some circles of friends demand greater conformity from their members than others. Some vocational groups regulate the behavior of their members closely, others far less. As a result, the social setting determines what kinds of new ideas emerge by setting limits on *both the amount and also the kind* of divergence that is permitted, or by guiding creative thinking into particular channels. One way it can do this is *via motivation*. There is little incentive to produce novelty or surprise that no one else is willing to support, or that is actively discouraged, or even punished. Despite this social determination of the amount and kind of creativity, some exceptional individuals who *swim against the current* are still seen to emerge, and this tells us more about both the *Person* and the *Press*.

The Effect of the *Amount* of Creativity

Apparently, the *amount* of deviation from the customary is decisive for public acceptance of novelty. Very large departures from what the group in question is used to may be socially unacceptable and even labeled in a variety of pejorative ways as previously described. This suggests that an important ability for the creator (whether individual or organization) is to learn how to link novelty to an existing framework—in other words, to understand how to introduce *just enough* novelty so that the amount of deviation is tolerated by the social group. Failing to do this risks the rejection of the creativity, not because it is inherently flawed or ineffective, but simply because it is too different! A real example of this question of judging just how much deviation can be tolerated is found in the example of the Hungarian doctor Ignaz Semmelweiss (1818–1865). He discovered that the incidence of *puerperal fever*—a serious, often fatal, infection to which women are prone immediately after childbirth—could be cut drastically by simply by having doctors wash their hands before delivering babies. Unfortunately, this idea conflicted with the prevailing views of the day, and Semmelweiss was also unable to explain why it worked. Worst of all, this proposed deviation from a norm was seen as insulting to doctors, implying that they were dirty. Here was a case of a level of deviation intolerable to a social grouping suppressing a creative solution (literally—it was novel and effective).

The Effect of *Kind* of Creativity

The Semmelweiss example was a case of *too much* deviation. The *kind* of deviation may also be important. The socially derived distinction between *kinds* of creativity can be regarded as involving, on the one hand, socially *radical* and, on the other, socially *orthodox* effective novelty. From a social point of view, radical novelty arises out of willingness to venture into the area of socially frowned-upon ideas or actions. Orthodox novelty involves generating effective novelty while remaining within socially prescribed limits. This distinction is similar to the one Millward and Freeman (2002) made between change that stays *within* the existing social system (what I am calling *orthodox creativity*) as against change that *challenges* the system (i.e., *radical creativity*). In a recent paper, Sternberg (2006) also linked his own cognitive approach to creativity with social factors when he divided the processes of his propulsion model (Chapter 4) into those that *accept current paradigms* (i.e., orthodox creativity), and those that reject current paradigms (i.e., radical creativity). Sternberg also suggested the existence of a third variant, creativity that synthesizes current paradigms.

The distinctions between *amount* and *kind* of creativity, and the labels I have attached (orthodox and radical), suggest a 2-by-2 matrix as a mechanism for classifying examples of creativity in particular social settings (Table 7.1).

TABLE 7.1 Domains and the *Kind* and *Amount* of Creativity

		Kind of creativity	
		Orthodox	**Radical**
Amount of Creativity	Small	Physics, Math, Engineering	History
	Large	Art History	Drama, Art

Differences among domains (engineering, art, history, etc.), and also among other social groupings (e.g., teachers, students, managers, organizations), can all be classified using this grid. Some domains, for example, are frequently open to high levels of radical creativity (e.g., drama or art), whereas others are mainly restricted to low levels of orthodox creativity (physics, mathematics). Thus, placement of the domain *Engineering* in the *orthodox/small* quadrant means that this domain is rated as *typically* involving lower levels of orthodox creativity, whereas the placement of *Art* in the *radical/large* quadrant means that this discipline is rated as typically involving higher levels of radical novelty. It is important to note that where I have placed domains *does not mean* that other *amounts* and *kinds* of creativity are not possible within them. Rather, this is an attempt to compare the different domains to each other in terms of general trends. Of course, there can be radical/large creativity in engineering, just as there can be orthodox/small creativity in art. What the table does suggest is that radical/large creativity in engineering is *harder to achieve*—relative to the social setting—than it is in art or drama. In simple terms, engineering is probably more socially conservative than art or drama, and the risk of the rejection of radical/large creativity is greater.

Glück, Ernst, and Unger (2002) showed that differences along these lines exist among other social groupings as well—for example, art and physics teachers. As a group, the former tolerate or encourage originality, risk taking, impulsivity, and nonconformity (in our terms, radical creativity), whereas physics teachers as a group prefer convergent problem solving, responsibility, and reliability (orthodox creativity or no creativity at all). In a similar way, some *students* display radical creativity; some, orthodox creativity; and some, little of either.

This classification system can be expanded to take account of a further distinction. Research such as that by Simonton (1997) suggests some societies are *product* oriented (they focus on producing novel works such as art, literature, machines, and gadgets, preferably high-tech gadgets, etc.), whereas others are *process* oriented (they focus on techniques, production, and management procedures, etc.). Different communities of experts or specialists may also reflect this difference.

Engineers, for example, may place greatest value on *product-oriented novelty;* while philosophers, on process-oriented novelty. The same may well be true among teachers in different disciplines. An example is the difference between engineering faculty who emphasize the development of imaginative and original solutions to problems (a *product orientation*) and those who emphasize procedures, algorithms, and the like (a *process orientation*). We can also see this as contrasting the achievement of the end result with how it was achieved. These considerations suggest an extension of Table 7.1 by adding a third dimension: *orientation* (process vs. product). It may be that engineering education has developed an *overemphasis* on a process orientation. This is a fact that, coincidentally, seems to militate against teaching divergent thinking and creativity in favor of analysis and convergent thinking (i.e., lots of math and other quantitative knowledge).

This classification can also be applied to other kinds of social settings, including *organizations*. From the point of view of fostering creativity, the ideal situation seems to be an alignment between all *players* in the social setting—in other words, they all occupy the same quadrant. This would occur when manager/teacher, individual, and discipline/area of operations were all located in the same quadrant, especially in the case of creative individuals. Harrington (1999) argued that there is no single "best" set of environmental circumstances that is favorable for everybody's creativity, but that the decisive factor is the *goodness of fit* between the characteristics of the environment and those of the individual. Table 7.1 can thus be regarded as providing a starting point for developing a schema for *diagnosing* goodness of fit in terms of demands of the domain or area of operations, teachers' or managers' orientation to novelty, and students'/engineers' production of novelty.

The Social Influence on the Content of Creative Behavior

The social environment is not simply a passive recipient of whatever novelty people generate, with its function confined to being surprised or not, and applying or withholding the seal of approval. As well as influencing—at least to some extent—what kind of novelty and how much novelty is produced, the society affects the fields in which people become active, the novelty generating tactics they employ, and the contents of their creativity. These effects are not only personal—affecting the creativity of a particular individual—but also *general*. In other words, environmental factors influence not just the novelty produced by individuals but also that produced in the *society* more generally. For example, there are documented cases of different people in a domain all adopting a similar novel approach, or coming up with the same novel idea at about the same time. An example of this is the simultaneous, but independent, invention of calculus by Newton and Leibnitz, or the controversy surrounding the invention and patenting of the telephone by Bell and others. However, the phenomenon is more general than this. Research (Simonton, 1994) has shown convincingly that in times of economic prosperity or depression; before, after, and during political and social upheavals; or following a successful or an unsuccessful war, differing patterns of creativity occur. This involves not only the number of creators who emerge, but also the domains in which creativity occurs and the kind of novelty that is generated (think here of the Propulsion model—Chapter 4—i.e., amending what exists, finding new lines of attack, generating radical and unprecedented novelty). To take one concrete example, Simonton (1998, p. 105) showed that "melodic originality" is higher among composers in wartime.

Such phenomena are commonly referred to as reflecting the *zeitgeist*—literally, the spirit of the times. The internal explanation of the *zeitgeist* is that a domain possesses its own pattern or pathway of growth that is inherent in the domain in question; one thing, so to speak, leads almost perforce to another. Closely related is the idea that each domain has its own system of internal logic and that growth in the field must follow this logic. Related to products, this would mean that novel products can occur only in a relatively fixed order and at the right time, after earlier events have opened up the field in a new way. Electronic devices, for example, obviously could be invented only after the discovery and harnessing of electricity. Of course, the discovery and harnessing of electricity did not guarantee that television would be invented, but was a necessary precondition for its invention. This is a generalization of the argument in Chapter 4 that individual novel products are very frequently extensions of what already exists, rather than unprecedented breakthroughs.

Another wartime example serves to illustrate this point. In the Second World War, Great Britain first deployed the radar countermeasure nowadays called *chaff*. This consists of small pieces of aluminum, or other reflective materials, that cause spurious contacts on a radar screen. Britain developed this in order to help protect Allied bombers from German radar-controlled anti-aircraft gunfire and fighter aircraft during mass bombing raids over Germany. Curiously, however, Britain did not initially use chaff, despite having developed it, because they feared that if they did so, this would reveal the technology to Germany, who might then use it themselves in bombing raids on Britain. When the British did introduce it, in July 1943, it was initially highly effective (although the Germans quickly learned how to disregard the spurious radar contacts—demonstrating the decay of novelty leading to a decay in effectiveness; see Chapter 5). In the end, it turned out that the

Germans had already developed the same technology themselves, at about the same time as the British!

A contrasting, external model of the appearance of effective novelty in various domains sees the influences on creativity as lying outside the domain itself, in the *broader social environment*. In this case, fluctuations in people's production of novelty are linked to *broad social conditions*, such as tolerance of variability by those who wield power in that society. A good example of this broad, social environmental influence—as distinct from a narrower, more domain-specific influence—is the influence in past centuries of the Catholic Church with regard to scientific opinions on the solar system. The prevailing view for many centuries—one that dominated thinking in entire nations—was the heliocentric model of the solar system. Deviating from this view could prove costly, as scientists like Galileo discovered. In more recent times, we have seen similar whole-of-society effects on deviations from the norm in countries like the former Soviet Union. Societies seem to need certain kinds of creativity (or lack of creativity) at certain times in their social, economic, and political development and to transmit this need to creative people in a global manner.

An interesting question for educators is how this occurs. An early answer was the proposal that all societies, as a kind of natural law, oscillate in long waves between a "sensate" and an "ideational" orientation (Martindale, 1990). The sensate orientation is empirical and deterministic; the ideational, intuitive and based on feelings. This is somewhat akin to a societal-level type indicator rather like the MBTI discussed in Chapter 6. Differences between empirical and intuitive novelty production would hardly be surprising; however, this pattern in the character of societies seems to be something that changes relatively slowly, and is of limited value to us in examination of the more day-to-day issues surrounding fostering creativity. For the purpose of understanding how creativity is helped or hindered, it is sufficient to understand that societies directly influence people's creativity through various social mechanisms that influence behavior. These have effects in both the long term and the short term.

MOTIVATION: THE SOCIAL NATURE OF THE CREATIVE IMPULSE

Why do people produce novelty at all? There are motivating factors, as I have already outlined, that are focused on the psychology of the individual. There are also factors that are individual and biological in nature. Berlyne (1962), for instance, argued that novelty and uncertainty act on the central nervous system to help people maintain an optimal level of neural activity.

A certain amount of exposure to novelty production is *biologically necessary*. Studies of sensory deprivation (Zuckerman, 1969) and of the effects of monotony on children raised in orphanages (Dennis, 1973) have shown the effects of denial of novelty in a dramatic way. The effects include anxiety, hallucinations, bizarre thoughts, depression, and antisocial behavior. In the case of young children denied novelty, effects include apathy, dullness, and stunted emotional and intellectual development. Heron (1957) called this the "pathology of boredom." Of great interest is that the nature of the novelty is important. Unexpectedness, incongruity, and the like (surprise) are more effective in promoting *normal* development than simple fluctuations in the usual (i.e., mere variability); in other words, the novelty must be relevant and effective! There is also a social component to the biological driver just mentioned: there is an *interaction* between the biological makeup of the individual and some aspect of the environment in which the person operates.

Social Motivation and Creativity

The impulse to be creative is also some or all of the following:

- economic (to make money);
- professional (because it is part of the job);
- personal (e.g., because someone is curious);
- social (e.g., because it brings status and acclaim).

Thus, in addition to affecting the kind and amount of novelty produced, as well as determining which novelty is judged to be effective, the social environment also plays a substantial role in determining whether people are *inclined to produce* novelty at all, i.e., in motivating (or not motivating) the production of novelty. Many creative products are developed "to satisfy the needs of ... social groups" (Sosa & Gero, 2003). The *needs* may be *concrete* and down-to-earth, such as cheaper power, or a cure for a particular disease. They may also be more *general*, such as better educational methods. Equally, they may be more *abstract*, such as improved ways of expressing feelings through music. Generally, the "social groups" consist of people who are knowledgeable in a domain—specialists or experts—and users of the domain. In much the same way that someone with no knowledge in a given domain is unlikely to *produce* creativity in it, someone who does not know that a particular domain even exists cannot experience a *need* for creativity in it. The people who are *motivated* to solve the problems of a domain are most commonly people active in the domain.

The idea that creativity is linked with meeting the needs of social groups means that the problems creative people seek to solve are at least partly *socially determined*. Where there is no social awareness that a problem exists, there may be no drive to produce solutions and no creativity. A simple example is the area of the design of common objects. An artifact may be awkward to use and inefficient, or possibly even dangerous. However, it may be so familiar to so many people that they have become accustomed to its disadvantages and may be able to use it very effectively, despite the disadvantages and inconvenience. They may even be incapable of imagining that things could be different. In this case, there is *no social pressure* to introduce effective novelty and, in a sense, no problem, no matter how bad the design may be, because *society has decided* there is no problem.

A good example is the automobile. The internal combustion (gasoline) engine is fairly inefficient, cars are very dangerous, and they pollute the environment. However, novelty in automobile design is limited to tinkering with details, and little genuinely radical originality has been seen since the introduction of the horseless carriage about 100 years ago. In fact, the motor car is only a coach or wagon with a motor replacing the horses! The basic design of a rectangular box, with a wheel at each corner, into which people climb, was well known *thousands of years ago*. Even the hybrid car is nothing more than a standard automobile with a different fuel system. This is, at least in part, because car manufacturers prefer the certainty of selling, for example, 100,000 standard (i.e., traditional) cars per year, rather than the uncertainty of trying to sell jet-powered hover cars! This is not necessarily because jet-powered hover cars *cannot* be built, but because society prefers the familiarity of the traditional design and doesn't see cars as inherently problematic. As a result, there's no social pressure motivating change.

Problem awareness in the individual, as against in the social group, has already been discussed (see Chapter 3). There may even be a tension between the society's problem awareness and that of individuals. The problem may be apparent only to experts in an area, or perhaps only to one such person, and may not provoke a publicly perceived need for novelty. If only the insiders, or even a single insider, are dissatisfied and experience the urge to produce relevant, effective novelty, then the social motivation may trump the individual motivation. In this case, the society's lack of problem awareness may inhibit motivation to introduce novelty and thus block creativity. This suggests that a *culture of problem awareness* would foster creativity.

Society, the Individual, and Creativity as a System

Fredrick Winslow Taylor, the father of modern studies of work and work training, started his own work career as a machinist on the shop floor of the Midvale Steel Works, and advanced through supervisory positions to become a member of senior management. He learned about scientific methods of observation and systematic drawing of conclusions when he studied engineering while an employee, and later transferred these to analysis of work practices in the steel works. Taylor's suggestions for scientific management came at a time when the new technology of fast steel cutting made new management practices possible. Because of its highly organized and systematic nature, the steel cutting technology permitted the new style of management. At the same time, fast steel cutting could not be organized within existing management practices. Because of their highly organized and systematic nature, the new management practices made fast steel cutting possible. The conclusion from this is that there is a *reciprocal relationship* between the introduction of effective novelty and the environment into which the novelty is introduced. Importantly, the dynamics of the relationships do not all go in the one direction: as has been shown, the environment permits or calls forth and directs or guides creativity, but creativity changes the environment.

Another variant of this interaction is to be seen in the way creativity not only is determined by social criteria, but itself shapes the criteria. Among other things, especially among domain insiders, effective novelty may

- push thinking about how to solve certain problems into a particular pathway. This may later become a corset, possibly acting as a source of tension for those active in the area and, paradoxically, blocking the emergence of further effective novelty. Earlier creativity blocks later creativity;
- alter the way other solutions in the area are judged (sometimes causing them to be judged as uncreative, the added value of the effective novelty pre-empting less value-rich novelty);
- provide new criteria for judging later solutions (with the danger of the corset effect mentioned above);
- expand the way the domain is conceptualized in the society, thus opening up new possibilities for creativity (this is the idea of *seminality;* see Chapter 4);
- suggest new issues not previously noticed (i.e., *germinality;* see Chapter 4); or
- suggest new ways of solving problems in the area.

Creative products thus not only reflect social forces, but may themselves alter those forces or even influence the way societies see the world. Sosa and Gero (2003) argued that the Sydney Opera House not only provided a solution to the problem of an opera house for Sydney and changed architecture and building techniques (there seems to be little doubt about this), but has also, as I discussed in Chapters 3 and 4, become part of the Australian consciousness and has become "an emblematic icon" of the Australian identity. The social aspects of creativity thus exert their influence in both directions: the environment influences generation of novelty and judges its effectiveness, and the novelty influences the society's willingness to tolerate novelty and how it judges effectiveness.

This interaction is also seen at the level of the individual. To give some examples:

- remote associates (see Chapter 5) arise out of deep domain-specific knowledge; broad, open perception; and networking in the processing and storing of information;
- resistance to group pressure is necessary for nonconformist behavior and autonomy of thinking—at least at certain times and in certain settings (such as the workplace);
- readiness to take risks permits remote associations;
- playfulness and willingness to experiment go with fluency and flexibility; and
- tolerance of ambiguity is supported by passion.

In fact, the influence pathways run in both directions: As Shaw (1989) put it, there are "loops" (I will discuss these more in Chapter 8).

The systems-like character of *interacting* factors—especially individuals and their environment—involved in the production of effective novelty is shown in Figure 7.3.

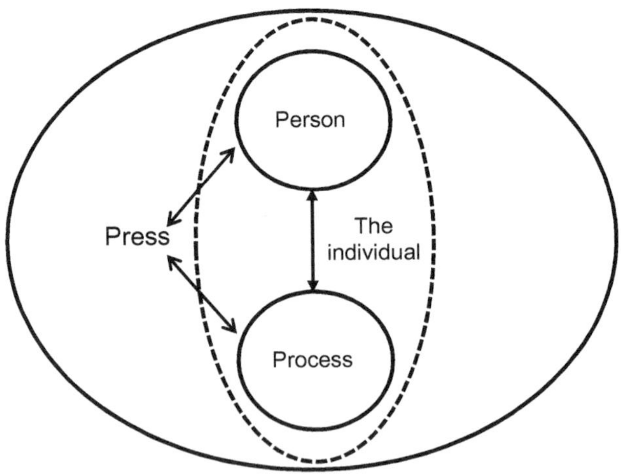

FIGURE 7.3 The interaction between properties of the individual and the environment.

Society's Ability to Tolerate Novelty

I have already discussed the fact that society defines *what is* creative, in part, as a result of its ability to *accept* the creativity. Consider the following example. An unusual design for a contemporary automobile would be to arrange the seats so that all passengers except the driver face the rear. This design would also be effective in dramatically reducing deaths and injuries among occupants of motor vehicles involved in serious collisions. However, it has never even been tried by auto manufacturers, because it is clear that car buyers would not accept it. Even more peculiar is the fact that infant car seats are made to face the rear, specifically because it is known that this is significantly safer. I have seen one figure that suggests it is five times safer! Despite the value placed on creativity and innovation in many contemporary discussions, not all deviations from the commonplace are equally acceptable to a society.

This phenomenon goes beyond acceptance or not of specific pieces of novelty, and extends to generation of novelty itself. Pseudo- and quasi-creativity (see Chapter 4) are often treated as harmless dreaming, letting off steam, etc., even if they are regarded as having no social value. However, some behavior that deviates from the social norms awakens anger, resentment, or rejection; the example of Andres Serrano's controversial art makes this point. In fact, only certain deviations will be tolerated by a particular environment. Some people even have a vested interest in maintaining the *status quo*. For instance, scientists who have invested a lifetime's work in a particular paradigm are understandably likely to resist novelty, even if it is effective.

Of course, many of the behaviors that lie outside a society's norms and are labeled *criminal* really are unacceptable in anyone's terms (think back to the discussion of *malevolent creativity* in this chapter). However, other proscribed behaviors are guilty only of deviating *too much* from what the society will tolerate at the present time. A society's reaction to levels of novelty that exceed the limits—that introduce *intolerable* levels of surprise—is closely connected with the age, occupation, or social role of the person involved. For example, in typical Western society, an artist is allowed to be more outrageous than an engineer, or a brain surgeon, while society is more tolerant of deviation in children than in adults. Indeed, as discussed in following sections, there seems to be an inverse relationship between age and tolerance of deviation: the older people are, the less they are expected to generate effective novelty!

Social Mechanisms that Encourage or Discourage Creativity

In the course of human development, we learn specific behaviors for specific situations: how to obtain food or move about in safety; rules for living harmoniously with other people; cultural skills such as language; a concept of the good person; and techniques for dealing with stress, anxiety, and so on. These are acquired through interactions with the various element of society: our family, schooling, our peers, our role models, various forms of media. Furthermore, they are strengthened by having an affective (i.e., emotional) or evaluational component. Thus, children learn not only that we *do* something in a particular way, but that those who do it *differently* are ignorant, naughty, or evil. For example, a child in one social group may be taught to eat with his fork held in his left hand and his knife in his right, placing food in the mouth with the fork, but always with the tines curving downward. This child may also be taught that people who eat differently are ignorant and/or of a lower social class (this was certainly still the case when I was a teenager in the United Kingdom in the 1980s). There was a time when, on first encountering the North American way of transferring the fork to the right hand and spooning up food with the fork held upside down, a child from Britain, for example, might be amazed at how many *ignorant* and *lower class* people there are in North America (according to their upbringing)!

In effect, generating variability involves breaking the social rules. All people are capable of a wide range of responses to life situations, but in the process of growing up, they learn that most ways of behaving are actually forbidden, and usually restrict their responses to a narrow range of socially tolerated behaviors—the preceding example makes this point. Regardless of how the British child viewed North American eating habits, it is likely that she would quickly learn to change her behavior, in order *to fit in*, if the child moved to the United States. A. J. Cropley (1967) studied the reactions of schoolchildren in social situations in which a number of alternative courses of action were possible, of which one was highly socially desirable (e.g., "You have promised to visit your grandmother but are tempted to go to the movies instead"). He concluded (p. 46) that the children were guided by "stop-rules" that forbade most of the wide range of possible reactions in a particular situation in favor of the socially approved, correct one. As Fromm (1980) put it, societies have "filters," and these inhibit divergent behavior or even discourage thinking about different possibilities. I am not only talking about obvious rules—in Australia, we drive on the left. If you want to try driving on the right, you can, but you will probably be arrested, hopefully before you cause an accident. There are rules not only about behavior but also about which *opinions* are correct, indeed about the right way of thinking and the contents of correct thought. Societies conduct "surveillance" (Amabile, Goldfarb, & Brackfleld, 1990) to detect people who deviate.

A simple example of the way society controls what novelty is generated can be seen in the phenomenon that is nowadays referred to as *political correctness*. Ideas that can be interpreted as criticizing certain groups, or as showing lack of respect for them, or as denying a particular status of that group are not discussed (or discussed at great risk). Political correctness is a form of social filter, but in some countries, it may actually be *illegal* to discuss certain ideas. This often prompts heated discussions about the extent of free speech. We even see this social filter in relation to issues such as the debate over climate change. A society's openness for novelty is frequently very limited. A result of this is that introducing novelty requires a special form of *courage*—a willingness to challenge prevailing views, even if this means being made a pariah.

There may be good reasons why societies have developed in this way. I have already discussed the fact that not every act of undisciplined, disruptive, or ignorant behavior, or every case of defiance, aggression or nonconformity should be acclaimed in the name of creativity. Knowledge, accuracy, speed, good memory, and the like are obviously important, most obviously for relevance and effectiveness. A society makes a substantial effort to train its members in its ways because this means that they can function effectively in a social environment. Indeed, acquisition of the social rules has an important survival value. To take a simple example, if city children do not know how to cross the road safely in high traffic areas, many of them will be killed or injured. Society has a strong interest in preserving most of the achievements of the past, as well as in limiting the extent to which people's behavior deviates from the well established. Some writers equate introduction of effective novelty exclusively with evolutionary change and imply that it cannot be introduced where the forces of preservation are strong. It is certainly true that most people would probably prefer the engineers who build the aircraft in which they fly to stick to the tried and trusted, rather than introducing too much novelty into the situation! Nonetheless, even such areas are not completely static. *Caution* is not the same as total absence of change, and this is another aspect of the constraints under which engineers operate when they engage in creative design.

A high level of conformity to social norms has the advantage that life becomes predictable, since it is more or less known what can be expected in everyday situations. However, the disadvantage is that unusual, unexpected behavior may become rare. In some societies, dislike of deviation may penetrate the public consciousness and become part of everyday, normal attitudes and values to such an extent that generation of novelty is subjected to extremely strong and widespread everyday sanctions. Gribov (1989) reported that the former Soviet Union was marked by a widespread public resentment of, and hostility toward, individuals who deviated from narrowly prescribed social

TABLE 7.2 Opposing Forces in a Society

Force	Effect	Change mode	Benefits
Conserving	Change: is relatively slow; builds on what already exists; may appear to be blocked.	*Evolutionary*	Despite change: the world remains orderly and understandable; existing knowledge and skills remain useful; people's feeling of security is not threatened; experts' self-image of competence is preserved; power structures and the like remain intact.
SLOGAN: "If it ain't broke, don't fix it!"			
Renewing	Change: is rapid (paradigm shift); sweeps away what already exists.	*Revolutionary*	As a result of change: novelty is obvious; progress is often rapid; problems are often solved quickly; people are encouraged to introduce novelty; existing structures are threatened.
SLOGAN: "Altius, citius, fortius!"			

norms and generated novelty. I mentioned this in Chapter 6, in relation to the individual, but Burkhardt's (1985) concept of *Gleichheitswahn* (sameness psychosis) also applies to a *societal* mass psychosis involving a drive to resist change.

There seem, in fact, to be two opposite forces in any society: *forces of conservation* and *forces of renewal*. The nature of these forces is summarized in Table 7.2.

Because these forces are logical opposites, they are often treated as being in opposition to each other. In fact, both are capable of producing change, and they can also work in a complimentary fashion. From a practical point of view, the biggest difference between them is that the *conserving* pressures in a society allow only slow, gradual *evolutionary* change, whereas the reforming pressures encourage more dramatic changes that are larger and occur quickly; i.e., they lead to *revolutionary* change. This is usually the kind of change people have in mind when they talk of creativity.

Degree of Openness of the Society

What we learn from this is that openness to the new (or lack of openness) is a characteristic not only of individual people (see Chapter 6), but also of *societies*. People who produce novelty in societies that are not open for it are likely to suffer various kinds of sanctions. The situation of such people is exacerbated by the fact that some traits associated with creativity may lead to disorganized, even chaotic behavior or to behavior that is regarded as antisocial or arrogant (e.g., impulsiveness, lack of concern about social norms, lack of interest in making a good impression, tendency to lose themselves in their work; again, see Chapter 6). Cognitive characteristics such as making remote associations that are *too remote* for most observers worsen the situation. The result may be that observers concentrate on the deviant or unpleasant, even antisocial behavior of the people concerned, and the link between their behavior and the production of effective novelty may become difficult for others to recognize.

In general, there are rules about breaking the rules! People publicly acclaimed as creative break the rules but succeed in staying within acceptable limits. If they do not, they are likely to be regarded as eccentric, immoral, mentally disturbed, or criminal rather than creative, with the possibility of being criticized, shunned, or even locked away.

Socially Assigned Roles and Creativity

Another way that society exerts a *control* on creativity is demonstrated by research on creativity and *age*. There is a well-documented relationship between the two. Focusing on people who actually became famous for their creative achievements (i.e., producers of acclaimed novelty), the early classic study of Lehman (1953) reported that peak performances occurred most frequently between 30 and 40. This view is still widely supported (see Simonton, 1988a, for a detailed analysis based on case studies of famous people). Lehman's findings indicated that the age at which peak performances occur varies from discipline to discipline, mathematicians tending to become famous particularly early. Nonetheless, there is agreement in the research literature that, allowing for differences in definitions and methodology, somewhere around 40 is the most productive age. Despite this, one should bear in mind that many famous creative individuals continued to produce until well into later life: Darwin, Freud, and Einstein became famous in their 20s and remained active into their 70s. Those who start youngest seem to continue longest!

However, a major weakness of many case studies of age and acknowledged creativity is that they often include creative people who died young. Naturally, such people made their creative achievements at an early age! The picture becomes somewhat different when people who have lived into their 70s and longer are studied separately. Lindauer (1993) showed that the average age of peak performance of long-lived artists was not 35, but 50. Focusing on individual people, he reported substantial differences from person to person in patterns of creativity and age. In general, longer-lived famous artists maintained a high peak of creativity over a period of three decades, usually in their 30s, 40s, and 50s. Peak creativity was unusual in the 20s (although not nonexistent) and fell off in the 60s and later, although it was more common after 50 than before 30. A high level of creativity even in old age was quite common. Two members of the group of artists studied by Lindauer reached their peak in their 20s, to be sure, but six reached it in their 60s. There were noticeable gender differences: famous female artists produced more creative work in their 20s, and men more in their 60s or even later, although both genders experienced their peak years between 30 and 50. This suggests a social effect—in this case possibly sex-role expectations—on the development of creativity. I will return to the issue of roles and creativity, as well as gender and creativity, later in this chapter, and comment on this in relation to engineering.

Turning from artists to scientists, Root-Bernstein, Bernstein, and Garnier (1993) studied the productivity of 40 men who had all made enduring high-impact contributions in physics, chemistry, biochemistry, and biology, including several Nobel Prize winners and a number of men who had been nominated for the Nobel Prize without actually winning it, some of them on more than one occasion. The contributions of these men were studied over a period of 20 years. The authors concluded that a fall-off in creativity after early achievements is common but by no means necessary: many of the people they studied went on producing into their 50s and 60s. Of particular interest for the present discussion is that those who made a single achievement early and then ceased to be creative tended to have moved into *management*, whereas those who continued to be creative avoided administrative work! We see this occurring in engineering very clearly, in part, because the pathways to higher salaries are often only those that involve management. Dudek and Hall (1991) showed that architects who resisted retiring were creatively productive for many years more than those who retired early.

The existence of a negative correlation between creativity and age (older means less creative) may be due, at least in part, to social factors, and not to disappearance of the psychological potential for creativity at all. These factors include the social convention of ceasing to work at about 60–65; the expectation that older members of a craft, trade, or profession will concentrate on training the next generation and supervising their work; the expectation that such people will focus on exploring novelty generated by younger people; or that they will become guardians of the *status quo*. These are all examples of the mechanisms through which social factors may lead to reduced creativity at older ages, and the mechanisms are more closely connected with socially defined career patterns and the behavior expected in certain social roles than with ability, personality, or motivation.

THE INSTITUTIONAL ENVIRONMENT AND CREATIVITY

At the start of this chapter, I explained that there are two contexts to our discussion of *Press*. I have examined the broader *social* environment, and I now want to turn our attention to the more specific *institutional*, or organizational, environment. Many of the same principles apply; however, some factors are more active in this narrower context. As engineers, we are also concerned with the organizational press because of the fact that engineering—whether education or work—usually takes place within a well-defined institutional context. The environment in which people learn or work plays a major role in encouraging or discouraging generation of effective novelty. Moreover, this environment goes beyond physical structures and equipment and includes the people in it, their attitudes, values, goals, and the like, and the way these are perceived (in other words, the organizational *climate* or *culture*). Working with other people in groups and teams is another aspect of this press, and has both favorable and unfavorable effects on production of effective novelty—largely through *roles* people play, *power*, *group processes*, the system of *rewards*. *Thought leaders* (professors, managers) can affect an organization's climate through the role models they offer, the way they acknowledge performance, and their effects on communication.

The Problem of Facing Organizations

Turning to business, Higgins (1994) described 10 challenges that he anticipated organizations would have to master with the help of innovation (and creativity) in the first decade of the 21st century. These challenges describe difficulties and potential constraints, but they also offer opportunities. The challenges include an accelerating rate of change, increasing competition, globalization, and the transformation of First World economies from industrial to knowledge-based economies. These factors mean that business is operating in an environment that is not only highly competitive, but also unpredictable. Indeed, economic theory suggests that returns on investments in rich countries should have been lower during the second half of the 20th century than during the first half, because the stock of capital was rising faster than the workforce. However, the fact is that ROIs were considerably higher. How was this possible? The decisive factor that defeated the law of diminishing returns was the *addition to the system of new knowledge and technology*, i.e., creativity and innovation. In fact, *innovation* currently accounts for more than half of economic growth (Thanksgiving for innovation, 2002, p. 13).

Pilzer (1990) describes this phenomenon through the concept of *economic alchemy*. He argues that a society's wealth is not limited by the availability of physical resources but that "technology controls both the definition and the supply of physical resources" (p. 2). It follows, therefore, that "[i]n fact, for the past few decades, it has been the backlog of unimplemented technological advances, rather than unused physical resources, that has been the determinant of real growth" (p. 2). I take this a step further by suggesting that it is a *lack of innovation* (the exploitation of creativity), or *poor* implementation, that has prevented growth from reaching its real potential. Looking to business in the future, writers such as Oldham and Cummings (1996) concluded that innovation is a key factor in the prosperity of organizations exposed to the conditions that exist today.

Unfortunately, psychological analyses of organizations show that they resist the introduction of novelty (Katz & Kahn, 1978), even minor change. Florida (2002) refers back to the work of Olson (1982), who discussed the way organizations resist change. Olson described the phenomenon that once an organization has prospered as a result of functioning in a certain way, it is difficult or even impossible for it to adopt novelty, no matter how effective it might be. Olson called this "institutional sclerosis"—a kind of organizational hardening of the arteries. Cultural and attitudinal norms become so powerfully ingrained that the organization rejects new ways of doing things. This frequently acts to stamp on creative people and stamp out the introduction of effective novelty. How can this be prevented, and how can an organization's natural resistance to change be broken down?

The processes involved in change can be mapped onto the three steps of innovation: *generation* of novelty, *exploration* of the novelty (including *evaluation* of effectiveness), and *exploitation* of the novelty. Organizational sclerosis first *discourages* the generation of novelty—*we've always done things this way here, and it's worked for us*! It then uses the process of exploration to *discredit* and *belittle* the novelty—*those ideas are no good; our focus groups rejected them*! Finally, sclerosis *blocks* exploitation—*sorry, but we are prioritizing project X instead of yours*! Naturally, as I have already discussed, I am not advocating change simply for the sake of change. *Not* all change is good, so that even in an organization open to creativity and innovation, the exploration phase is of great importance. Consider, for instance, how well the Coca-Cola Company *explored* its own decision to introduce *new* Coke in 1985. Ultimately, the company had to back down in the face of consumer resistance and re-introduce *classic* Coke.[3] Was this a case of unexplored change that should have been rejected? Unexplored (or *blind*) change or incorrectly explored change can cause problems, as with the Coke example, while even *successful* change involves risks such as overconfidence.

Organizations as the Site of Creativity

In relation to organizations and creativity, there are two broad strategies for approaching the task of fostering of creativity. Sosa and Gero (2003) called these "bottom-up" (focused on dispositional characteristics of the *individual* such as intelligence, personality, interests or motives) or "top-down" (focused on leadership, roles, group pressure, distribution of power, system of rewards, and so on, in the environment in which novelty is to be generated, explored, and inserted). As Sosa and Gero (2003) pointed out, behaviors that seem to be remarkable and to require a complex explanation in terms of personal dispositional factors (i.e., bottom-up) may seem unremarkable when looked at top-down. The approach in this book has been somewhat bottom-up in nature, in the sense that we have looked at different aspects of the 4Ps in turn, but also top-down, in the sense that we understand there is a system-level interaction of the 4Ps that results in creativity. If we isolate just the *Person* and *Press* for a moment, the point of the present discussion is to shift our thinking about this interaction and to give temporary precedence to the top-down, *Press* factors.

One reason for taking this shift in focus is that the extent of the influence of the environment is often underestimated, especially in psychological discussions of creativity. This is probably due in part to the so-called fundamental attribution error identified by social psychologists Ross and Nisbett (1991): when explaining behavior, observers tend to *over*estimate the importance of characteristics of the individual and underestimate the effects of the environment. This tendency is known to be particularly marked when attempting to explain *unusual* behavior. Since creativity is by definition surprising, its study would be particularly susceptible to the fundamental attribution error.

However, as I have already emphasized, the appearance of creativity in a given *Press* depends on an *interaction* between personal properties

[3]Of course, this may have been simply an innovative approach to marketing. If that is the case, then the example still serves as an important case study of creativity in a particular domain.

and the environment. For the purposes of the present sections, I am interested in the way in which institutional settings in particular enable (or disable) people to become creative. One way of conceptualizing this enabling/disabling function without disregarding personal properties was suggested by Harrington (1999): he suggested that it is a matter of *goodness of fit* between the organizational environment and the properties of individuals.

A Broad Understanding of "Organization"

Drilling down to this more specific level of *Press* (look back at Figure 7.1), it is helpful to define more precisely what I mean by *organization*. For present purposes, an organization incorporates

- *material institutional structures and facilities* such as work stations, laboratories, information-processing facilities, libraries, classrooms and workshops, etc. These are found in businesses, factories, and the like, but also in schools and universities;
- *people*, not only managers or instructors, but also fellow workers or students;
- *immaterial institutional factors* influencing the interactions between material structures and people, such as traditions, standards, norms, and customs;
- *psychological institutional factors* influencing these interactions, such as roles, relationships, social hierarchies, interaction rules, communication pathways, and the like.

Figure 7.4 shows this organizational Press in more detail.

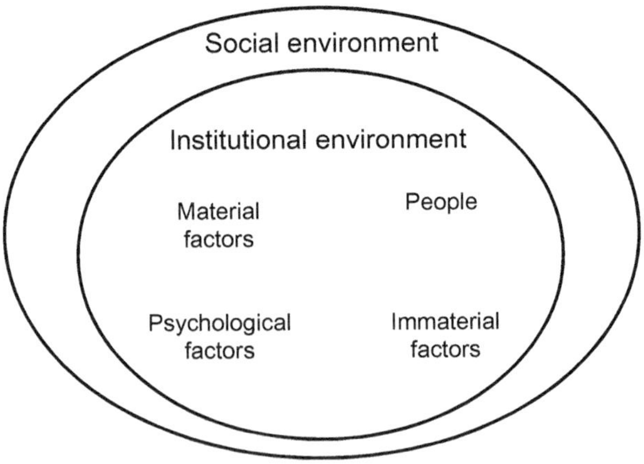

FIGURE 7.4 Factors of the institutional environment.

The Congenial Environment

Earlier in this chapter, I referred to an environment that encourages the production of effective novelty. The previous emphasis was on the role of the social environment in providing source material that can be manipulated to generate novelty, in deciding what is creative, or in determining effectiveness. In this section, a *congenial* environment can be thought of as providing more specific conditions that permit, release, encourage, or foster the creativity of individual people or of groups. These include

- the *amount* of divergence or risk taking that is tolerated/encouraged;
- the *kind* of variability that is tolerated/encouraged (for instance, routine extensions of the already known versus radical deviations);
- the *resources* that are made available (not only material, but also human) to support production of novelty; and
- the *rewards* (or punishments) that are offered to people who diverge from the usual.

These are not substantially different from the factors that I discussed as social environmental factors; however, the way in which they play a role is somewhat different in the context of the organization. The quality, quantity, and timing of these factors affect production, exploration, and insertion of novelty by people functioning within the organization. For instance, in an institutional environment, the combination of encouragement of risk taking, tolerance of radical novelty, provision of ample time and other resources (funding, lab equipment, access to information), and high levels of reward for those who depart from the usual (promotion, bonuses) would be highly *congenial* and, not surprisingly, would be expected to encourage generation of novelty.

In fact, the beneficial effects of such circumstances are not restricted to encouraging divergent thinking (i.e., *cognitive* aspects of novelty production), but also promote

- a positive attitude to generation of variability in general;
- positive social status of creative individuals in teams, work and social groups;
- a positive self-image among divergent thinkers;
- appropriate motivation (e.g., the urge to innovate, willingness to take risks, tolerance of ambiguity); and
- willingness to express personal characteristics such as openness, nonconformity, independence, and flexibility.

One of the better-known examples of the deliberate creation of an environment that we would call *congenial* in the context of creativity is the Lockheed Martin *Skunk Works*® concept. Responsible for the company's Advanced Development Programs, the Skunk Works traces its origins to the development of advanced fighter aircraft in World War II. The term has become synonymous with any small and loosely structured team of people engaged in the development of a project in which a primary consideration is innovation. Typically, a *Skunk Works* operates with a high degree of independence from normal managerial and reporting constraints, and is protected from interference by a *champion*. By operating outside the *normal* rules of product development, physically separate from the main site of its parent company, and with minimal communication overhead and strong autonomy, a Skunk Works seeks to achieve rapid development of novel product concepts that subsequently may be reinserted into a normal product development cycle.

Some of the *rules* of Skunk Works that are relevant to the present discussion are

- teams must have a high degree of *autonomy*;
- project teams must be kept *small* (typically only 10%—25% of normal teams);
- reporting requirements must be kept to a minimum;
- a high degree of *cooperation* and *communication* between developer and customer must be maintained; and
- *reward* systems must be based on outcomes achieved.

Environments that are simultaneously challenging and supportive have been shown to sustain high levels of creativity in individuals and teams (West & Rickards, 1999). According to Mathisen and Einarsen (2004, p. 119), creativity-fostering organizations are characterized by

- ambitious goals, which promote dissatisfaction with the *status quo*;
- freedom and autonomy in (a) choosing tasks and (b) deciding how they are carried out;
- encouragement of ideas;
- time (for creating ideas);
- feedback, recognition, and rewards;
- lack of threat of sanctions when brave attempts go wrong;
- interest in excellence;
- expectation and support of creativity;
- permission to take risks;
- tolerance of errors;
- loosely specified objectives (or clearly specified objectives plus opportunities to challenge them).

The Bell Labs is another excellent example of how a highly creative organizational climate can be created and nurtured (Gertner, 2012). There is also increasing evidence of the role that simple physical factors play in fostering creativity. Dul, Ceylan, and Jaspers (2011), for example, have studied the impact of the physical environment (and the interaction of that with the social work environment) on creativity. Some of the physical factors that positively impact on creativity include plants, décor, visual access (to the outside environment), and lighting.

Assisters and Resisters

It is not just the organization of work or study that is important in encouraging creativity. Individual people also play an important role. Diaghilev is remembered as the father of Russian ballet, Stravinsky as one of the most famous composers born and bred in that country, yet both were *law* students at university in Saint Petersburg around 1900! While still law students, both came under the influence of the revered musician Rimsky-Korsakov, who advised each of them to give up law and focus on music, although he strongly advised Diaghilev against trying to become a composer as he intended. Diaghilev then focused on ballet. The result of Rimsky-Korsakov's influence on Stravinsky and Diaghilev was perhaps a loss to the law, but a vast enrichment of world music.

Another curious example is the development of novel submarine technology in Germany both prior to and during the Second World War. In 1934, an engineer at Germaniawerft in Kiel, Hellmuth Walter, developed a novel, closed-cycle propulsion system for submarines, based on hydrogen peroxide. The advantage of such a system is that it does not require a source of atmospheric oxygen for combustion, meaning that submarines using this system could remain underwater for longer periods, at much higher speeds. Unfortunately for Walter (and fortunately for the Western Allies), the idea was rejected by the German Naval High Command—they found his claims for an engine that could power a submarine at a submerged 30 knots, rather than the conventional 7 knots, *fanciful*. Walter persisted, and found a champion in Captain Karl Doenitz in 1937. With Doenitz's help, Walter won a contract, in 1939, to develop a prototype submarine using the new technology. Once this prototype demonstrated its performance, the Naval High Command were impressed and awarded Walter a contract to build six. The twist in the story is that Doenitz, now head of the German U-Boat service, did not want industrial resources taken away from the program building Germany's standard submarine, and killed the project.

The facilitating effect of people such as Rimsky-Korsakov or Doenitz is well known: Treffinger (1995) referred to the presence in organizational environments of "resisters" (people or circumstances that inhibit production of novelty) and "assisters" (people or circumstances that facilitate it). Although *assisters* can be resources and other abstractions, the focus here is human assisters: people who foster the generation and insertion of effective novelty by merely tolerating appropriate behavior, actively encouraging it, or even functioning as models of it.

A comprehensive study of 20th century British novelists (Crozier, 1999) concluded that differences in their productivity were largely attributable to the influence of "social support factors." Csikszentmihalyi (1988) postulated that "social support networks" are vital determinants of creativity in the lives of individual creators: these include parents and teachers, mentors, colleagues, and managers. In a discussion of introduction of novelty into an organization, Mumford and Moertl (2003) emphasized the importance of, among other things, a "persuasive and effective advocate" (p. 264)—in other words, what we would call a *champion*. These people are not always single individuals; they may also consist of groups or networks.

Assisters seem to be important among other things for development and maintenance of the intense *motivation* that is needed to generate, explore, and apply high levels of novelty. Petersen's (1989) study of hobby authors showed that support from other people—assisters—was vital in her participants' ability to avoid writer's block and maintain their motivation to write, thus demonstrating the importance of the social support system not only in acclaimed but also in everyday creativity (recall the 4Cs model of creativity discussed in Chapter 5). As Bloom and Sosniak (1985) showed, human creativity assisters need not be powerful figures like Rimsky-Korsakov or Doenitz, but in certain settings can be humble and unsung people such as a grade school teacher or perhaps a manager or colleague. Thus, some assisters energize, activate, or release creativity in others without necessarily producing effective novelty themselves. This is a major responsibility for instructors and managers: to facilitate the creativity of their students and colleagues.

An important function of such people is to offer creative individuals a safe space where they can break the rules without punishment, thus protecting them from social or other sanctions (such as having research funding cut off, being fired, or, in the case of students, failing a course). Another is to offer them a positive perspective on themselves, for instance, the view that their ideas are not crazy but creative. This recognition can help to foster the courage to deviate from what everyone else is doing, among other things, by offering an opportunity to test the limits of the acceptable without risk or feelings of guilt. Human assisters can also help creative people to communicate their ideas to others by acting as an advocate.

Institutional Climate

In addition to relatively concrete and specific assisters and resisters, however, it is possible to speak of a more general element of a creativity/innovation-friendly environment: its *climate*. I defined climate earlier in this chapter as the particular conditions *in* the environment. Siegel and Kaemmerer (1978), citing Litwin and Stringer (1968) focused on concrete aspects of climate by defining it as "a set of measurable properties of the work environment that are perceived by those working in the environment and influence their motivation and behavior" (p. 554). These include

- *recognition* of the value of generation of variability through promotion or raises or, in the case of colleges, praise, high grades, scholarships and assistantships, and the like;
- *decision-making processes* that do not stifle change by bogging it down in a quagmire of discussions and procedures;
- contact with *models of creative behavior*;
- provision of *appropriate opportunities* to generate novelty; and
- presence of *people who encourage generation* of variability.

As Ekvall (1996, p. 105) explained, an institutional climate is also a percept in the mind of the people working in it: "a conglomerate of attitudes, feelings, and behaviors that characterize life in an organization." These include a feeling that generation of novelty is welcome and that people who generate it are respected. It also involves factors like feelings of tolerance and safety. These also define a *congenial* environment.

CREATIVITY AND GENDER

Two particular aspects of the *Press* deserve some additional attention: gender and teams. This is true of the relationship between creativity and *any* environment—social and organizational—but I want to focus on these in the context of the *engineering* Press. Perhaps the most interesting social issue in the discussion of institutions and creativity is the question of possible gender differences. Are women more creative than men? Are men more creative than women? Is there no difference in the creativity of men and women? Does the environment play a role in support of one gender or the other? It cannot be denied that, historically speaking, far fewer women than men have been acclaimed for what in Chapter 6 I called *sublime* creativity. Unfortunately, this remains true despite the many contributions in the past of creative women and the fact that women's opportunities for receiving acclaim have undoubtedly been severely limited by, for instance, a refusal of those (men) who had the power and status to give socio-cultural validation to the novelty they generated. How is this to be understood? More importantly, if our focus is on *fostering* creativity, what must we do to ensure that the opportunity for creativity is maximized, both across society as a whole,

TABLE 7.3 Examples of Motivational, Personal, and Social Factors Associated with Creativity

Psychological domain		
Motivation	**Personality**	**Social skills**
• goal directedness • persistence • curiosity • risk taking (courage) • drive to ask questions (even "uncomfortable") • unwillingness simply to carry out orders • desire to do things differently • drive to reveal one's own unusual ideas to others • mastery drive • desire for acclaim	• openness to the novel • flexibility • independence • acceptance of things that are "different" • self-image as innovative and daring • tolerance for ambiguity • sensitivity to problems • mental toughness • autonomy • self-centeredness • intuitiveness • playfulness	• team work • willingness to go it alone • willingness to risk looking foolish • communication skills • confidence in a group • willingness to admit not having an answer • low level of respect for "sacred cows" • willingness to be disrespectful to authority • willingness to risk hurting people's feelings

and in organizations more specifically? Finally, is the engineering profession missing out on untapped creative potential through its failure to attract more women into the profession (and is that failure, at least in part, a *Press*-related issue)?

Characteristics Needed for Creativity

I will start by looking at creativity and gender in a general, individual way, and then relate this to Press. A. J. Cropley (2002) carried out a psychological analysis of the issue of creativity and gender based on concepts outlined in the preceding section. He started by listing some of the personal characteristics thought to be linked to creativity (see Chapter 6). These are summarized in Table 7.3. These characteristics were then used as the basis for an analysis of the relationship of gender and creativity (see the next section).

Stereotypes of "Male" and "Female"

Without disputing the existence of biological differences between males and females, we can say that there is a great deal of disagreement about the nature and extent of social and psychological differences

TABLE 7.4 Stereotypes of *Male* and *Female*

Area	Stereotype	
	Female	**Male**
Cognition	– concrete – narrowly focused – convergent – intuitive	– abstract – broadly focused – divergent – logical
Motivation	– irresolute – seeks security (avoids risks) – seeks to avoid failure – reactive – pursues long-term goals	– persistent – takes risks – seeks success – proactive – pursues short-term goals
Personality	– cautious – empathic – timid – sensitive – oriented toward feelings – lacking self-confidence – responsible	– daring – egocentric – aggressive – insensitive – oriented toward ideas – self-confident – adventurous
Social Properties	– people-oriented – wants to be liked – communicative – slow to come forward – allows herself to be dominated – gives in to authority – fears criticism	– task-oriented – wants to be respected – taciturn – seeks limelight – tries to dominate others – challenges authority – fights back when criticized

between them. It is also obvious that *male* and *female* are not discrete categories: some biological females display some characteristics traditionally socially labeled *male*, and some biological males display socially defined *female* characteristics. In fact, it is unclear to what extent gender is a matter of biological destiny, to what extent a social construct, and to what extent a psychological disposition. Thus, as *psychological* categories, *male* and *female* are probably best regarded as *stereotypes;* they describe common and general patterns but are neither all-encompassing nor exclusive. If we accept these as reasonable descriptors, then it is possible to see that typical differences, with regard to creativity, are thought to exist. Lipman-Blumen's (1996) distinction between male and female "achieving styles," for example, suggests that there are characteristics of cognition and personality that are stereotypically male or female. Table 7.4 summarizes some of these. The contents of the table are based on discussions in Millward and Freeman (2002), Powell (1993), and

Schein (1994), and strongly reflect the classic analysis of Maccoby and Jacklin (1974), but I have mapped these onto the psychological dimensions used to structure the discussion of creativity in this book.

When the requirements for creativity in Table 7.3—general characteristics of creativity—are juxtaposed with the *stereotypes* of male and female in Table 7.4, it becomes apparent that the *male stereotype* fits the requirements for creativity outlined in earlier chapters and summarized in Table 7.3 much better than the female stereotype! Before my female readers throw this book away, let me explore what this means in practice!

I have already pointed out that the gender profiles in Table 7.4 are *stereotypes*. The key point here is that these are heavily influenced and determined by the social and, in the case of engineering, organizational press. Remember the role that the Press plays in *what is* creative and *who is* creative. These stereotypes exert a strong influence on aspects of experience such as the way boys and girls are educated or treated by their parents and by society. To take a simple example, it is commonplace in Australian TV reports to see the fathers of adult female athletes who have just won a world title or an Olympic gold medal refer to their daughter as "my little girl." Apparently, the fathers are astonished that their cuddly little baby in pink ribbons has become a highly focused, competitive, high-achieving adult. By contrast, it is almost unimaginable that the parents of a similar *male* athlete would talk about their "little boy." Millward and Freeman (2002) linked society's stereotypes of male and female directly to management by drawing attention to evidence indicating that the stereotypes have consequences for the way female managers are regarded by their seniors (and thus for factors like authority and promotion), as well as for females' actual management behavior. In fact, Schein (1994) concluded that the stereotypes dog female managers from the very beginning of their careers. Indeed, in engineering, my personal view is that these stereotypes play a negative role even *before* a female ever embarks on an engineering career, and may be the single most important factor in the poor participation rates by women in engineering degrees and subsequent engineering careers. An important mechanism through which stereotypes affect the behavior of females and males is also *role expectations*. Scott and Bruce (1994) showed that these expectations have direct effects on creative behavior. For instance, not only do male managers expect their female colleagues to avoid risks, but the women too are familiar with the stereotype and the associated role expectations, and often tend to behave accordingly.

Lipman-Blumen (1996) carried out an extensive analysis of male–female stereotypes and the way males and females are shaped into different achieving styles during the process of psychological development. There are a number of psychological mechanisms that could lead people to acquire existing stereotypes:

- imitation (Bandura, 1962);
- identification with the same gender parent who conforms to the stereotype (Hoffman, 1971);
- differential reinforcement by parents, teachers, and the like of what are perceived as gender-appropriate behaviors (Fagot & Leinbach, 1993); or
- the view that acquisition of clear gender roles is vital for healthy psychological development (Kohlberg, 1966).

Thus, even if they are no more than stereotypes, a society's ideas on gender can affect not only what others regard as normal in men and women, what duties women are assigned, and so on, but also, through internalization of the stereotypes by women themselves, what ambitions they develop, what kind of management behavior they exhibit, and what careers they choose. More recent research has also addressed this issue through the construct *stereotype threat* (Spencer, Steele, & Quinn, 1999). What, then, is the answer to the question about men, women, and creativity?

The Paradoxical Personality Revisited

A. J. Cropley (1997a) drew attention to the relevance for answering this question of what he called the *paradoxes* of creativity. I already discussed this concept earlier in the book in relation to *Process* and *Person*. These *polarities*—apparently mutually exclusive or contradictory properties or states that have to be simultaneously present for effective novelty to be generated—help us not only to understand how to foster creativity, but also give us some insight into issues that may affect the participation of women in a domain such as engineering. Some examples of the paradoxes that we have already encountered are that creativity requires

- openness for multiple solutions versus drive to find the best possible solution;
- openness to material from the subconscious versus a strong orientation to reality;
- a critical, almost destructive attitude versus constructive problem solving;
- openness to a variety of possibilities versus passionate engagement for a particular solution;
- a self-centered attitude versus consideration for others;
- emotional involvement versus cool self-control;
- a *loner* work style versus capacity for teamwork;

- a high level of expertise versus ability to think like a novice;
- playfulness versus strong sense of responsibility; and
- low levels of need for feedback versus high levels of willingness to communicate results to others.

Striking about these paradoxes is that at one pole they involve what seems to be stereotypical *male* characteristics (e.g., coolness, independence, secretiveness, preference for logic); at the other, stereotypically *female* properties (e.g., emotional involvement, team attitude, communication skills, intuitiveness). Crucially, therefore, it appears that to achieve fully effective creativity, across the sequence of phases that I have already outlined, requires *both stereotypically male and stereotypically female properties*. In the same way that we need to be able to think divergently in some phases and convergently in others, this may be part of a more general pattern of oscillating between creativity-favoring styles that society has, for better or for worse, labeled *male* and *female*. What is necessary, therefore, to promote creativity in organizational settings is thus an "integration of opposites" based on "the art of balancing" (Urban, 1997). This point of view has been extensively elaborated by Lipman-Blumen (1996). She called for a fusion of stereotypically male and female styles to yield *connective* leadership.

From the point of view of engineering creativity in particular, I think that the preceding discussion also reveals a deeper issue that may be contributing to the poor participation by women. The tendency of engineering programs to focus on convergent, analytical material, as I've already discussed, not only damages the preparation of fledgling engineers by denying them the skills and abilities for divergent synthesis that we know are a vital part of creative design and engineering problem solving, but also deemphasizes the more stereotypically female elements of the creative problem-solving process. My concern is that young women look at prospective engineering programs and only see something that looks stereotypically *male*. By failing to sell the integration of opposites that in fact represents engineering problem solving, we may be making engineering both less appealing and less accessible to women. This is a shame, not only in terms of equality, but because it is just plain dumb if engineering—and creativity—is as important as I tried to explain in Chapters 1 and 2. Any profession that is making itself available only to about half of the population is going to struggle as a result.

In fact, there is mounting evidence that shows a link between women and innovation performance in organizations. The recent report by the Anita Borg Institute[4] titled *Innovation by Design: The Case*

[4]http://anitaborg.org/

for Investing in Women discusses evidence that having greater numbers of women at all levels in an organization is linked to improved performance and competitiveness. Specific outcomes include *increased innovation, better problem solving,* and *better group performance.* The mechanisms delivering these outcomes appear to be related to the greater diversity of teams with a better gender balance, an improved sense of psychological safety in balanced teams, greater self-confidence and willingness to experiment in balanced teams, and other factors that contribute directly to many of the creativity-enabling factors that I have already discussed.

GROUPS AND CREATIVITY

Another important aspect of the Press that resonates with engineering is question of groups (or teams) and creativity. As Harrington (1999, p. 333) put it, groups can provide a "responsive" or "nourishing" audience. Paulus (1999) listed a number of beneficial effects of working in a group on creativity: They can

- provide broader and more varied information than that possessed by a single person;
- motivate creative activity;
- provide models; and
- give feedback.

One of the most frequently cited benefits of working in groups is the positive effect they are thought to have for idea production. VanGundy (1984) pointed out that groups usually

- possess more knowledge;
- arouse intense interest in group members;
- develop a broader perspective on the problem;
- generate more and higher quality ideas;
- come up with more candidate solutions;
- make riskier decisions;
- explore candidate solutions, to use our terms, more effectively;
- enhance acceptance of novel solutions (in our terms, are more open); and
- lead to greater satisfaction with solutions.

Groups also tend to enhance implementation of solutions, which is not surprising, since a solution offered by a group solution will have the support of a greater number of *persuasive advocates* (the members of the group) than a solution worked out by a single person. All in all, VanGundy concluded that groups are most effective for solving problems that can be solved through division of labor, either because the problem can be broken into separate areas on which subteams can work simultaneously or because it can be broken into sequential stages that can be worked on and solved sequentially.

However, groups can also *inhibit* creativity. Larey and Paulus (1999) listed a number of creativity-inhibiting tendencies in groups:

- free riding (individuals reduce their effort and leave it up to the group);
- evaluation apprehension (fear of negative reactions from the others);
- production blocking (one person dominates and blocks others);
- social comparison (people make sure that their ideas conform to the group tendency);
- matching down (out of, for instance, solidarity, the standard drops to that of the weakest member of the group);
- focus on shared information (special knowledge of individuals is ignored or kept hidden because of the factors already listed);
- premature closure (to keep the peace or because of the urge to be democratic or respectful, group members agree too quickly); and
- fixed roles or fixed power structure (in the group there are leaders and followers or bosses and minions; the former possess authority, and the others do what they are told).

In numerous studies, group brainstorming has been shown to produce *fewer ideas than the same number of individuals brainstorming alone* (Paulus, 1999). Apparently, some of the group members hold back on ideas, possibly because of the factors just listed. This is particularly the case where the individuals who come together to form the group all possess much the same knowledge base: the group does not broaden knowledge or add new perspectives, but just has more people working on the same knowledge. As Puccio (1999) pointed out, research also shows that the effectiveness of brainstorming groups in generating *effective novelty* (as against producing a large number of ideas) depends strongly on the number of highly innovative people participating in the brainstorming: groups without innovators do not generate much effective novelty, even if they do brainstorm and produce ideas. Thus, brainstorming makes use of what is available in the group, rather than adding some new element that transcends the individuals in the group: the group is simply the sum of the links in the chain, not a new entity. There is a growing body of work on groups/teams and creativity, including, for example, Baer et al. (2008) examining aspects of team composition and personality, while Paulus and Nijstad (2003) look at a range of issues surrounding groups and creativity. Most recently, Mumford (2011) covered a wide range of topics relevant to organizations, including group/team factors in creativity.

Teamwork

If the group/team is the physical structure—the collection of people—then teamwork is a feature of what they do together. Thus, a group can fail to work together, or a group can work in such a way that they achieve far more than they can as individuals. In other words, teamwork describes the outcome of a group, when the whole is greater than the sum of the parts. This is what many of VanGundy's (1984) *advantages* describe. Abra (1994) showed that achieving spectacular breakthroughs often requires cooperation with others. Sir Harold Kroto (winner of the 1996 Nobel Prize in chemistry for the discovery of Fullerenes) and William Phillips (winner of the 1997 Nobel Prize in physics for development of methods to cool and trap atoms with laser light) are two examples of contemporary Nobel Prize winners who emphasized *teamwork* when discussing their own processes of innovation. Kroto (Frängsmyr, 1997) argued that competition must be avoided at all costs. In the 1997 Nobel Lecture, Phillips emphasized that he always worked in a team. He gave as examples of the team's function "testing out ideas," "getting other people's feedback," "getting their suggestions," "asking questions," and "answering questions."

Despite the benefits listed for working in teams, actions like taking a strong stand in favor of an innovation are risky, and people's willingness to be publicly creative/innovative is thus affected by, among other things, fear of being publicly wrong, of exposing themselves to criticism, or of looking foolish. Production of novelty also depends on people's dissatisfaction with the *status quo*, which may require standing against the team, or not accepting existing situations that the team sees as perfectly satisfactory. As a result, considerable courage and willingness are required to stand alone in a team. This is a situation in which a human *assister* can be of great value. What seems to be clear is that there are great advantages to be had for creativity when groups/teams are used, but that these must be managed with care and understanding (of what can inhibit creativity). A great deal of engineering work is done in teams, and it is therefore incumbent on teachers and managers to understand the interaction of groups and creativity.

Downstream Consequences

Earlier discussions emphasized that the introduction of novelty into a functioning system must build on what already exists. Thus, creative people must operate on the basis of a foundation of knowledge. In an organizational setting, this means that they not only must know about the material the team is working on, but also must be able to take account of factors such as the *goals of the organization*, and to calculate an innovation's *downstream consequences*. This is as true of institutions of higher education as it is of business. There is, for instance, no point in teaching students to learn in ways that will cause them to fail in other professors' classes, or make them incapable of getting a job after university. Thus, successful innovation also requires knowledge of the organizational *status quo* and the ability to make accurate decisions based on existing knowledge. This in turn requires familiarity with the facts, good memory, rapid recall, logical reasoning, and the like—aspects of what I have called "convergent" thinking. Once again, the message I am pushing is for a careful, informed integration of the principles of creativity and innovation, whether into a university setting or into a business organization like an engineering firm.

ASSESSING THE ORGANIZATIONAL CLIMATE

The last piece in our discussion of creativity and the Press is the question of measuring and diagnosing. In previous chapters, I have examined the different scales and instruments available to help us put our knowledge of creativity on a firm, quantitative footing. Whether as managers or teachers, we can draw on a range of instruments that have been developed for different aspects of the social and organizational environment. In the following sections, I summarize some of the better known of these and suggest other sources for finding reputable, evidence-based scales for other dimensions of the Press.

Tests of Organizational Conditions

Mathisen and Einarsen (2004) reviewed four organizational climate inventories and summarized their psychometric properties (see Table 7.5). The scale *KEYS: Assessing the Work Environment for Creativity* (Amabile & Gryskiewicz, 1989) has been used in organizations in many fields, including electronics, high tech, pharmaceuticals, manufacturing, and banking. It consists of 78 statements about the organization, such as "The tasks in my work are challenging," " I feel challenged by the work I am currently doing," "A great deal of creativity is called for in my daily work," and "I believe that I am currently very creative in my work." Respondents rate their own organization by answering "never or almost never in this organization," "sometimes," "often," "always or almost always." The scale rates the organization on 10 dimensions: organizational encouragement, supervisory encouragement, work group support, sufficient resources, challenging work, freedom, organizational impediments, workload pressure, creativity, and productivity.

TABLE 7.5 Overview of Tests of the Workplace Environment

Test	No. of items	Level	Aspects tested
Situational Outlook Questionnaire (English-language version of Creativity Climate Questionnaire—Swedish) (Isaksen et al., 2001)	50 statements: Participants agree/disagree on 4-point scale from "not at all applicable" to "applicable to a high extent"	Engineers and scientists	**Nine scales:** challenge; freedom; idea support; trust/ openness; playfulness/humor; debates; conflicts; risk taking; time for ideas
KEYS: Assessing the Work Environment for Creativity (Amabile & Gryskiewicz, 1989)	78 statements: Participants respond on 4-point scale: "never or almost never in this organization," "sometimes," "often," "always or almost always"	Many organizations including electronics, high tech, pharmaceuticals, manufacturing, and banking	**Ten scales:** organizational encouragement; supervisory encouragement; work group supports; sufficient resources; challenging work; freedom; organizational impediments; workload pressure; creativity; productivity
Siegel Support for Innovation Scale (Siegel & Kaemmerer, 1978)	61 items: Participants respond on 6-point Likert scale: "strongly agree" to "strongly disagree"	Schools, engineering firms, university school of nursing	**Five dimensions:** leadership (support of ideas, diffusion of power, support of workers' individual development); ownership (of ideas, procedures, and processes); norms for diversity (being different is accepted, workers choose ways to solve problems, creativity is rewarded); continuous development (fundamental assumptions of the organization are constantly questioned, its goals change, and its methods change); consistency (people work together toward goals)

(*Continued*)

TABLE 7.5 (Continued)

Test	No. of items	Level	Aspects tested
Team Climate Inventory (N. R. Anderson & West, 1994)	38 items: Sometimes responses on 7-point Likert scale from "not at all" to "completely," sometimes 5-point scale from "strongly disagree" to "strongly agree"	Health services University staff Oil companies TV production company	**Four dimensions:** vision (clearly defined goals, shared goals, attainable); participative safety (safe to present new ideas); task orientation (shared concern with excellence); support for innovation (approval and practical support of attempt to introduce novelty)
Lipman-Blumen Organizational Achieving Style Inventory (Lipman-Blumen, 1991)	45 items: Participants respond on 7-point Likert items from "never" to "always"	Many different organizations	**Three broad domains, each with three specific styles (9 styles in all):** relational style (vicarious, contributory, collaborative); direct style (intrinsic, competitive, power); instrumental style (entrusting, social, personal) Can be administered to assess organization and individuals, and assess the degree of match between relational styles of individual people and those of the organization.

The subscales have reliabilities (alpha coefficient) around .70–.85, and their validity is supported by factor-analytic studies, as well as some applications in real organizations.

A second scale is Isaksen and colleagues' (2001) *Situational Outlook Questionnaire,* an English-language version of a rating scale originally published in Sweden. It consists of 50 statements about the organization.

People rating the organization agree or disagree with these statements on a 4-point scale from "not at all applicable" to "applicable to a high extent." The scale has been applied mainly in scientific and engineering organizations, as well as large manufacturing and business firms. It yields scores for the organization on nine scales: challenge, freedom, idea support, trust/openness, playfulness/humor, debates, conflicts, risk taking, time for ideas. Alpha coefficients of .62–.90 are reported for the various subscales, with most of them being above .80. Factor analytic studies have confirmed that the scale measures nine dimensions.

Siegel and Kaemmerer's (1978) *Siegel Support for Innovation Scale* consists of 61 items with which respondents agree or disagree on a 6-point Likert scale ranging from "strongly agree" (1 point) to "strongly disagree" (4 points). It has been used in organizations, schools, engineering firms, and a university school of nursing. It assesses five dimensions of the organization: leadership (support of ideas, diffusion of power, support of workers' individual development); ownership (of ideas, procedures, and processes); norms for diversity (being different is accepted, workers choose ways to solve problems, creativity is rewarded); continuous development (fundamental assumptions of the organization are constantly questioned, its goals and its methods change); and consistency (people work together toward common goals).

The *Team Climate Inventory* (N. R. Anderson & West, 1994) has been used to assess, among others, health services, university staff, oil companies, and a TV production company. It consists of about 40 items, depending on the version being used. On some items, the people rating the organization respond on 7-point scales ranging from "not at all" to "completely," sometimes on 5-point scales from "strongly disagree" to "strongly agree." This scale yields scores on four dimensions: vision (the organization has clearly defined goals, shared goals, and attainable goals), participative safety (it is safe to present new ideas), task orientation (members of the organization have a shared concern with excellence), and support for innovation (within the organization there is approval and practical support of attempts to introduce novelty).

The idea of achieving a good fit between the characteristics of the people in an organization and the psychological characteristics of the organization has already been mentioned. Lipman-Blumen (1991) has specifically incorporated this into her assessment procedure, the *Achieving Styles Questionnaire* (ASI), for which there are two versions: one for organizations and one for individuals. The two forms can be administered and then used to make a diagnosis based, in essence, on goodness of fit. The individual version consists of 45 statements (e.g., "Faced with a task I prefer a team approach to an individual one," or "I achieve by guiding others toward their goals") to which participants respond on a 7-point Likert-type scale ranging from "never" to

TABLE 7.6 *Press* and the Phases of Problem Solving

		Invention					Exploitation	
Dimension	**Phase** **Poles**	**Preparation** Knowledge, problem recognition	**Activation** Problem definition, refinement	**Generation** Many candidate solutions	**Illumination** A few promising solutions	**Verification** A single optimal solution	**Communication** A working prototype	**Validation** A successful "product"
Press (Organizational climate)	High Demand vs. Low Demand	High	Low	Low	Low	High	High	High

"always." These define nine achieving styles such as "collaborative," "competitive," "vicarious," or "personal." Lipman-Blumen (1991) reported reliabilities (alpha coefficients) ranging from .82 to .91 for the nine subscales, while construct validity was demonstrated by means of factor analytic studies as well as correlations with data on variables like task accomplishment, gender roles, or leadership style. Puccio and Cabra (2010) review a wider range of studies and instruments, spanning nine different studies and many dimensions.

SUMMARY

Table 7.6 now maps the *Phases* of creative problem solving against the changing nature of the *Press*. Like the Product, Process, and Person, the characteristic of a favorable Press changes as the problem-solving activity unfolds.

CHAPTER

8

Innovation: Exploiting Creativity

"But remember this, Japanese boy... airplanes are not tools for war. They are not for making money. Airplanes are beautiful dreams. Engineers turn dreams into reality." **Hayao Miyazaki, 1941–, Film Director**

Back in Chapter 3, I showed how it was possible to blend together models of the creative problem-solving process and engineering design. I presented engineering design as a special case of creative problem solving and explained that four factors—the *4Ps*—help or hinder creativity (and therefore engineering design) at each stage of the process. I then spent the next four chapters studying each of those 4Ps—*Person, Product, Process,* and *Press*—in more detail in order to understand how they can act either to foster or inhibit creativity. I am now in a position to bring all of that information together in a single model that describes the impact of the 4Ps on the process of generating and exploiting effective and novel solutions. We can now also put a new label on that process: *innovation.*

The purpose of this chapter is to draw these threads together. I will briefly describe why the term *innovation* is an appropriate one for us to use now, and I will then give you the single model that serves as a high-level summary of all factors, phases, and their creativity-favoring poles. Equipped with this information, you will then know what aspects of the Product, Process, Person, and Press favor creativity, and therefore innovation, at each stage of the engineering problem-solving process, and be ready to drill deeper as required in order to get the best result at every step. D. H. Cropley and Cropley (2014) give a detailed discussion of the *Person* in the context of innovation. Finally, in keeping with my emphasis on measurement as the mechanism for turning ideas into useful tools, I will describe a measuring instrument—the *Innovation Phase Assessment Instrument* (IPAI). The IPAI has been developed to assist managers of innovation in diagnosing how well aligned a team or organization is to the conditions that we know favor creativity and innovation, at each step of the problem-solving process.

DEFINING INNOVATION

Early in this book, I discussed the drivers that are connecting creativity and engineering. These drivers stem from *change* and a pervasive need to react to that change with novel technological solutions. Over the preceding four chapters, I have built up the idea that creative technological solutions—*engineering solutions*—are designed and developed by individuals, frequently working in teams, and within the environment of specific organizations. Furthermore, these solutions are creative because they are, as a minimum, both *effective and novel*. Crucially, this activity takes place across a number of *phases*, each of which has particular requirements for creativity to be realized successfully. *Creativity*, at its core, is the *generation of effective novelty*.

The more we discuss the *overall activity*—developing novel technological solutions to problems—the more helpful it seems to be to have a term that is a little broader. In many ways, however, this is dictated by the different disciplines that have an interest in this activity. It is clear that engineers have an interest in creative problem solving—that is the reason for this book. However, economists are also vitally interested in the activity for what it tells them about how organizations and entire nations function economically. These economic approaches generally trace their development back to Schumpeter's (1942) Theory of Economic Development and are concerned with the commercial, financial, and organizational aspects of creative problem solving. They also prefer the term *innovation*. At the highest level of abstraction—a *macro* level (for instance, national innovation policy)—innovation is understood to be vital to meeting the challenges of the early 21st century arising from technological advances, social change, globalization, and now the global financial crisis. At an intermediate, *meso* level (the individual organization), innovation is "a key to organizational effectiveness and competitive advantage" (Davis, 2009, p. 25) and thus ultimately to commercial success and creation of wealth.

If creativity is the generation of effective novelty, then *innovation* implies a particular focus on the *exploitation* of that *effective novelty*. The term *value innovation* (W. C. Kim & Mauborgne, 2004; T. A. Dillon, Lee, & David, 2005) is somewhat more explicit: it focuses on innovation as a process through which organizations find novel and effective ways of serving their current customers and identifying new markets, thus linking innovation to what customers value. This terminology makes it clear that at the level of organizations' innovation is not just a matter of coming up with a new idea but requires a valuable product. However, *product* is not confined to technological devices or even tangible objects, but covers the full value chain, including marketing, market research, sales, advertising, distribution, and customer service.

Bledow et al. (2009) defined innovation as the development and intentional introduction into practice of new and useful ideas by individuals, teams, and organizations, while Luecke and Katz (2003) illustrated the process as shown in Figure 8.1. In the context of innovation, they preferred to think of creativity as invention, emphasizing the generational aspect of creativity in the overall process. Mokyr (1990) also distinguishes between invention and innovation, drawing on the work of Schumpeter and noting that "invention does not imply innovation" (p. 9). Mokyr (1990) further noted that "Without invention, innovation will eventually slow down and grind to a halt ... Without innovation, inventors will lack focus and have little economic incentive to pursue new ideas" (pp. 10–11). In other words, creativity and innovation are highly interdependent—two sides of the same problem-solving coin.

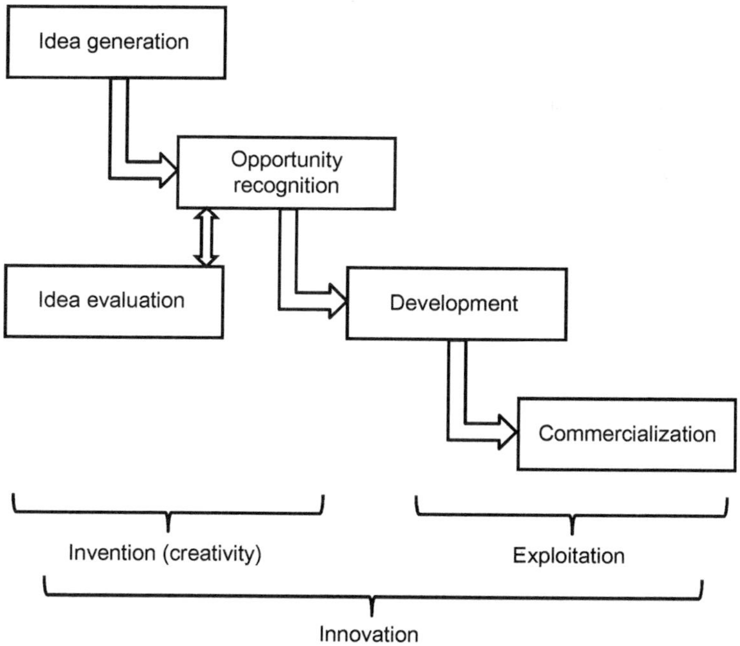

FIGURE 8.1 The innovation process (Luecke & Katz, 2003).

COMPETITION AND INNOVATION

Another benefit of thinking of creativity in the broader context of innovation and economics is that we develop a deeper understanding of the impact of a competitive business environment. In earlier chapters, I indicated that creativity, and especially engineering problem solving, is a competitive endeavor. I also suggested that the core characteristic of creativity—novelty—confers competitive advantages on the creating organization. Two types of competitive environment exist, and creativity/innovation plays an important role in each. *Scramble competition* (F. M. Hruby, 1998) describes a competitive landscape in which new players emerge with no notice with a new product. *Contest competition,* by contrast, involves established players battling for a share of existing markets. In both cases, the competitive advantage stems from effective novelty. This may, in the case of the *scramble* competitor, be the application of a new technology or the identification of a new need (or both)—in other words, driven by either technology push or market pull (see Chapter 2). In the case of the *contest* competitor, the drivers remain the same; however, the scale and pace of the innovation may be lower. The difference between scramble competitors and contest competitors therefore lies *not* in the nature of what they do—finding solutions to problems driven by supply (of technology) and (customer) demand—but in the scale and pace of how they do it. Scramble competitors are more likely to disrupt the competitive environment with *wholly* new technologies meeting, or creating, *completely* new needs, whereas contest competitors are more likely to make incremental changes to existing technologies and needs.

Christensen (1999) described these two classes of innovation as *sustaining*—"they give customers more and better in the attributes they already value" (p. 8), and *disruptive*—they "introduce a very different package of attributes to a marketplace than the ones that mainstream customers have historically valued" (p. 9). For the purpose of the present discussion, the key point is that both forms of competition draw on the generation of effective novelty—creativity—as the determinant of success. The question of how these different forms of innovation interact is also important in understanding the value of creativity.

UNDERSTANDING INNOVATION

Whatever term we used to describe the process of exploiting effective novelty—*value innovation* seems a good fit to the engineering problem-solving context—the key remains understanding what factors determine the success or failure of this activity. I have endeavored to build up a framework for understanding *creativity* across the stages of engineering problem solving, based on the 4Ps and a recognition that what is good for creativity in one phase may be bad for creativity in another.

The idea that there is a complex relationship between the 4Ps and the phases of creative problem solving is not without precedent in the literature of innovation. For example, Bledow et al. (2009) pointed out (p. 306) that a "pervasive theme in research on organizational innovation is that innovation is characterized by tensions ... paradoxes ... contradictions ... [and] dilemmas." These generate apparently "conflicting demands" and call for what seem to be "conflicting activities." Dealing with the conflicts necessitates a "dialectical perspective." Lewis et al. (2002) identified "tensions"; Miron, Erez, and Naveh (2004), "paradoxes"; and Benner and Tushman (2003), "dilemmas." Haner (2005) gave a good example of a paradox revealed by research on groups and innovation: innovation requires "simultaneous agreement and disagreement" (p. 291) among the members of the group, consensus, and yet absence of consensus. Looking at the individual, Hulsheger, Anderson, and Salgado (2009) gave another example: the need both to do things your own way and yet also rigorously implement other people's ideas.

These tensions, paradoxes, contradictions, and dilemmas are exactly what I have described in the context of engineering problem solving. Critically, the "key management issue is to integrate and balance ... complementary processes" (Bledow et al., 2009, p. 364) so that the process of developing and exploiting creative solutions to complex technological problems can be as successful as possible. Haner (2005, p. 297) called for research on "principles according to which organizational innovation ... can be conceptualized." Although they were not writing directly about innovation but indirectly via a discussion of creativity and personality, Batey and Furnham (2006) made a highly relevant call for "a comprehensive taxonomization" (p. 410). They went further and suggested that such a taxonomy should be based on the 4Ps approach.

THE INNOVATION PHASE ASSESSMENT INSTRUMENT (IPAI)

One response to the various calls for addressing paradoxes of innovation has been developed using the foundation of psychological concepts of creativity described in earlier chapters. In Chapter 3, I spelled out a model of the phases involved in creative and engineering problem solving. They map closely to the general model of innovation represented in Figure 8.1. For each one of these phases I showed, in Chapters 4–7, how each of the 4Ps—Product, Process, Person, and Press—must oscillate between two fundamental poles as the phases unfold. In some phases, for example, divergent thinking is necessary—or *active*—in creativity, whereas in other phases, convergent thinking is active. In the course of those four chapters, I have built up model of the innovation process—the exploitation of effective novelty—that provides an answer to the question of how to conceptualize innovation and how to integrate and balance seemingly conflicting requirements.

The Innovation Phase Model (Table 8.1) was first described by A. J. Cropley and Cropley (2008) in the context of creativity education, and later in D. H. Cropley and Cropley (2010b) for the first time in the more general context of organizational innovation. The model was then developed as the basis for *diagnosing* the *alignment* of organizations to conditions ideal for creativity and innovation in D. H. Cropley and Cropley (2011), and this became formalized as an instrument—the Innovation Phase Assessment Instrument (IPAI)—in D. H. Cropley and Cropley (2012) and D. H. Cropley et al. (2013).

The IPAI consists of a set of 168 questions that tap into the matrix of 7×6 (42) *nodes* represented by the intersection of each phase and the 4Ps (Table 8.1)—*Person* was given greater weight in the IPAI by expanding it to include motivation, personal properties, and feelings, as described in Chapter 6. For any given *node*—for example, the intersection of *Preparation* and *Process*—the IPAI defines four dichotomous (yes/no) questions that examine the respondent's view on which pole of the relevant paradox is representative of his organization. When the responses are analyzed, the IPAI is looking for differences between the ideal response for a given node—for example, for *Preparation/Process*, the ideal is convergent thinking—and the actual response given. Thus, while the theoretical ideal might be convergent thinking, a respondent might feel that divergent thinking is favored in the particular organization. By aggregating responses, one is able to build up a picture of an organization's overall alignment to the conditions that favor innovation at every stage. For example, if the members of an organization generally feel that people think divergently when that is appropriate, and convergently when that is appropriate, and so forth for the other Ps, then the organization is well aligned to the conditions necessary for success. Conversely, if members of an organization report a mismatch between the ideal conditions and actual conditions and behaviors, then the organization is poorly aligned.

Table 8.2 gives a hypothetical example of the data for an organization assessed with the IPAI. The score in each node ranges from "0"—*none* of the given responses corresponded to the ideal responses for that node—and "4"—*all given* responses corresponded to the ideal responses. Responses are aggregated across the entire sample and

TABLE 8.1 The Innovation Phase Model (IPM)

		Invention					Exploitation	
	Phase	**Preparation**	**Activation**	**Generation**	**Illumination**	**Verification**	**Communication**	**Validation**
Dimension	**Poles**	Knowledge, problem recognition	Problem definition, refinement	Many candidate solutions	A few promising solutions	A single optimal solution	A working prototype	A successful "product"
Process	Convergent vs. Divergent	Convergent	Divergent	Divergent	Convergent	Convergent	Mixed	Convergent
Person (Motivation)	Reactive vs. Proactive	Mixed	Proactive	Proactive	Proactive	Mixed	Reactive	Reactive
Person (Properties)	Adaptive vs. Innovative	Adaptive	Innovative	Innovative	Innovative	Adaptive	Adaptive	Adaptive
Person (Feelings)	Conserving vs. Generative	Conserving	Generative	Generative	Generative	Conserving	Conserving	Conserving
Product	Routine vs. Creative	Routine	Creative	Creative	Creative	Routine	Routine	Routine
Press	High Demand vs. Low Demand	High	Low	Low	Low	High	High	High

TABLE 8.2 Hypothetical IPAI Data

Phase dimension	Preparation	Activation	Generation	Illumination	Verification	Communication	Validation	Row sums
Process	3	3	2	3	3	3	4	**21**
Motivation	3	4	3	4	4	4	4	**26**
Personal Properties	3	3	4	3	2	4	4	**23**
Feelings	4	3	4	3	3	3	4	**24**
Product	4	2	2	3	3	4	1	**19**
Press	3	4	2	3	3	4	3	**22**
Column Sums	**20**	**19**	**17**	**19**	**18**	**22**	**20**	**135**

would normally result in more complex data; however, I have used integers in this example for simplicity and clarity.

If an organization was perfectly aligned to the ideal conditions, as indicated by the respondents, then the maximum aggregate score on the IPAI is "168." Similarly, for each Phase (e.g., Preparation) or Dimension (e.g., Process), it is possible to compute an aggregate. In Table 8.2, the aggregate score for the Preparation phase is "20," while for Process it is "21." This means that there are three levels of analysis possible in the IPAI: Node, Phase/Dimension, and Total.

The diagnostic rationale of the IPAI means that an appropriate approach to the interpretation of data is to examine *relative* strengths and weaknesses for a given organization. In addition, this interpretation must be made against an understanding of the aims and objectives of the organization. The total score is therefore of less interest than Phase, Dimension, and Node scores, with the possible exception of benchmarking across similar organizations. Once an organization has been assessed, the first level of analysis is to examine relative strengths and weaknesses for each Phase and Dimension. This is best done by computing a mean Phase score (19.29) and standard deviation (1.6) and a mean Dimension score (22.5) and standard deviation (2.43). Each individual Phase and Dimension score is then ranked as a strength, opportunity, threat, or weakness according to its score relative to the mean (Tables 8.3 and 8.4).

Finally, individual nodes can be examined to locate specific areas for attention. In the example given, analysis might focus, for example, on

TABLE 8.3 Interpretation of IPAI *Phase* Scores

Classification	Score band	Preparation	Activation	Generation	Illumination	Verification	Communication	Validation
Strength	>20.89							
Opportunity	19.29–20.89							
Threat	17.68–19.29							
Weakness	<17.68							

TABLE 8.4 Interpretation of IPAI *Dimension* Scores

Classification	Score band	Process	Motivation	Properties	Feelings	Product	Press
Strength	>24.93						
Opportunity	22.50–24.93						
Threat	20.07–22.50						
Weakness	<20.07						

the *Generation* phase (Weakness). Studying Table 8.2, we can see that there are some areas of strength (*Personal Properties* [4] and *Feelings* [4]), but notable areas of relative weakness (*Process* [2]; *Product* [2]; and *Press* [2]). The diagnostic role of the IPAI is then to provide guidance for improving the weaknesses as a means for maximizing the potential for successful overall innovation.

Guidance for improving weaknesses is therefore targeted and specific. A misalignment in Process means that, in the phase of Generation, where divergent thinking is most appropriate for a successful outcome (Table 8.1), there is a perceived misalignment in the organization; in other words, people in the organization perceive that divergent thinking is either not always possible or is hindered in some way in this phase. Further insight is then obtained by examining the specific questions that were misaligned for this node. For example, one of the misaligned questions for this node is "Staff like to link disparate ideas." The ideal response for the Generation phase is that staff *do* like to link disparate ideas; i.e., they are able to engage in making remote associations. An approach to improving the organization relative to this node might therefore be to provide employees with improved skills in divergent thinking and better guidance for when these skills are appropriate.

SUMMARY

The IPAI is a response to the call for better integration and management of the range of paradoxical factors that influence the successful execution of the engineering problem-solving process. Its purpose is not to classify organizations, but to assist managers and employees in understanding how each of the 4Ps affects the process of innovation. It chief strength is that, unlike other instruments, it does not assume a one-size-fits-all approach to creativity and innovation. The IPAI is based on a theoretical model that acknowledges the paradoxes that are inherent to creativity and innovation. What is good for creativity in one phase, for example, may inhibit creativity in another phase. The IPAI helps managers to understand where they are in the process and whether their organization is aligned to the conditions needed for success in each phase.

CHAPTER

9

Creativity Training

"You can't use up creativity. The more you use, the more you have."
Maya Angelou, 1928–2014, Author, Poet

The theme of reconnecting creativity and engineering now sits on a framework that is summarized by the Innovation Phase Model (IPM) presented in the preceding chapter. This connects the *4Ps—Person, Product, Process,* and *Press*—to the *Phases* of engineering problem solving. It tells us what is needed at each stage of the problem-solving process to maximize the opportunity for successful creativity and innovation. In earlier chapters, I also described the many scales and instruments available to us that make it possible to measure the 4Ps, while in Chapter 8, I presented the IPAI as a mechanism for managers of creativity and innovation to diagnose the alignment of their organization to the conditions that favor creativity and innovation.

A question that now arises concerns *what we do with this knowledge*. If, for example, a manager determines that her organization is misaligned based on an IPAI assessment—with a particular weakness suggested in divergent thinking, for example—she might drill deeper into this issue by testing her employees' capacity for divergent thinking using the Torrance Tests of Creative Thinking (see Chapter 5). Whether that reveals that the misalignment is due to poor skills in divergent thinking or that the real problem is an unfavorable workplace climate (Chapter 7), what is she to do about this? In other words, it is one thing (and a necessary prerequisite) to know that the problem exists, but another thing to *fix it*. This is where I turn to a discussion of creativity training and education. In this chapter, I will focus on this topic in a more general sense—blurring somewhat the lines between *training* (in the sense of gaining proficiency in a process or method) and *education* (in the sense of a more holistic, intellectual exercise). The technical differences were once explained to me as follows: how would you like your teenage daughter to learn about the birds and the bees—training or education? In relation to creativity, I will look at both the more general questions of creativity education, and the more specific issues of creativity training. I will follow this by discussing education in the specific case of engineering creativity in the next chapter.

CAN CREATIVITY BE TAUGHT?

A critical, underpinning premise for our discussion of training and education—one that makes change and improvement possible—is the assumption that creativity is *not* a fixed, inborn trait. Olken (1964), however, identified the belief, even among engineers, that creative people are born, not made, as the major factor blocking efforts to establish creativity training in the immediate post-Sputnik years. Even relatively recently, Edwards (2001) confirmed that numbers of creativity researchers still do not believe that creativity, as a psychological characteristic of individual people, can be inculcated where it is not already present. Indeed, since personality is known to be partly dependent on biological factors and is, in any case, substantially formed early in life, it is reasonable to question the ability of training to make people more open or flexible or to enjoy risk taking.

Fortunately, there is now widespread agreement that creativity is *not* the expression of a divine spark—bestowed on a few poets and artists by their muses. On the contrary, for the practical purpose of teaching people how to do it, creativity is best thought of as an accessible, although statistically uncommon, characteristic of people and products. Pioneers of the field such as Torrance (1972) were adamant that creativity can be taught. This is also the view of, for instance, Amabile (1983). Authors such as Richards, Kinney, Bennet, and Merzel (1988) or Runco and Richards (1997) showed that in the course of everyday life *ordinary* men and women frequently produce effective novelty in fine arts, the sciences, the humanities, crafts, and so on. A. J. Cropley (2001) gave a number of examples of "everyday" creativity, including generation of effective novelty in a women's weekly sewing circle or in a junior soccer team.

This position is optimistic in that it implies that many people can produce effectively novel products (i.e., be creative), if the circumstances are right. Recall also our discussion of the 4Cs model (Kaufman & Beghetto, 2009). The idea of *mini-c*, *little-c*, and *Pro-C* creativity, as distinct from the *Big-C* creativity of geniuses, suggests that it must be possible for *ordinary* people to generate effective novelty. If creativity is a statistically uncommon trait but is regularly exhibited by many people, then it must be something that can be learned. Nicholls (1972) went further. He argued that creativity is, in fact, a normally distributed trait like intelligence, and emphasized the importance of creativity in people who have never received public acclaim and never will—the little-c creatives. This is not to say that everybody can be expected to become a Big-C creative genius, just as not everybody performs astonishing feats of intelligence. It does suggest, however, that even in ordinary settings—engineering, for example—people can be *encouraged* to generate effective novelty consistent with their level of knowledge, ability, skill, talent, and experience.

Sternberg (2007) characterizes creativity as a "habit" that can be either encouraged or discouraged through *opportunity*, *encouragement*, and *rewards*. It should be clear to readers that the key factors of creativity that characterize the individual—*Person* and *Process*—are teachable in the sense that they can be trained into a person or developed through education. For example, I can *train you* to become more proficient at divergent thinking by getting you to spend one hour every day coming up with alternate uses for ordinary objects. I can also *educate you* in the definition of creativity so that your alternate uses training is enhanced with knowledge that divergent thinking requires fluency, flexibility, and originality. I can also train you to become more tolerant of ambiguity, for example, by presenting you with problems that are open-ended, and have unclear or incomplete information. I can also educate you as to the value of creativity as a job skill, thus increasing your intrinsic motivation. The list goes on. Once again, the 4Ps and the link to problem-solving phases provide the basis of our understanding of what is important to creativity, why it is important, and now, what to do about it.

Nevertheless, the question still is asked—*can we teach creativity*? The real question is not *can it be taught?* but *are we teaching it right*? This is an issue of *effectiveness*.

THE EFFECTIVENESS OF CREATIVITY TRAINING

There has been considerable debate over the *effectiveness* of creativity training in the research literature. From early in the modern, post-Sputnik era, doubts have been expressed about whether training actually achieves what it sets out to do, which is to foster creativity (Mansfield, Busse, & Krepelka, 1978). Wallach (1985), for example, argued that the effects of creativity-facilitating programs are very *narrow* and *specific*, and scarcely generalize to behavior in settings other than those closely resembling the training procedure itself. Treffinger, Sortore, and Cross (1993) concluded that it has *not* been shown that

- there are clearly definable effects of creativity training on specific cognitive or personal characteristics;

- particular programs foster specific aspects of psychological development;
- people with one particular psychological profile benefit from a specific program, whereas other people need a different one.

Whether this is a failure of the training or a question of *misalignment* of training and desired outcome is a question that Baer (1996) touched on and the explored further (Baer, 1998). One possible explanation for these apparent deficiencies is that creativity training has concentrated overwhelmingly, in the past, on the *cognitive* (i.e., *Process*) aspects of creativity, even if factors such as self-concept or positive attitudes to problem solving are sometimes considered. D. H. Cropley and Cropley (2000b) criticized this narrowness in the conceptualization of creativity inherent in many programs—they called this *fast-food creativity*—and called for an integrative, holistic (or *spinach*) approach. Urban (1997), likewise, saw training programs as too narrowly focused. An integrative approach is consistent with the position adopted in this book that creativity should be looked at in a differentiated way, and the wide range of components leading to production of effective novelty—the cognitive, personal, motivational, and social (i.e., 4Ps)—must all be taken into account in the design *and* assessment of training.

At least some of the evidence suggesting poor training outcomes may also be more specifically connected to the question of domain-specificity and creativity (see, for example, Chapters 3 and 4). This has been addressed, in part, through the Amusement Park Theory (Baer & Kaufman, 2005). I will talk more about this issue later in the chapter. Further evidence supporting the specificity of creativity training is found in Dow and Mayer (2004).

Despite the concerns just raised, there is, in fact, convincing empirical evidence that creativity training, *properly conceived and implemented*, is effective (Scott, Leritz, & Mumford, 2004a). Scott et al. identified 70 studies published in or after 1980 in which the effects of creativity training were tested empirically. The studies had to meet strict methodological criteria that eliminated some of the criticisms of earlier research reporting favorable effects of creativity training. For example, there had to be a specific focus on creativity training, the procedure employed in the training had to be clearly described, the measures used to assess the effectiveness of the training had to be clearly identified and described, and statistical data on effectiveness had to be provided. There also had to be some kind of control condition, either a control group or at least a test–retest design. The effects of creativity training were tested statistically by calculating *effect strengths* (in statistical terms, the strength of the influence of the independent variable—i.e., the training provided). Ma (2006) extended this work, conducting an extensive analysis of creativity training, covering various aspects such as divergent thinking as well as attitudes to creativity, and found a large effect size overall, confirming the findings of previous studies.

There is also considerable evidence of the positive benefits of specific creativity training in school and college curricula. A review by Hunsaker (2005) and specific examples (McGregor, 2001) support the beneficial effects of creativity training in this particular context. DeHaan (2009), for example, studied "specific instructional strategies" to promote creative problem-solving in science and engineering, while McFadzean (2002) discussed "paradigm stretching" techniques that may encourage creative ideas. Convincing evidence described by Diamond et al. (2007) showed that specific aspects of mental ability associated with problem solving (cognitive flexibility, working memory, and inhibitory control) were boosted by programs that focused on finding alternative ways to solve a problem. My own research, specifically with engineering students (D. H. Cropley & Cropley, 2000b), showed significant gains on the TCT-DP (see Chapter 5) following a six-week program of creativity-fostering activities.

Debate will no doubt continue regarding what *aspects* of creativity can be taught and trained, *how long* any benefits persist after training, and *who benefits* most from these interventions; however, there seems to be ample evidence that creativity can be fostered and developed through specific activities and with appropriate guidance.

The Effect of Creativity Training

The strongest effects found by Scott et al. (2004a) were obtained when the criterion was cognitive *processes* (i.e., improvements in people's divergent thinking and problem solving after training). Within the cognitive domain, the single largest effect of creativity training was on *originality* of thinking (see Chapters 4 and 5), although training also enhanced fluency, flexibility, and elaboration. Thus, after training, people produced a greater number of surprising ideas; in other words, their capacity to generate novelty improved. The second strongest effect of training was on creative performance (i.e., the creative *products* people generated after training). There were also noteworthy effects on *attitudes* (i.e., *Person*). The effects of creativity training were strong in both children and adults in both educational and noneducational organizations, and were found in both gifted and nongifted samples. There were sizable effects for both males and females, but the effects were larger for males, especially with regard to divergent thinking. Thus, we see that training is able to have a measurable effect on aspects of the *Person*, the *Process*, and the *Product*.

Effectiveness of Different *Kinds* of Training

Evidence is also available that some forms of creativity training work *better than* others: there are differences between training procedures in the *strength* of the effects obtained (Scott et al., 2004a). When cognitive, social, personality, motivational, and combined training procedures—different *kinds* of training—were compared, it was found that the cognitive (i.e., *Process*) approach had by far the largest effect. Scott et al. (2004a) divided *cognitive* training procedures according to the particular process that each procedure emphasized, and found that training in *problem identification* (corresponding to the *Activation* phase; see Chapter 3) and *idea generation* (in the *Generation* phase), contributed most to the success of the training.

The best way to foster these processes was to give participants opportunities to analyze novel, ill-defined problems. Mere unfettered expression of unexplored ideas was negatively related to the effectiveness of training. Scott et al. (2004a) also found that highly organized and systematic training, based on realistic examples and involving substantial periods of structured, focused practice (i.e., relevant to a field or domain), was most effective. Finally, training that started by introducing specific relevant concepts and basic principles and then moved to targeted practice aimed at acquiring specific skills achieved stronger effects than *holistic* training. In connection with the intrinsic versus extrinsic motivation debate (see Chapter 6), it should be noted that provision of evaluative feedback positively affected improvements in problem-solving and relevant performance criteria, but *inhibited* improvement in divergent thinking. The two studies by Scott et al. (2004a, 2004b) are enormously informative with regard to a wide range of factors that influence creativity training. The analysis extends to the effect of different forms of instructional media (lecture, video, etc.), forms of exercise (written, group, etc.) and types of assessment, and should be used to inform the design of curricula for engineering creativity in university programs (I will discuss this issue in more detail in the next chapter).

WHY DO WE NEED TO TEACH CREATIVITY?

It is reasonable to wonder why we need creativity training in the first place. Are we not naturally creative? If so, what is going wrong that requires a discussion of what is, in effect, *remedial* training? This is another hot topic in creativity research, especially in an educational context, and one which has received some renewed impetus with the publication (Kim, 2011) of a study strongly suggesting that, over a period of nearly 50 years, creativity scores have decreased significantly.

Among factors suggested as responsible for this "creativity crisis" is an increasing emphasis on high-stakes, standardized testing (Kim & Coxon, 2013; Sternberg, 2007). It is suggested, furthermore, that the emphasis on *convergent, one-right-answer* problem solving that this testing environment encourages may inadvertently (or unintentionally) suppress the opportunity to engage in divergent thinking, with a consequent impact on creativity. Unfortunately, there are good reasons to believe that this problem is also occurring in university-level engineering education.

However, Kim (2011) has suggested that the age range experiencing the largest declines in creative thinking—kindergarten to grade 3—is less impacted by school-related factors and more influenced by *domestic* factors. The blame therefore cannot be laid solely at the feet of elementary and secondary schooling. In fact, statistically significant declines are also present among older age groups, suggesting that the training and education—and indeed, the *environment*—that encourage and foster creativity are required at home, in pre-school education, at all levels of regular schooling, and also at college and at work.

WHAT ABILITIES NEED TO BE TRAINED?

The evidence presented so far tells us that there is a need for creativity training/education and that training/education can be designed to deliver specific improvements. The Scott et al. (2004a) data suggest that targeted, specific training is more effective than holistic approaches; however, it seems that this fails to account for some of the more qualitative aspects of fostering creativity. In any discussion of training versus education, we must examine the extent to which we merely wish to develop a skill in contrast to developing understanding. Is the need in engineering for people who are simply better at generating many ideas, or do we need engineers who also understand when and why the ideas need to be generated?

If our focus is only on *how* to generate ideas, then training may well be sufficient. If our focus is not only how, but also *what*, *when*, and *why*, then we need a more holistic framework as an umbrella to the specific tools and techniques. The 4Ps is such a framework, and I will discuss how this informs the development of a creativity *curriculum* in the next chapter. Thus, even though empirical evidence may not support holistic approaches to training, the reason is that the assessment of training effectiveness is focused *only on how*, not on what, when, and why. If we assess the effectiveness of a creativity training course by looking at *fluency* before and after the training, then it is no surprise that the greatest gains in fluency will come from practising generating lots of ideas.

It will be no surprise that the same training fails to deliver any measurable change to a person's ability to describe why creativity is important to organizations.

A bridge between the *quantitative focus* of training and the *qualitative focus* of education is provided by Sternberg (1985), Sternberg and Lubart (1995), and Sternberg and Williams (1996), who describe the link between creative *work* and three *abilities*. Each of these abilities can be developed through training and/or education:

- *Synthetic Ability:* Most commonly associated with the concept of *creativity*, synthetic ability is the ability to generate novel, relevant ideas, link disparate concepts, etc.
- *Analytic Ability:* Most commonly thought of as an ability to think critically, this ability supports the generation of ideas by evaluating and analyzing their value and potential.
- *Practical Ability:* This is the ability to implement ideas and move from theory to practice. This ability can be thought of as the transition from creativity to innovation.

Creativity—or perhaps it is better to say engineering problem solving—is more than just the ability to generate ideas. It requires a balance of these three abilities, some of which are developed through more quantitative training, and some through more qualitative education. Furthermore, embedding creativity in an organizational or educational setting requires finding a balance of all three.

Developing the Creativity *Habit*

How is a balanced approach to be achieved? Sternberg (2007) points to "twelve keys for developing the creativity habit in children" (p. 8) as one mechanism for addressing the problem (Table 9.1), while Kim (2011) outlines a range of similar steps necessary to encourage creativity among children and adolescents. Kim and Coxon (2013), furthermore, outline approaches to blending a standards-based curriculum with creative activities as a means for offsetting the decline in creativity even within the constraints of the current educational climate. In fact, there is no reason that this approach cannot be used to guide the development of creativity at all levels. Despite these helpful guidelines, many parents, teachers, faculty, and managers may lack the ability to translate these guiding principles into concrete actions and activities that provide relevant *opportunities* for creativity that can be suitably *encouraged* and *rewarded*. In advocating, for example, for a restoration of "free, uninterrupted time ... so that children can engage in reflective abstraction" (Kim, 2011) (p. 293) it may not be clear to parents and teachers what this is seeking to achieve and how it benefits creativity. Similarly, directing

TABLE 9.1 Sternberg's 12 Keys for Developing the Creativity Habit

#	Summary of key
1	Redefine Problems
2	Question and Analyze Assumptions
3	Do Not Assume that Creative Ideas Sell Themselves: Sell Them
4	Encourage Idea Generation
5	Recognize that Knowledge Is a Double-Edged Sword and Act Accordingly
6	Encourage Children to Identify and Surmount Obstacles
7	Encourage Sensible Risk-Taking
8	Encourage Tolerance of Ambiguity
9	Help Children Build Self-Efficacy
10	Help Children Find What They Love to Do
11	Teach Children the Importance of Delaying Gratification
12	Provide an Environment That Fosters Creativity

(Sternberg, 2007)

parents and teachers to "encourage idea generation" (Sternberg, 2007, p. 10), while absolutely correct, may be difficult for nonexperts to achieve without a suitable vehicle around which to build these habit-forming activities and behaviors. The material I have presented in this book provides the rationale, and our present discussion of training offers practical methods to addresses the means. Integrating real-world problem solving into the curriculum, as a means of fostering creativity, or indeed constructing similar meaningful activities for children at home, is likely to present challenges but has clear benefits (Kim & Coxon, 2013).

What may be lacking therefore, both at home and at school, is an understanding of *how* to generate suitable "opportunities" around which a structure of encouragement and reward can be built. Benson (2004), for example, presents some evidence to suggest that primary (elementary) school teachers are unclear on what creativity means, and believe that it is simply a matter of letting children "do their own thing" (p. 138) and that creativity, at its core, is "developed mainly through art and music" (p. 138). For parents, teachers, faculty, and managers to implement successfully the principles that underpin the creativity habit, more specific guidance is needed. What, they might ask, is suitable as a problem requiring divergent thinking that can serve as the basis for allowing children, students, and workers to develop good judgment, question assumptions, engage in idea generation, surmount obstacles, take sensible risks, tolerate ambiguity, build self-efficacy, and so on?

DOMAIN-SPECIFICITY AND CREATIVITY TRAINING

A recurring theme of this book is the development of a differentiated understanding of creativity revolving around the 4Ps. I have already pointed out that creativity-training procedures have tended to focus only on a cognitive (*Process*) approach. This narrowness is compounded by a second common assumption—that a single, general approach will foster all kinds of creativity in all kinds of people.

To what extent is such an undifferentiated approach useful? Dow and Mayer (2004) looked at the issue in a limited way (solving insight problems) but still made an interesting point. They argued in effect for a differentiation into *verbal*, *mathematical*, and *spatial* creativity. They concluded in three studies that training in solving spatial insight problems (which they specifically referred to as "creativity training") was the only form that consistently fostered creativity. More importantly, as I discussed in a previous section looking at the *effectiveness* of creativity training, spatial training only clearly improved the solving of spatial problems. Thus, Dow and Mayer showed that not all approaches to training creativity are equally effective and that the effects of training are *domain-specific*.

Baer (1998, p. 174) pointed out that "domain" can be understood in terms of "cognitive content domains" such as linguistic, mathematical, or musical, but also in terms of "task content domains" such as poetry writing or collage making. He summarized findings of a number of studies and reported that individual people's levels of creative performance in different domains (e.g., verbal, mathematical, mixed) were highly variable, with correlations between scores in different domains not uncommonly being *negative*. This was true for both cognitive content and also for task content. He also reported that creativity *training* showed similar domain-specificity, with benefits found only on tasks resembling the original training.

Both these reports suggest that

- creative performance is domain-specific (i.e., a person may show dramatically different levels of creativity in different domains), and
- creativity training is narrow in its effects.

This supports the call for differential diagnosis of creativity and subsequent differentiated training. At the same time, it is clear that a substantial diagnostic system accompanied by a wide range of specific programs would no doubt be more complex to administer. Once again, creativity presents us with a paradox. For economy and efficiency, a one-size-fits-all training approach that covers a number of domains and teaches a range of common creativity skills is desirable. This approach would achieve effects that transcended a single domain, and thus deliver general benefits, even if the effects of creativity training really are specific. Yet much evidence suggests that there are strongly domain-specific elements to creativity. What is the answer? Baer (1998) strongly supported this view, advocating use of a wide variety of tasks from different cognitive content domains as well as different task content domains.

Creativity in Different Domains

The idea that creativity can be trained by means of a general procedure implies that creativity is the same in all content domains; i.e., that creativity in, let us say, engineering is the same as creativity in philosophy. To what extent is this true? On the basis of an analysis of more than 1,000 eminent creators in various fields, in which he analyzed creativity in terms of the *psychological demands of the field* (in other words, he looked at the problem's effects on creativity), Ludwig (1998) concluded that the relationship between creativity and field of endeavor is governed by the mathematics of fractal geometry. This is the geometry of "self-similar" objects that are characterized by a structure that repeats itself again and again at progressively smaller scales.

Ludwig identified four central dimensions for describing these demands. Different fields differ by being

- impersonal versus emotive,
- objective versus subjective,
- structured versus unstructured,
- formal versus informal.

He then divided fields of activity into two kinds: "investigative," on the one hand, and "artistic" on the other. The characteristics of these two broad categories are summarized in Table 9.2.

The "examples" in the table are arranged hierarchically, with mathematics the most *impersonal, objective, structured,* and *formal;* medicine

TABLE 9.2 Differences among Academic Fields

Field			
Investigative		**Artistic**	
Characteristics	**Examples**	**Characteristics**	**Examples**
Impersonal	Mathematics	Emotive	Art history
Objective	Engineering	Subjective	Writing
Structured	Biology	Unstructured	Composing
Formal	Medicine	Informal	Visual arts

and art history in the middle; and visual arts the most *emotive, subjective, unstructured*, and *informal*.

Ludwig went on to show that the four-dimensional framework just outlined applies not only to the broad division into investigative versus artistic, but that it also applies *within* fields. Thus, it is possible to use the same four criteria to divide investigative fields once again into *natural* sciences, such as mathematics or physics (which are more impersonal, objective, structured, and formal), versus *social* sciences, such as economics or psychology (which are more emotive, subjective, unstructured, and informal). According to Ludwig, this subdivision can continue, always based on the four dimensions, yielding ever-finer differentiations. For instance, natural sciences could be divided into *physical* sciences that are more impersonal, objective, structured, and formal than *life* sciences. Similarly, physical sciences could be subdivided into *theoretical* versus *applied* sciences, the former being more impersonal, objective, structured, and formal than the latter. An applied field such as engineering could be divided into different subfields ranging from those that are extremely impersonal, objective, structured, and formal (e.g., *structural engineering*) to those that are (comparatively speaking) more emotive, subjective, unstructured, and informal (e.g., *environmental engineering*).

Ludwig's analysis suggests that if we wanted to train people to be creative in fields that are *investigative* (i.e., structured, formal, objective, and impersonal), a different approach to training would be necessary compared to fields that are *artistic* (i.e., personal, subjective, unstructured, and informal). I have already argued—and there are a number of recent discussions of this dichotomy (e.g., Baer, 2010; Baer, 2012)—that fostering creativity has both general aspects, connected mainly with *quantity* of novelty generated, and domain-specific aspects, which are more relevant to *quality* of novelty. Ludwig's research yields guidelines for the nature of domain-specific training that may be particularly helpful within a domain such as engineering.

GENERAL APPROACHES TO CREATIVITY TRAINING

Plucker (1998), however, defended a general approach, and made the interesting point that claims of domain-specificity, which seem to be highly differentiated, may in fact be too sweeping! Plucker's review suggests that

- *amount* of originality may be general, and
- *quality* may be domain-specific.

This view is consistent with one of the earliest distinctions I made about the nature of creativity—namely, that creativity has two core components:

- *generation* of novelty; and
- *exploration* of the novelty in order to identify *effective* aspects.

Although *generation* may be nonspecific (i.e., general), *exploration* may require knowledge, skills, ability, attitudes, values, a self-image, and social support that are (a) specific to particular tasks; (b) affected by the specific setting (i.e., *Press*); and (c) dependent on convergent factors (i.e., analysis).

This *general-yet-specific* approach is the concept that I have developed throughout the book. Think of the Innovation Phase Model (Table 8.1). The *Process* oscillates between convergent and divergent thinking, suggesting that training will need to address both forms. That one of these—divergent thinking—is more domain-general in nature, while the other is more specific to particular domains does not prevent us from formulating an approach to training. Similarly, the *Person* oscillates between various poles for motivation, properties, and feelings. Some of these may be more domain-general, while others are specific to particular domains. The same can be said for the remaining dimensions (*Product* and *Press*). This is also consistent with the division of the creativity of products into *amount* of creativity (more general) and *kind* of creativity (more specific). This suggests that general creativity-facilitating techniques (such as those described later in this chapter) will help to encourage increases in the *amount* of novelty generated (as well as other of the more domain-general aspects of creativity), but that more specific approaches are needed in order to improve the *quality* of this novelty. Perhaps this is the main difference between creativity training and creativity education. The former is more domain-general and addresses those aspects of the 4Ps that are common to all creative endeavors, whereas the latter builds on that by integrating the domain-specific elements necessary for a fully effective problem-solving process.

FOSTERING CREATIVITY IN INDIVIDUAL PEOPLE

I turn now to the fundamental question of how to help individual people generate effective novelty—domain-general creativity training. Before I launch into some detail, however, I draw attention to the fact that the boundaries of *domain-general* versus *domain-specific* and *training* versus *education* are still somewhat hazy. As Ludwig's (1998) taxonomy suggests, factors that are more specific in one broad area may be more general in another. For this reason, keep in mind that training methods and programs discussed in the following sections tend toward the domain-general, but this is not set in stone. This is why it is useful to consider these training options against a background of the Innovation Phase Model (Table 8.1), and with a specific domain (e.g., engineering) in mind.

In general, there are two approaches to training individuals for creativity:

- helping people *acquire* knowledge, skills, methods, attitudes, values, and personal properties that are necessary for creativity, but *which they do not already possess*; and
- *removing blocks*—e.g., fear of the unknown, lack of self-confidence—that prevent people from using the creativity *that they already possess*.

The first approach assumes that something needs to be added to people that is not yet there. Guilford (1950) promoted this approach. This leads to a concentration on *assisters* or *facilitators* of creativity. I discussed this in Chapter 7 mainly as an aspect of the *Press*, but assisters can be *anything* that promotes the acquisition of knowledge, skills, motives, and personal characteristics necessary for generation of effective novelty, and anything that enables the effective use of these acquired qualities. Nickerson (1999) gives a wide-ranging analysis of many of the core issues of enhancing creativity through training, while Pfeiffer and Thompson (2013) examine a number of aspects of creativity training in an educational context, including question of *when* and *how* the enhancement should take place. There are many methods, tools, and processes available for this purpose, and some of them will be discussed more fully in this chapter.

By contrast, the second approach is based on the argument that all people already possess creative ability, and that creativity will emerge naturally, provided that it is not inhibited or *blocked*—for instance, by conditions in organizations. The idea is that people know how to think divergently and make remote associations because they *already possess* the cognitive skills necessary for generation of variability, possibly, because these are part of the functioning of the central nervous system. It is also argued that people are open to new ideas, flexible in thinking, and have the courage to take risks, and so on, *at least as potentials*. However, in the otherwise highly desirable and necessary process of socialization (see the discussion about how *Press* influences creativity in Chapter 7), most people learn to suppress or hide these properties. Creativity is thus *blocked*, rather than lacking, possibly to the point where the ability is close to being completely lost. Fostering creativity is seen, in this second case, as a matter of eliminating blocks, principally in the areas of motivation and personal properties (e.g., fear of taking a risk, desire to avoid uncertainty, intolerance of ambiguity, fear of complexity).

Like much of what I have discussed about creativity, the reality is probably a complex blend of factors. Creativity arises as an emergent property of the interacting system of Person, Process, Product, and Press (Figure 1.2), across a series of phases. Any given person will have a blend of preexisting enabling qualities in varying degrees, some of which need to be added or strengthened (the first approach). In addition, that person will also have a set of currently active blocking factors that need to be reduced or eliminated (the second approach). I will discuss these two approaches in a little more detail and then move on to discuss specific tools, techniques, and programs that aim to achieve the goal of fostering creativity in individuals.

Inculcating What Was Not Previously There

The first approach to fostering individual creativity involves helping people develop skills and qualities that they do not already possess (or that they possess only weakly). In fact, this is the dominant approach to fostering creativity. Treffinger, Sortore, and Cross (1993) reported in excess of 250 published sources that can be regarded as resources for training people in how to generate novelty. Huczynski's (1983) encyclopedia of methods listed dozens, including "buzz groups," "flexastudy," "lateral thinking," and "mathetics." The *Mycoted* website[1] lists more than 200 procedures. A striking feature of these programs is that they focus overwhelmingly on training the cluster of thinking skills (*Process*) that I described in Chapter 5 as belonging to the family of *divergent thinking*. This is neither surprising, nor unwise, in view of Scott et al.'s (2004a) conclusion that divergent thinking has been shown by 50 years of research to contribute to "many forms of creative performance" (p. 363). Similarly, Scott et al. (2004b) note that "idea production training is clearly the most common approach" (p. 168), while Clapham (2003) reviews a number of common techniques.

This approach—*positive encouragement* of creativity—can go further than just the cognitive skills such as divergent thinking. Even without leaving the realm of the domain-general, the approach that underpins this book suggests that effective training should not confine itself only to teaching people thinking skills that they did not possess before. Instead, creativity training based on the *acquisition approach* should focus on

- encouraging people to *develop* their existing ability to generate novelty;

[1]http://www.mycoted.com/creativity/techniques/

- giving them *practice* in doing so;
- encouraging them to place a *high value* on generation of novelty;
- encouraging them to *feel good about themselves* when they do it; and
- encouraging a respect for creativity in others.

In other words, a more holistic yet domain-general approach gives more emphasis to *noncognitive* aspects. Domain-specificity can also be catered for drawing on Ludwig's (1998) concepts. Training aimed at getting people to generate novelty in a particular field (as opposed to continuing to generate orthodoxy in that field) would focus on the Ludwig's contrasting four dimensions of impersonality versus emotivity; objectivity versus subjectivity; structure versus lack of structure; and formality versus informality. Individuals in highly impersonal, objective, structured, and formal fields (perhaps *requirements engineering*, for example) would benefit from creativity training that helped them work in a more emotive, subjective, unstructured, and informal way. Conversely, individuals in a highly emotive, subjective, unstructured, and informal field (e.g., visual arts) would benefit from training in the opposite facets. Such an *acquisition* approach to noncognitive factors, based on Ludwig's (1998) four-dimensional model, would need to diagnose not only the characteristics of individual people but of individual fields too. I am advocating an *acquisition approach* that is a blend of the traditional domain-general focus on cognitive skills *and* a focus on domain-general noncognitive qualities.

Eliminating Blockers

The second approach described earlier—removing blocking factors—recognizes that these are partly properties of the environment (*Press*), both physical (such as facilities and resources) and social (e.g., organizational status afforded people who generate novelty, approval/disapproval offered by other people), but also personal (i.e., *Person*: personality and motivation). *Blockers* prevent or frustrate creativity not because they involve lack of ability, but because they *inhibit* people from doing what they already can. This occurs, for instance, because blocking factors make people *frightened* to be creative (e.g., an organizational block like a hostile boss) or channel people's thinking into unproductive pathways (e.g., a cognitive blocker). Some typical blocks in the *social climate of an organization* could include

- exaggerated success orientation;
- strict distinction between work and play;
- intolerance of questioning;
- crushing conformity pressure;

- rigid maintenance of strict sex roles; and
- intolerance of diversity.

However, because environmental and social factors are not *part of* the individual person, I will not discuss them in any further detail. This does not mean that they are not important! They may have profound effects on creativity. Our current focus on the individual, and especially on creativity training, means that the blocking effects of *Press* factors are outside the scope of this chapter. Training relevant to the intersection of creativity and the *Press* is a question of *how to manage enterprises for creativity*, rather than training individuals to be more creative. The material in Chapters 7 and 8 is the key.

Returning to the individual, typical blocks *located in a person's own mind* include

- inability or unwillingness to relax control and let ideas flow;
- inability to handle a flow of ideas when it occurs;
- fear of letting the imagination loose; and
- fear of giving the wrong answer.

Blocks also have *cognitive* (*Process*) aspects. Some of these are *more general*, for instance:

- excessive emphasis on speed;
- almost total reliance on external evaluation;
- one-sided emphasis on analytic thinking;
- inability to break an existing set; and
- heavy reliance on verbal communication.

Other cognitive blocks are *more specific*, such as

- assuming that new problems must be attacked from existing perspectives;
- imposing limits on the way a problem is looked at, more or less from habit;
- assuming that there is always a single, best answer, if we can only spot it; and
- assuming that the solution can only take a certain form.

It seems that many of these blocks can be eliminated with the help of simple, general procedures. A. J. Cropley (2001) summarized studies showing, for instance, that simply giving people examples of unusual responses seemed to increase their scores on a divergent thinking test. Other researchers reported higher scores on creativity tests brought about simply by allowing people (admittedly, children) to play with test materials or to watch a video of a comedian, or to watch a film of a person solving a creativity test. Indeed, even telling someone to *be more creative* can have a positive effect.

There are also activities more specifically designed for releasing blocked creativity. Cropley (1992b) gave a number of examples of simple games and game-like activities for breaking blocks. These include "bridge building," "idea production," or "creative connections," activities that are based on simple *thinking techniques* such as (a) reversing the problem; (b) considering the end result; (c) focusing on the dominant idea; and (d) discarding irrelevant constraints. Procedures such as these aim, in essence, at loosening people up. They seek to encourage expression of what is already there. This is why creativity trainers often start training sessions with such games. The idea is to get what is already there flowing. This is done at the personal level by breaking through unwillingness to relax control and let ideas flow, reducing fear of letting the imagination loose, encouraging people to break existing mindsets, encouraging their confidence in their own ability to handle a rush of ideas when it occurs, or reducing fear of giving the wrong answer. At a more cognitive level, the activities help to escape from stereotypical patterns of thinking, break down barriers against unusual ideas, avoid imposing self-generated limits on thinking that are not inherent in the task, and so on. The definitive term describing what is required in this area of creativity-facilitation is *flow* (Csikszentmihalyi, 1996).

In some respects, these activities resemble team-, trust-, communication-, and morale-building exercises that can be treated as *indirectly* related to creativity. In other words, some people use them directly, to build trust or morale, for example. For creativity training purposes, we use them to build trust or communication, not because that is our primary goal, but because those things then influence creativity (by removing blocking factors).

SPECIFIC CREATIVITY-FACILITATING TECHNIQUES

To finish this chapter, I will now review some specific training techniques. Although somewhat at odds with my arguments in this chapter—the need for a more differentiated, 4Ps + Phases approach—these techniques are generally more of the domain-general, cognitive type. They remain part of any differentiated approach and have many benefits. I will address the more holistic approach—educating for creativity—in the next chapter. The only necessary cautionary note is: if all you do is give people some domain-general, cognitive training, then be prepared for results that are consistent with that approach. It may (indeed, probably will) improve their capacity for generating novelty, but it may not equip them with the three abilities I mentioned earlier in this chapter. In other words, a narrow, training focus is likely only to develop a *limited synthetic ability*, and not analytic or practical ability that characterizes the fully effective creative, or engineering, problem solver.

The range of approaches to training, as discussed already, frequently emphasizes the need to learn special thinking *techniques* such as (a) reversing the problem, (b) considering the end result, (c) focusing on the dominant idea; and (d) discarding irrelevant constraints. Other procedures for training specific thinking skills are game-like, including (a) producing, (b) analyzing, (c) elaborating, (d) focusing, (e) associating, (f) combining, (g) translating, (h) breaking out, and (i) recognize the new. All of these are specific cognitive techniques that can be learned quickly and then applied in a wide variety of settings as a help to generating novelty. A. J. Cropley (1992b) reviews a number of these.

Some procedures are more formal and broader, and based on at least a rudimentary theory of the nature of creativity. An example is the *SCAMPER* procedure, originally developed by Osborn (1953) in the advertising industry. This approach assumes that the production of novelty involves changing what already exists, and includes seven *change techniques*: **S**ubstituting, **C**ombining, **A**dapting, **M**agnifying, **P**utting to a different use, **E**liminating and **R**earranging/**R**eversing. Another procedure in this category is *Bionics*. This involves seeking out instances in nature in which a solution to the problem at hand already exists, and transferring this solution to the human problem. For instance, a dirt-resistant paint was developed by noting that the underside of the lotus plant's leaves allow virtually nothing to stick to them and reproducing the method used by the lotus plant in a paint. This is now being referred to more and more, not as a specific creativity technique, but as a legitimate field of study in its own right: biomimetics.[2] As a creativity technique, the procedure involves three steps:

- analyze the essence of the problem to which a solution is being sought;
- find examples in nature where this problem has already been solved;
- transfer as much as possible of the solution to the human problem.

Some procedures are more elaborate still and may have a formal, organized set of steps, although they still consist of fairly specific procedures that can be learned and then applied in different situations, usually in order to get ideas, i.e., to generate variability (e.g., *Synectics*, *Brainstorming*, *Morphological Analysis*, *Imagery Training*, *Mind Maps*, the *KJ Method*, the *NM Method*). Most of these are described by Torrance (1992) or Michalko (2001), and in many original publications specific to the technique in question (e.g., Gordon, 1961). Although many

[2]See, for example, Harman (2013).

were developed for schoolchildren, they are also widely applied with adults—for instance, in business in areas such as product development or in advertising. A few of these are discussed in detail in the next section.

Creativity Tools

Brainstorming (Osborn, 1953): This is probably the best known of all idea-getting procedures, and has become the prototype for a number of related techniques. In the early 1940s, Osborn, an advertising executive, concluded that business meetings often blocked generation of new ideas. He proposed a set of rules for interactions among people in groups that would eliminate the blocking effect and encourage participants to come up with ideas for solving a specific problem on which the group was working. The core of brainstorming is unrestricted production of ideas. Although this can be done alone, as part of so-called *nominal* groups, *classical* brainstorming is a real group activity (i.e., members are present in the group together) in which all members of the group are encouraged to put forward ideas without any constraints, such as fear of looking foolish, no matter how implausible the idea. There are four key principles:

- *quantity* of ideas, rather than quality, is stressed—premature evaluation is discouraged;
- *criticism* is not permitted because of its inhibitory effect;
- *hitchhiking* by attaching one's ideas to those of others is encouraged; and
- wild or *exaggerated ideas* are welcome.

There are various procedures for recording, selecting, testing, and otherwise relating ideas to reality (i.e., although the approach emphasizes *generating* novelty, *analysis* in the stage of *Verification* and *Validation* is not ignored). Proponents claim that brainstorming has been used successfully in a vast range of different settings.[3] These include *planning advertising campaigns* and *developing marketing strategies* (as might be expected, given Osborn's background), but also in developing new products, making investment decisions, designing better insurance policies, developing government policy, and, particularly interesting in a higher education context, designing research projects and planning written documents and reports. There are now also many empirical studies of brainstorming. At least some of these question various aspects of the effectiveness of the technique. Useful overviews of keys issues are given in Runco (2010) and Sawyer (2010).

[3]http://www.brainstorming.co.uk

Synectics: This procedure emerged at about the same time as brainstorming (Gordon, 1961). Gordon worked in the Invention Design Group of Arthur D. Little Inc., a consulting firm that helped companies develop new products. He noticed that some groups were highly productive, others less productive, and concluded that the differences in productivity derived from the way members of a group interacted, not from ability, knowledge, etc. In 1960, he founded the firm Synectics. Synectics is available in a formal format.[4] Like brainstorming, it is based on unfettered production of ideas in order to solve a problem. However, the essence of Synectics goes beyond simply generating novelty. At its core is the principle of seeing connections between things not normally regarded as connected. Such unexpected links (I used the term *remote associates* in Chapter 5) generate novelty. The technique involves making analogies—seeing aspects of the *other* thing that are like aspects of the present problem. Analogies are frequently based on metaphors and are frequently made with living organisms (e.g., a vacuum cleaner is like a pig, because both ingest garbage—so, somewhat akin to the concept of *biomimicry*). Synectics is also a group procedure. Ideally, one person states a problem, the group helps to define the essential core of the problem, and group members begin generating remote associates by looking at what already exists. They do this, however, by seeing both the present object (let us say a new kind of vacuum cleaner) and the remote associate (in this case a pig) in a new way—what Gordon described as *making the strange familiar* and *making the familiar strange.* Seeing a vacuum cleaner as a garbage eater makes the familiar (a vacuum cleaner) strange. Seeing a garbage eater as a pig makes the strange familiar. Such analogies are constructed with the help of special techniques, including Subtracting, Adding, Transferring, Empathizing, Animating, Superimposing, Changing size, Substituting, Fragmenting, Isolating, and Distorting. Subsequently, the group identifies the link between the new way of looking at the situation and the desired solution to the problem, and works out the value of the analogy in practice (i.e., in our terms, they *explore* the novelty and verify and validate its relevance and effectiveness). For instance, the group might suggest the idea of a vacuum cleaner with teeth that grind dirt.

An example of a synectic-like approach to generating effective novelty can be seen in the way the renowned British engineer Sir Marc Isambard Brunel is said to have hit on the method of tunneling he adopted to build the Thames Tunnel, completed in 1843.[5] He saw his problem as the same as that of the saltwater *Teredo navalis* shipworm

[4]http://www.synecticsworld.com

[5]http://en.wikipedia.org/wiki/Thames_Tunnel

that bores through submerged timber and possesses a hard outer shell. As a result of his observations of the shipworm, while working in a shipyard, Brunel devised the *tunneling shield*, making it possible to tunnel through unstable soil.

Morphological analysis: Morphological analysis (MA), developed by Caltech astrophysicist Fritz Zwicky, is a technique for finding new combinations of conditions that define a novel product (Ritchey, 2006). Take the development of a novel kind of pencil as a simple example. The technique (done as a group or individual) begins with a listing of all the attributes of the product that is to be developed or improved. The attributes of a pencil would include the material it is constructed of, material that does the actual writing, hardness/softness of writing material, color, thickness, etc. The next of MA involves drawing a table with each attribute as the heading of a column. All the possible imaginable variations of the attribute in question are then entered into the respective column. These are typically derived from an idea generation method (e.g., brainstorming). For a pencil, the column *pencil material* might have entries *wood, plastic, glass, metal, clay*, and so on. The column *writing material* might have entries *graphite, ink, paint, mud*, and so on. When complete, the table will contain many possible variations of each attribute. In the final step of MA, a kind of automated *Generation*, one entry from each column is selected (i.e., one kind of pencil material, one kind of writing material, and so on), and these are combined: e.g., *metal body + ink*, plus one further attribute from each of the other columns. Importantly, these combinations of attributes/options in the columns can be done randomly, or on the basis of some kind of heuristic, rule, or other criterion. Each set of combined elements, one from each column, defines a potential new kind of pencil. The decision as to which are practicable designs (if any) is a matter for the analytical phase of *Verification*. I have also used MA in engineering problem solving where the listed attributes are, in fact, the core functions of the system or product that is being designed. Each function represents a problem to be solved—for example, if a particular function is *move vehicle* (recall my discussion of verb-noun pairs in Chapters 3 and 5), then the problem to be solved is how to move the vehicle—and the solution options can be generated by any suitable method. Once the MA table is populated, it can also be used to drive the next stage: analysis. As can be seen in Table 9.3, even a simple system with only five functions soon generates a large number of potential combinations. In the case shown, five functions, each with five solutions, means that there are 3,125 possible overall combinations! If we have done our *idea generation* correctly, and not evaluated any options prematurely, then even this small MA table generates far too many options for each one to be considered. One way to address this is to use the

TABLE 9.3 Morphological Analysis in Engineering Problem Solving

	Option 1	Option 2	Option 3	Option 4	Option 5
Function 1 (Move Vehicle)	A (5:4:3:4)	B (5:3:3:5)	C (3:2:4:2)	D (4:2:4:2)	E (2:5:3:5)
Function 2	A	B	C	D	E
Function 3	A	B	C	D	E
Function 4	A	B	C	D	E
Function 5	A	B	C	D	E

Creative Solution Diagnosis Scale (CSDS, see Chapter 4) to filter the solution options. For example, each option is given a score (e.g., out of 5), using the original CSDS, for (a) Relevance and Effectiveness; (b) Novelty; (c) Elegance; and (d) Genesis.

A rule can then be applied—for example, we are interested only in options with a minimum score of "4" for Relevance and Effectiveness, or we will consider only options that have the maximum score for Novelty—and the combinations from this reduced set of options are then considered. In Table 9.3, this could knock out two or three options per line, still leaving, say, 32 promising combinations (e.g., $2 \times 2 \times 2 \times 2 \times 2$).

KJ Method: The procedure was developed in Japan and is closely associated with modern quality management processes in that country. Somewhat like *brainstorming*, it is based on generating large numbers of ideas, and is also usually a group procedure. Individual members of the group write on cards their ideas about what constitutes the core of the problem at hand (one idea per card). The cards are then sorted into *sets*, i.e., groups of cards containing statements that define the nature of the problem in a similar way. The different sets are given labels that summarize the essence of the problem as it is defined in a particular set. For instance, a group applying the method to designing a revolutionary new form of public transport might come up with sets of cards that focus on the following categories:

- How could it be financed?
- What energy source could be used?
- How should routes be planned?
- How could safety levels be raised?
- What special staff training would be needed?

Sets of solutions can then be constructed via a similar procedure.

Mind Maps: This procedure was originally developed in the 1960s but has been substantially developed since (Buzan, 2003). It is usually carried out by individuals working alone. Mind maps retain the idea of unhindered generation of variability but go beyond simple *blind* generation of ideas. A central *theme* is written down, usually in the center of a blank sheet of paper or whiteboard, and then a *spray* of associations recorded, without pausing to evaluate, judge, or edit. Each association functions as the possible beginning of a new spray of further associations, rather like tree- or spider-diagrams that we use in engineering to represent hierarchies. Solutions to the central problem or question are found by identifying patterns or threads in the network of associations. An example related to public transport might result in a mind map with *public transport* at its center, and a spray of associations surrounding that, including *passengers, schedules, vehicles*, and so on. A second level of associations, for example, stemming from *passengers*, might be *complaints, peak hour numbers, fare evasion*, and *security*. *Schedules* might then result in a third tier of associations, such as *frequency, reliability*, and *routes*, and further associations for *vehicles* might include *comfort, safety*, and *power source*.

If a problem is now identified—for example, *how can we increase patronage?*—solution options are found by identifying patterns or threads found in associations of the mind map. An example of a possible solution derived for the mind map in Figure 9.1 might be to increase comfort and safety by providing better seating and improved lighting, as well as more frequent and reliable services. The mind map does not provide the answer, but helps us to structure our knowledge of the problem, and from this derive ideas for solving the problem.

Hierarchical Method (Butler & Kline, 1998): This method involves an even stronger element of organization and structure. Although the core idea of generating large amounts of variability (possible solutions) is retained, this approach is based on the idea that a hierarchical organization of ideas (rather than simple clusters of related ideas or

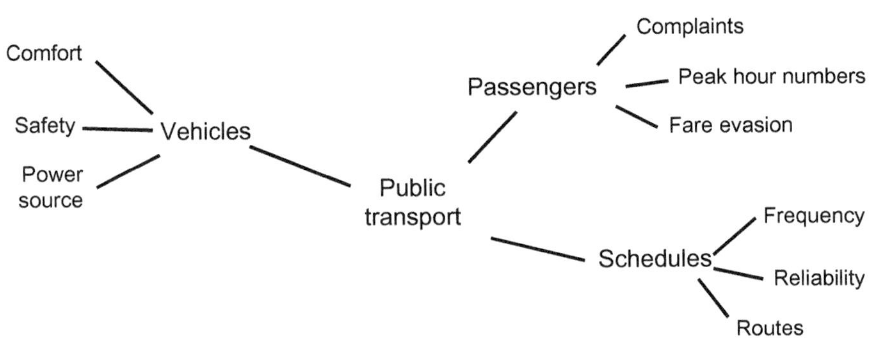

FIGURE 9.1 A mind map of public transport.

associational chains) produces solutions of better quality. Suggested solutions—quite possibly derived from a technique like brainstorming—are sorted into *classes* based on common content. Subsequently, hierarchies are formed by combining lower-level classes into superordinate classes or, by contrast, breaking down higher-order classes into lower-level categories. I discussed this in Chapter 5 in the context of remote associates, categories, and networks (see Figure 5.1). Once categories are identified, this often stimulates the identification of *new categories* that can then be populated with further solution ideas. This is often an effective way of extracting more value from brainstorming.

TRIZ (Altshuller, 1988): The TRIZ method is based on the idea that all inventions (i.e., creative solutions to problems) display the same pattern of emergence of ideas. These patterns were identified by Altshuller in an analysis of thousands of successful patent applications. TRIZ defines 40 *inventive principles* and describes various tools and techniques for their application. The 40 principles act as a set of *stimulators*, directing idea generation along predetermined but previously successful lines. Examples include

- *Segmentation* (divide an object into independent parts, or, make an object modular, or, increase the degree of segmentation);
  - For example, replace a mainframe computer with many individual personal computers and a network.
- *Extraction* (separate an interfering part or property from an object, or single out the only necessary part or property);
  - For example, put the compressor part of an air conditioner outside the building.
- *Asymmetry* (change the shape of an object from symmetrical to asymmetrical, or vice versa);
  - For example, change the shape of an O-ring from circular to a specialized, irregular shape, to obtain a better seal.
- *Merging* (merge identical or similar objects, or assemble identical or similar parts to perform parallel operations, or make operations parallel in time);
  - For example, combine many individual transistors on a single silicon chip to create an *integrated* circuit.
- *Reversing* (invert the actions used to solve a problem, or implement the action opposite to the requirement, or make fixed parts moveable and moveable parts fixed, or turn an object or process upside down).
  - For example, rotate the part to be machined, rather than the tool.

The *principle* itself is not a solution, but a characteristic of the way previous problems were solved. The 40 principles represent a catalog of broad categories (think remote associates and networks again; see Chapter 5) that can be reapplied to new problem situations. Thus, while

TRIZ *reduces* novelty at the level of the category, the goal is to increase novelty at the level of the individual solutions.

Savransky (2000) gave a detailed description of the application of TRIZ and also discussed the processes through which existing knowledge is used to develop effective novelty. He argued that inventive solutions to problems always involve a change in what already exists. While this represents only a subset of the pathways I described in Chapter 2, it nevertheless sheds useful light on the way that new solutions can be derived, to satisfy new needs (i.e., *redirection;* see Figure 2.4). Savransky described six ways in which this *technology push* can occur:

- *Improvement* (improvement or perfection of both quality and quantity of what already exists);
- *Diagnostics* (search for and elimination of shortcomings in what already exists);
- *Trimming* (reduction of costs associated with existing solutions);
- *Analogy* (new use of known processes and systems);
- *Synthesis* (generation of new mixtures of existing elements);
- *Genesis* (generation of fundamentally new solutions).

In fact, Savransky's sixth option, *Genesis,* describes not a modification of an existing solution, but a genuinely new solution, and is more consistent with a reaction to market pull.

Formal Training Programs

The tools that I discussed in the previous section are characterized by the fact that they are typically very specifically directed at one aspect of creativity and creative problem solving—for example, idea generation or idea evaluation. Their purpose is usually not to train the user to be more creative, but simply to provide a mechanism to cause the generation of novelty to occur. It is possible to be trained in their use, to become more proficient with the tool, and this may result in some ongoing improvement in the individual's creative ability.

Of great interest also are the more organized and formalized creativity-training programs that consist of a package of sequential tutorials or lessons and accompanied by exercises, practice sessions, and the like, often supported by print materials such as handbooks and/or by audiotapes and video, etc. It is true that such programs are available for some of the tools I have already described—e.g., brainstorming and mind maps—however, I will now outline some other examples of programs. Some of these can be described as having a *formal* basis in the sense that they have evolved from a more theoretical and research-driven starting point, whereas others are more *popular* in nature. There are a number of analyses of these training programs, for example, Rose and Lin (1984), in addition to the studies of effectiveness that I have already mentioned in this chapter.

Purdue Creative Thinking Program (PCTP): This program (Feldhusen, 1988) is based on Torrance's divergent thinking concepts. It is intended for use with elementary-age schoolchildren, but is included here because it illustrates very clearly the degree of organization and formalization of what I refer to as *programs*. It is also true that the general approach seems to be readily transferable to higher education and organizational settings. The third edition of the program, dating from the 1980s, consists of 36 lessons of 14 minutes duration on audiotape. In each lesson, a principle for enhancing one of the aspects of divergent thinking is presented for 3–4 minutes and then demonstrated in action through a case study of an individual person (e.g., Henry Ford, Alexander Bell, Guglielmo Marconi) lasting about 10 minutes. Finally, each lesson is accompanied by three or four exercises, in print form, that provide practice in applying the principle demonstrated in the lesson.

CoRT Thinking Lessons: Devised and marketed by Edward de Bono,[6] the CoRT Thinking lessons are based squarely on the view that *thinking* skills (in general, not just for creative problem solving) cannot be left to chance, or assumed to be the inevitable result of academic study. Instead, these skills must be taught directly and explicitly. The program consists of six sections, each containing 10 lessons; the lessons consist of teaching materials, teacher's notes and students' notes, and are available in print form or as videos. The section most directly relevant to this book is the section on *creativity*. This is based directly on de Bono's concept of lateral thinking (see Chapter 5), and teaches principles for changing the way people look at things. These principles include FIP (first important priorities), CAF (consider all factors), and APC (alternatives, possibilities, choices). It is suitable for use with children and adults.

Creative Problem Solving (CPS) is based on Wallas's stage model (see Chapter 3). In its *classical* form (Parnes, 1981), it involves five steps which can be applied in a systematic way to finding, investigating, and solving problems. Treffinger, Isaksen, and Dorval (1995) added a preliminary stage at the very beginning to give six steps: Mess finding; Fact finding; Problem finding; Idea finding; Solution finding; Acceptance finding. In his book, Parnes (1981) went through a large number of problems with readers in order to make the steps automatic, so that they can be reapplied over and over again with new problems. Treffinger (1995) extended understanding of CPS by emphasizing that it is not a purely cognitive exercise: he drew attention to the role of other people

[6]http://www.edwdebono.com/cort/index.html

in the Acceptance finding phase, both assisters and resisters. I also described CPS in Chapter 3, as a model of the stages of creativity, and the latest version of the program—the Thinking Skills Model—is described by Puccio and Cabra (2010). The program is also well supported online, and further details are available from The Creative Problem Solving Group.[7]

Popular Training Programs

In addition to more scientific work on fostering creativity, represented by the *formal* training programs in the preceding section, there is a substantial number of semi-scientific or *popular* programs. Typically, these are more commercial in nature and are aimed at organizations (business/commerce, the armed forces, government) and individuals (adults interested in self-help, teachers, and parents). Many of these were developed by *practitioners*, and not necessarily researchers or even traditional educators. Although they frequently lack empirical support, this is not to say that the programs are not effective. However, a characteristic of such programs often seems to be that they focus firmly on cognitive elements of creative problem solving and very little on other aspects of the 4Ps—especially *Person* and *Press*. Probably the best known are Edward de Bono's publications/programs, in which he elaborated on the concept of "lateral thinking" (de Bono, 1993). Originally a medical practitioner, he developed not only a graphic and picturesque terminology (e.g., *water* and *rock* logic, and the colorful *six thinking hats*), but also published the *CoRT Thinking Program* (de Bono, 1978) mentioned in the previous section. Another popular example is the work of Michael Michalko (1996, 2001), a former officer in the U.S. Army, who has become prominent in the United States, with programs such as *ThinkerToys* (aimed at nurturing business creativity) or *Cracking Creativity* (self-training).

Such books/programs are often based on scholarly findings, even if the connection is sometimes loose. Michalko, for example, cites work by familiar creativity scholars, including Guilford, Sternberg, and Barron. These popular programs are also frequently technically well produced, extremely readable, easy to understand, and plausible. In addition, they often contain sensible and cogent advice with which very few people would disagree, while many of them are undoubtedly capable of bringing benefits in creative thinking. However, there are problems with much of this popular literature—some of which I have alluded to already—and these were summarized by Hruby (1999). Although he

[7]http://www.cpsb.com

was reviewing a specific book, his comments are very pertinent and can be applied here in a more general way.

Hruby (1999) complained that enthusiasm for fostering creativity can "run away with itself" (p. 326). Among other weaknesses he identified in popular approaches are

- presenting speculations, conjectures, and hypotheses as established facts;
- confusing correlations with causal relationships;
- making unjustified sweeping generalizations that are either not unequivocally supported by research or are even contradicted by some findings;
- drawing unwarranted conclusions about the implications of research findings for practice;
- failing to understand the factors that inhibit conversion of admirable recommendations into practice.

Some popular books proclaim incompletely digested research findings as containing a *revolutionary* panacea that can be applied in a set way in all situations, without taking account of the individuals involved, the special characteristics of the situation, or the personal or structural factors facilitating or impeding implementation of good practice (i.e., each of the 4Ps). As I have already indicated, there is no harm in these programs, provided we use them judiciously, and with the benefit of the deeper knowledge of all the factors that are involved in successful creativity.

CHAPTER

10

Embedding Creativity in Engineering Education

"Thank goodness I was never sent to school; it would have rubbed off some of the originality." **Beatrix Potter, 1866–1943, Author**

I began this book by making the case for the importance of creativity (and therefore innovation) in engineering. Now, more than ever, society depends on the ability of engineers to develop novel and effective technological solutions to the flood of problems that result from an ever-increasing rate of change in the world. However, I also lamented the fact that there is *not* a stronger connection between creativity and engineering. Regrettably, this disconnect is probably strongest in engineering education.

There are many reasons for this state of affairs. At a very general level, there is the inertia and resistance to change that constrain many entrenched systems. Engineering education has done a reasonable job for many decades, and there is a natural reluctance to risk changing what *seems to be* a successful formula. This is compounded by a trend toward ever-greater specialization in engineering. As programs proliferate, it is almost inevitable that the only way they can be differentiated is to drill ever deeper into narrow fields. The danger, as Gandhi warned, is that "The expert knows more and more about less and less until he knows everything about nothing." The casualties of this over-specialization are *general* graduate attributes, skills, and abilities: *design, creative problem solving*, and *thinking*. At a more specific level, there is also the problem—a core idea in this book—that many engineering faculty, managers, and decision makers do not understand creativity and innovation sufficiently well to do anything to change the system, even if they are motivated to do so.

In this chapter, I will first discuss this problem in more detail. I will then attempt to provide some practical help in tackling the problem by doing two things. First, at the general level, I will supplement the content of the previous chapters with some *guiding principles* for engineering programs. In other words, a set of *creativity requirements* that would, in an ideal world, drive the design of engineering programs. Second, at a more concrete, specific level, I will develop an *exemplar curriculum* for a course on engineering creativity.

THE PROBLEM

The failure of education in general, and engineering education more specifically, to adequately address the need for creativity is reflected in the 1996 report of the Alliance of Artists' Communities (1996) which concluded, "American creativity" is "at risk." The problem is not confined to the United States of America, and goes beyond artistic or aesthetic areas. Employers surveyed in Australia in 1999 complained that three-quarters of new university graduates there show "skill deficiencies" in creativity, problem solving, and independent and critical thinking. Still in Australia, in 2013, the annual *Graduate Outlook Survey* conducted by Graduate Careers Australia[1] indicates that "Critical reasoning and analytical skills/Problem solving/Lateral thinking/Technical skills" is third on the list of top selection criteria for employers. Of greater concern, when asked to rate the employability skills of graduates *actually hired* in 2013, employers indicated that only 57.3% exceeded average expectations in problem solving—the lowest figure since 2009! Tilbury, Reid, and Podger (2003) also reported on an employer survey in Australia which concluded that Australian graduates lack creativity.

In the United Kingdom, Cooper, Altman, and Garner (2002) concluded that the education system discourages innovation. The British General Medical Council, for instance, recognized that medical education is overloaded with factual material that discourages higher order cognitive functions such as evaluation, synthesis, and problem solving, and engenders an attitude of passivity. Bateman (2013), meanwhile, reports on results of UK employment survey data in the area of computer science and IT, suggesting that graduates in this domain miss out on employment opportunities due to a lack of creativity.

[1]http://www.graduatecareers.com.au/wp-content/uploads/2014/03/Graduate_Outlook_2013.pdf

A similar picture is reported widely in the United States in various sources. Articles in *Time* and *Forbes Magazine*, for example, suggest that employers are frustrated by the fact that new graduates are emerging from universities lacking skills in creativity and problem solving.

It is worth noting in passing that the situation is much the same in schools. Despite the fact that research has shown that more than 25 years ago most teachers already claimed to have a positive attitude to creativity, even today in classrooms in many different countries, properties and behaviors actually associated with creativity are frequently frowned upon. The evidence summarized by A. J. Cropley (2001) is that teachers discourage traits such as boldness, desire for novelty or originality, or even actively dislike children who display such characteristics. Thus, although there are calls for creativity, there may be limited efforts to foster its emergence, or even dislike of people who display it.

The situation in *engineering* education seems to be no different. The United Kingdom's Royal Academy of Engineering, for example, published the report *Creating Systems that Work: Principles of Engineering Systems for the 21st Century* in June 2007 (Elliott & Deasley, 2007). Among six principles that the report presents as necessary for "understanding the challenges of a system design problem and for educating engineers to rise to those challenges" (p. 11) is an ability to "be creative." The report further recognizes the key role that creativity plays in successful engineering and defines creativity as the ability "to devise novel and ... effective solutions to the real problem" (p. 4)! Baillie (2002) similarly noted an "... increasing perception of the need for graduates of engineering to be creative thinkers..." (p. 185).

D. H. Cropley and Cropley (2005) reviewed findings on fostering creativity in engineering education in the United States, and concluded that there is little support for creative students. It is true that there has been some effort in recent years to encourage creativity in colleges and universities: for instance, in 1990 the National Science Foundation (NSF) established the Engineering Coalition of Schools for Excellence and Leadership (ECSEL). This had the goal of transforming undergraduate engineering education. However, a subsequent review of practice throughout higher education in the United States (Fasko, 2001) pointed out that the available information indicated that deliberate training in creativity was rare.

Kazerounian and Foley (2007) restate the fundamental problem: "If creativity is so central to engineering, why is it not an obvious part of the engineering curriculum at every university?" They suggested the reason is that it is "not valued in contemporary engineering education" (p. 762), but the problem runs deeper than that. Why is the compelling pressure for creativity in engineering education largely ignored?

STAKEHOLDER PERSPECTIVES

Before I discuss the reasons why I think creativity is not part of the engineering curriculum, I would like to turn to the other key stakeholder in this situation: the student/graduate. What do students and graduates think of the question of creativity and engineering? Some of the issues facing universities in relation to creativity and innovation in engineering curricula have been eloquently articulated by Wilbur (2013).

I have also had many discussions with undergraduate engineering students regarding creativity in engineering, both in the United States and in Australia. I have reproduced here salient comments from discussions in 2013 and 2014.

"As engineers we are supposed to be the innovators of the world, inspired by creativity and a passion for problem solving. However, many curricula drain students of excitement for challenges. Students are graduating unprepared."

What is striking about these comments is that the student in question has a clear appreciation of the connection between engineering, creativity, and problem solving, yet an intuitive sense that her degree is not preparing her for those activities. This mirrors the employer survey data I reported earlier.

"I feel that engineers need to have open discussions and team projects, rather than weekly homework that addresses only theoretical problems. I understand that a strong basis in the fundamentals is a necessary start, but it should not need to span four years of undergraduate studies, with no additional hands-on learning. Students forget why they even had a passion for engineering in the first place."

These comments reflect a sense that the current curriculum is heavily skewed toward convergent, analytical work. The comments also mirror a concern that too much emphasis is placed on narrow specializations. The comments also suggest an appreciation that knowledge underpins creativity.

"The same kids who had such excitement for a subject are stuck in a classroom, being told 'In the real world none of this applies.' How are we supposed to trust our education system, when it admits how much it is failing us?"

These comments suggest that students are beginning the process of engineering education highly motivated and primed to be creative, but are losing this in programs that they feel are not preparing them properly, and are disconnected from the real world.

"Encouraging creativity while teaching the fundamentals is a balance the schools have yet to learn. As engineering students, we take a couple of English courses and dabble in the humanities. Instead, what about a drawing class? By learning to draw, we can more clearly express our ideas. Da Vinci certainly

couldn't have been as creative as he was without this talent. I am involving myself in an extracurricular program where I will take a drawing course, a business course, and a project course, where real local companies ask for each group to come up with an applicable solution to one of their problems. Most of the companies end up using the students' ideas. These types of classes should be mandatory for engineering students, not a program that often doesn't work with our schedules."

This comment is one of the most telling, in my opinion. I see it reflecting a major reason why creativity is not more strongly embedded in engineering—where opportunities to develop some of the requisite skills are available, they are add-ons to engineering programs. In other words, engineering programs tolerate them, provided they are somebody else's problem, and do not detract from the core purpose of the engineering curriculum—which is to drill ever deeper into narrow specializations. Creativity should *not* be a remedial action that employers, or other parts of the university, have to add to correct the deficiencies in engineering students and graduates!

"It's amazing how many students don't even bother to show up to class, end up dozing off or fiddling with their phones (myself included) because the subject has lost its sparkle. In fact, in one class today the professor cut class short because he was losing so much attention from the students. This past year I have found myself becoming more and more discouraged by the program I am in. While I will stick with Engineering until I graduate, I see myself taking it a different direction, one that at this point does not include graduate school."

Can we really afford to discourage our students in this way? Engineering already struggles to attract women, and can ill afford to discourage students from pursuing graduate careers. As easy as it would be to dismiss these comments as outliers, it is hard to deny the evidence we see with our own eyes. Creativity is not just a necessary component of engineering education, but it also offers the means to revitalize engineering programs, making them far more motivating for students.

"Students forget why we are actually here—to learn to become engineers; to see a new and different perspective of the world; to look at a telephone line and think, 'How does that actually work?' rather than never pausing to wonder or ask questions. Students often take what their professors say as the truth without sitting and pondering why it is so, or maybe suggesting another vantage point. We need to be creative in the classroom and creative with our dreams, not always accepting the status quo. I really do feel that more people need to hear the message that engineering should be creative. The curriculum in school should be fun and exciting and teach us to embrace the power that we hold to make a difference."

The student's comments again show a natural willingness to embrace creativity, and a keen desire to engage in the fundamental engineering problem-solving process as I have described it in this book. If this is the raw material that we, as engineering educators, are working with, then we can be confident that any efforts to reconnect creativity and engineering will be embraced with enthusiasm by our stakeholders.

The Overspecialization Problem

Employers, industry bodies, and students see the value of creativity in engineering. The *need* has been clearly articulated. What is stopping the reconnection of creativity and engineering in higher education? Buhl (1960) summarized the underlying problem facing engineering education, whether in relation to creativity or not, when he highlighted that "Until the present day we have sought to expose the student to every conceivable situation he might encounter after he leaves the university" (p. 10). It is important to understand that Buhl did not mean this in a good way. Rather, he was drawing attention to the fact that engineering programs in 1960 suffered from the problem of *breadth at the expense of depth*. The issue that this created was that students and faculty, because of the sheer volume of topics, could only hope to cover those topics in a relatively superficial manner. The lack of depth occurred in two senses: a lack of coverage within any given topic, but also a lack of opportunity to develop deep understanding of any given topic. In other words, students were learning an awful lot of relatively superficial material, in a very superficial manner. Biggs (1999) referred to this as "the inevitable tension between coverage and depth of understanding" (p. 44). Buhl (1960) made it clear, in engineering, why this approach was a chimera. "The present growth of technical knowledge has placed this goal [exposing students to every conceivable situation that might be faced as a professional] beyond the reach of a four-year college education. The student may now be assured that ten or twenty years after graduation many of the problem solutions and 'facts' presented to him will have changed" (p. 10). Nowadays this problem of the *half-life of knowledge* is even more severe. Not only did this situation result in broad, shallow knowledge, but it also left little room for creativity.

The reaction to this state of affairs seems to have been to try the opposite—depth at the expense of breadth. I have already described the modern trend toward a focus on narrower specializations, and this proliferation has the same basic impact—no room for creativity, design, thinking, and other soft skills. The solution is not to attempt to cram ever more technical content into the curriculum, but, as Buhl (1960) noted, ". . . schools must educate the student for change. Students must not only learn the fundamental ideas upon which the various subjects are based, but they must learn how to solve a problem in a creative way…" (p. 11).

The problem of overspecialization is compounded by weaknesses in engineering pedagogy. Problem-based learning, for example, may be highly effective, but if the problems that it focuses on remain convergent and analytical, then it will do no more to engender creativity than any other paradigm. Walther et al. (2011) suggested that the problem may lie in "persisting difficulties of the construct of outcomes-based education as the current paradigm of formal engineering education" (p. 704). Walther and Radcliffe (2007) earlier expressed this as a *mismatch between different kinds of learning outcomes and predominant teaching approaches*. In simple terms, a learning outcome framed around the development of declarative knowledge of, say, engineering mechanics may be amenable to a "traditional" teaching approach in a way that a more diffuse graduate quality such as "the ability to think creatively" is not. A fundamental dilemma faced by engineering educators in preparing students for the "…diversity of competency demands" (Walther & Radcliffe, 2007, p. 44) is "…whether to equip students with a broad (and arguably shallow) knowledge base in many domains, or prepare them for specific job tasks and a contribution to a narrow subject area (technical depth)" (p. 44). Creativity is, by its nature, a broad, generic competency. If poorly understood, and perceived as the antithesis of the "serious business" of engineering (Kazerounian & Foley, 2007), it is little wonder that it is not only undervalued, but absent in most curricula.

The Pseudo-Expertise Problem

Related to the issue of overspecialization is the *kind* of knowledge developed, as I indicated in the previous section. Factual, or *declarative*, knowledge is easier to teach, and easier to measure, but is it the right knowledge that students need to be successful, creative engineers?

I have previously discussed the importance of *domain knowledge* as a foundation to engineering creativity. DeHaan (2009), citing Bransford et al. (2000) and Crawford and Brophy (2006), similarly discusses differences between experts and novices, and the role of creative thinking, suggesting that *minimal levels* of expertise and fluency are needed for expertise. Sawyer (2006) describes the fact that experts typically are distinguished by their deeper knowledge, recognition of patterns, ability to see relations among disparate facts, capacity to organize content, and so on. An excessive focus only on factual knowledge—even at great depth—denies students these other qualities needed for expertise.

Such a focus risks creating *pseudo-experts* (Sternberg, 2003a). Students grappling with a new subject are taught to solve problems by the application of algorithms and *procedural* knowledge (Biggs & Tang, 2011). If they do this enough, it can become routinized and may be considered *expertise*. DeHaan (2009), however, contrasts this with the need to move up the scale of what Crowe, Dirks, and Winderoth (2008) refer to as Higher Order Cognitive Skills (HOCS), and which Biggs and Tang (2011) would describe as higher (or deeper) levels of understanding. In other words, the argument is that true expertise, or *adaptive expertise* (Hatano & Oura, 2003; Schwartz, Bransford, & Sears, 2005) is characterized by an ability "to draw on ... knowledge to invent or adapt strategies for solving unique or novel problems within a knowledge domain" (DeHaan, 2009, p. 175). The foundation for creativity in engineering, then, is the development of adaptive expertise, which can come about only as a result of the development of appropriate *relational* and *extended abstract* functioning knowledge (Biggs & Tang, 2011). Pseudo-expertise, namely expertise characterized by knowledge that is overly declarative and procedural, and which is more superficial (characterized by unistructural and/or multistructural levels of understanding) militates against the development of creativity in engineering.

DeHaan (2009) also goes into some detail on university level teaching and creativity. *Passive* teaching and learning approaches, for example, fail to engender active engagement and cognitive flexibility. Citing Ausubel and Paul (2000), he links this to poor outcomes in creativity and creative problem solving because of the importance of cognitive flexibility as a core mental executive function in creative problem solving. Furthermore, the *transfer* of knowledge that is critical to the ability to apply ideas creatively in new contexts is facilitated by active learning strategies (Freeman et al., 2007).

This suggests that there is a *threshold of adaptive expertise*, which is a necessary but not sufficient foundation of engineering creativity. Even a shift to *narrow and deep* will not achieve this if the depth is only *declarative* and *procedural* in nature. Figure 10.1 shows the relationships between *kinds of knowledge, levels of understanding*, and three basic forms of expertise. Engineering programs that fail to develop *conditional* and *functioning* knowledge, no matter what level of understanding is achieved, can only hope to produce pseudo-experts. Adaptive expertise requires the development of *all forms of knowledge*. Furthermore, the potential for professional-level engineering creativity (Pro-C creativity) is highest when a threshold of adaptive expertise has been reached.

If engineering programs are producing only pseudo-experts, their ability to apply creativity will remain constrained and limited. If adaptive expertise is absent, it is likely that no amount of training in creativity will compensate for the deficiency. When present, with the addition of the requisite processes, personal qualities, and press, Pro-C Creativity in the engineering domain can be realized.

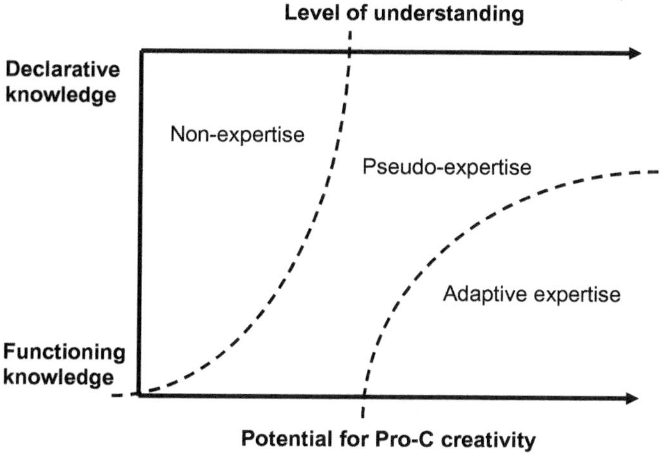

FIGURE 10.1 *Expertise* and creativity.

The Lack of Knowledge Problem

One of the most pervasive problems that is blocking the addition of creativity in the engineering curriculum is simply a lack of knowledge *about* creativity. Where discussions of creativity in engineering do take place in the literature, they typically follow a common pattern exemplified by Mishra and Henriksen (2013), and begin by propagating the myth that creativity is poorly defined, before offering their own definition. Even concerted efforts to explore creativity in engineering seem unaware of the existing body of knowledge. In 1998, a special issue of *European Journal of Engineering Education*, for example, began with the question "How does one implement creativity in engineering education?" (Ihsen & Brandt, 1998, p. 3). While their editorial is to be applauded for attempting to drive this issue to the forefront of thinking in engineering education, it also perpetuates some of the myths of creativity, in all domains, that retard progress. The most pervasive of these is the *creativity is hard to define* myth. Indeed, the authors point out that of the 13 papers selected for their special issue, 13 *different definitions of creativity* are given. This confusing state of affairs is compounded by the attitude that "we leave it up to the readers to think about their own definition of creativity in engineering education and to develop their own concepts and specific approaches..." (p. 3)! It is hardly surprising that creativity is not embedded in the engineering curriculum.

Another pervasive myth—not unique to engineering creativity—surrounds the question: *can creativity be taught*? Acar (1998), for example, argues that there is "no universal agreement on whether creativity can be taught or not," while Törnkvist (1998) starts his discussion in a more rhetorical manner, citing Evans (1991), who claimed that "[y]ou cannot teach creativity, but you can kill it." I have already discussed this question in Chapter 9. Benson (2004) reminds us why this lack of knowledge of creativity is problematic: "... unless misconceptions are identified and addressed, the development of creativity will almost certainly be hindered" (p. 138).

Amoussou, Porter, and Steinberg (2011) provide another interesting perspective on the question of the knowledge, or lack thereof, surrounding creativity in engineering and technology. Amoussou et al. surveyed computer science and engineering faculty in California's higher education system. On the surface, the study seems to suggest that faculty are doing well in promoting creativity; however, a number of weaknesses in the methodology are masking underlying problems. For example, it is unclear if the sample is representative of the wider population of faculty in engineering and computer science. Were respondents largely those who already have a favorable view of creativity? The survey also did not include items designed to check the honesty or social desirability of responses. For example, one question asked respondents the degree to which they explicitly instruct students to be creative. This needs to be balanced with a question such as: "I explicitly instruct students to be *analytical* in their designs." We would expect that if the pattern of responses was generally high for one, then it would be low for the inverse. Amoussou et al. make the point that items in survey are "based on psychological literature on creativity that is often unknown to computer scientists or engineers" (p. S2B-3), and yet include questions that could hardly be answered reliably by respondents lacking a knowledge of psychological concepts. A good example is the question: "Are your students taught about informational social influence?" Lacking a knowledge of psychology, a respondent is likely to answer in a socially desirable way—*this sounds like a good thing, so I'll say yes*. The study also failed to report reliability data, so there is no statistical evidence that respondents gave consistent answers. It is important that surveys such as this are conducted as part of addressing the problem of embedding creativity in engineering; however, they must be designed more rigorously, or they risk compounding the problem. Amoussou et al.'s (2011) points about how to encourage creativity are valid, but I suspect that this survey underreports the extent of the problem. The results suggest that the problem is not as extensive as I believe it really is, and that leads to a danger that decision makers take no action in support of it because they do not feel there is a problem.

Another recent study that illustrates a similar problem is that of Ahern et al. (2012) investigating "critical thinking" in engineering education. The study revealed that engineering faculty thought critical thinking was important, but found it hard to articulate what it was.

TABLE 10.1 Summary of the Problems

Problem	Symptom	Consequence
Overspecialization	Degrees focus on narrow specializations.	Focus only on technical content. No room for creativity.
Pseudo-Expertise	Teaching focuses on factual knowledge.	No threshold of adaptive expertise achieved. No room for creativity.
Lack of Knowledge	Faculty focus on "what is creativity?" and "can it be taught?"	Little real progress while the wheel is reinvented.

In technical disciplines, faculty equated critical thinking with problem solving and creative thinking, and "something a little more abstract and conceptual than simply learning facts" (p. 127). While this is promising, it reveals a lack of understanding of the topic. In the study, faculty also reported that subjects like engineering are "so content driven in the early years that the space for introducing critical thinking was minimal" (p. 128). Further findings from the study included the finding that "[l]arge sizes made teaching critical thinking skills harder" (p. 128), while "[t]here may be lessons that can be learnt by engineering from the humanities in terms of academics themselves becoming more aware of what critical thinking is" (p. 128). This is the same theme I have discussed in earlier sections—a lack of understanding of what creativity is, how to teach it, and how to embed it in the curriculum. At the same time, this study acknowledges that critical thinking (and creativity) is "an important attribute that universities can engender in graduates" and that "successful careers in these disciplines would usually require some level of critical thinking [creativity]" (p. 128).

Table 10.1 summarizes three major problems that hinder the reconnection of creativity and engineering.

BENEFITS OF CREATIVITY IN EDUCATION

If additional reasons are needed to embed creativity in the engineering curriculum, consider the value of creativity at the level of the individual. D. H. Cropley and Cropley (2000b) drew attention to the benefits of creativity in education: "modern research has demonstrated that although students with high IQs usually obtain good grades both at school and university, they are consistently outstripped by those with not only a high IQ but also high creativity" (p. 207). A. J. Cropley and Urban (2000) expand further on this point. Facaoaru (1985), studying professional engineers, determined that those rated by their peers as the *best* engineers were not only technically or conventionally better, but had more *characteristics* typical of creative people. A. J. Cropley (1994) suggests "creativity is indispensable for 'true' giftedness." In other words, the value of creativity to the individual is that it can be taught and developed (Torrance, 1972).

Fasko (2001) describes other examples of the benefits of creativity in an individual and educational setting, citing earlier work by Parnes and Noller (1972), who reported data on a study into the benefits of creativity courses. Fasko (2001) notes that "Parnes and Noller found that students who completed the sequence of creativity courses significantly outperformed comparable control students..." (p. 324) across a range of idea generation, evaluation, and problem-solving measures, and that their performance in other courses improved as well. A study by Mohan (1973) found a similar result for teacher training, while Mack (1987) discussed the perceived need for creativity training among teachers, and the perception of teachers of the importance of creativity training for children.

FIXING THE PROBLEM

Fixing the problem of creativity and engineering education requires many changes. These are explored also in D. H. Cropley (2014c, in press). Two concrete contributions relate specifically to the design of engineering programs and curricula. Even where program design guidelines exist—for example, the ABET (2011) accreditation criteria—these do not give enough explicit direction. Among the ABET criteria for accrediting academic programs, *Curriculum* talks about "carry[ing] knowledge further toward creative application" (p. 4). That, however, is the only use of the terms *creativity, innovation, creative,* or *innovative* in the criteria. The term *design,* however, is used frequently in the context of problem solving and meeting needs, suggesting that, while there is little specific guidance on embedding creativity and innovation in college-level engineering programs, the need for creativity is implicit in the specified student outcomes (e.g., "an ability to identify, formulate and solve engineering problems" or "an ability to design a system ... to meet desired needs within realistic constraints..." [p. 3]). If creativity is largely missing from engineering programs, it is not through a failure to articulate that need in the accreditation guidelines, although its specific role could be articulated much more explicitly. The problem remains one of how those guidelines are enacted in practice. The problem then returns full circle to issues of a lack of understanding of the role of creativity and innovation in engineering design and problem solving, and a lack of the requisite knowledge and skills needed to build creativity into the curriculum. While this may seem like a harsh indictment, this problem is by no means unique to engineering.

If I have painted a pessimistic picture, it is not to deny the many efforts that have been made to embed creativity in engineering programs. Acar (1998), for example, discusses features of an engineering curriculum that might be used specifically to foster creativity. For example, in describing a new curriculum approach for a master's degree in systems engineering in the United Kingdom, he highlights the importance of encouraging and rewarding creativity. At a more specific and concrete level, he makes explicit the link between a clear definition of the objectives in a system design activity and the fact that this will "ease finding alternative ways of looking at the problem" (p. 136). In other words, divergent thinking in the context of an engineering design activity will be facilitated by good problem definition. This can be mapped to the "Preparation" phase of the Innovation Phase Model (Chapter 8). Acar (1998) also notes the importance of defining student design projects in an open-ended manner, with problems selected that have no right answer. These two approaches deserve a deeper explanation. The former can be seen as an expression of the importance, both to design and to creativity, of a top-down approach (see Chapters 2 and 3). This is the difference between asking "what can I do with this brick?" and asking "what are all the ways that I can solve the problem of building a house?" The latter is an expression of the importance of first defining "what" a system must do (its function), in terms that are solution-free, followed by "how" the function will be implemented. Indeed, this whole issue of the definition of needs and requirements, and the relationship of this to creativity, is a topic of some importance (Hoffmann et al., 2005).

Other examples of work that is seeking to embed creativity in engineering and education, particularly in a more holistic and systematic manner, includes Baillie and Walker (1998); Chang, Hsu, and Chen (2013); and Liu and Schonwetter (2004).

I now offer 3 principles and 12 strategies that can help drive both program and curriculum design to enhance creativity. I offer these at a strategic level, as a set of guiding principles or requirements that should drive program and curriculum design. As a result, these are *not* simply statements such as *we need to teach engineering students how to brainstorm*. There is little value in a piecemeal approach unless it is placed inside a framework that supports all 4Ps of the creativity concept. Engineering students will develop creativity as a genuine graduate quality—as an emergent property of their education—only if these strategies permeate their programs and curricula as a *system*, and are not simply tacked on in a reductionist, remedial fashion.

General Principles

Sternberg (2007) outlines three things to promote the *habit of creativity* (p. 3). They should serve as general principles for curriculum and program design in engineering. First, students must have the *opportunity to engage in creativity*. This must be woven throughout programs and courses in an integrated and mutually reinforcing manner. Second, students must *receive positive encouragement* as they engage in tasks requiring creativity. Third, students must be *rewarded when they demonstrate the desired creativity*.

Sternberg (2007, pp. 8–15) then outlines 12 strategies that guide the development of the creativity habit. This is not to suggest that every aspect of engineering learning must be transformed. There will remain many areas of the curriculum that are best served by convergent approaches; there is, after all, still only one right answer to the question "what is 2 + 2?" However, wherever practical, these strategies should be used to guide the development of creativity as a desirable and vital graduate quality:

- **Redefine problems**—To make *good* choices, students need practice at making choices. When those choices do not work out, students need the opportunity to try again. To achieve this, engineering students need the opportunity to engage in projects that are presented as more open-ended and flexible. Highly constrained or overspecified projects do not allow students to develop this skill.
- **Question and analyze assumptions**—Students must be encouraged to ask questions and not just accept the problem as we as give it to them. This can be achieved partly through the way in which faculty respond to questioning, as well as the way in which faculty establish a *Press*, in which a questioning mindset is valued and modeled.
- **Sell your creative ideas**—Students need to learn how to persuade others of the value of their ideas, i.e., to justify their ideas. Team-based activities, as well as competitive elements to student projects, engender an environment in which the students must become adept at selling their ideas, both to each other, and to faculty.
- **Encourage idea generation**—We want students to get practice at generating ideas, but with constructive criticism. This should be encouraged *intrinsically*, as a necessary component of the activities students undertake, and *extrinsically*, by teaching students specifically how to engage in divergent thinking. In other words, students need to be taught (or perhaps, reminded) how to think divergently and must be given plenty of opportunity to use this rediscovered skill.
- **Understand the role of knowledge**—To be a creative engineer, you first need to be a competent engineer. The principle for the student

here is about broad preparation and *not overspecializing*. Students must be encouraged to see the value in developing other knowledge and skills. You never know when your knowledge of biology, for example, might give you an idea for solving a mechanical problem. Indeed, the growing field of biomimetics suggests that biological sciences may prove to be an exciting and valuable area for *broadening* education for engineering students. This principle also supports the value of diverse internships and work experience during the engineering students' time at college.

- **Identify and surmount obstacles**—We must present students with challenging tasks so that they build resilience. We need to give them opportunities to fail and try again. Certainly, in project work, but even in other courses, students need to develop an understanding that engineering, for example, is usually *not* simply a matter of rolling out a predetermined solution. Every problem is unique and bound by a unique set of constraints. What worked in another situation may not work in this one. Students who understand this are able to focus their energies on finding a new solution, rather than trying to puzzle out why the old solution does not work (and may never work).
- **Encourage sensible risk taking**—Students need the opportunity to try something, even though it might not work. They need to learn how to assess risks and judge that the risk is acceptable. This can be encouraged as simply as making it clear to students that they will not be punished for mistakes, both in terms of their grades and in real terms (for example, if they damage an integrated circuit in the course of a practical class). Clearly, they need to be taught which risks are okay and which are not. Connecting some electronic components on a breadboard in a new way is a sensible risk; forgetting to wear safety goggles when operating a drill press is not. However, if we overreact to the former, we encourage an extremely risk-averse mindset in which students never take sensible risks.
- **Encourage tolerance of ambiguity**—We need to present students with ill-defined problems. Creative people recognize that ambiguity gives them more space to be creative. This can be as simple as breaking away from a familiar lab paradigm of *here is the sheet for today's lab class. Follow the instructions.* Instead of giving students a highly structured menu for a lab class, for example, in electronics, give them a more open-ended problem statement that requires them to deal with the ambiguity and think more independently. Rather than a set of instructions along the lines of *put component X on your breadboard; now connect component Y to component X, touch your voltmeter probe to point Z, and write down the number on the voltmeter,*

we can achieve the same outcomes by saying to students: *Today I want you to find out as much as you can about how transistors work. You have everything you need in the lab, so go for it!* This approach may make students uncomfortable at first, but with the right encouragement, they will begin to take this uncertainty in their stride. When faced with ambiguity, some people close down and do nothing, whereas others see the ambiguity as an opportunity to try different things. We want our creative engineering professionals to be of the latter mindset.

- **Build creative self-efficacy**—Allow students to see that they *can* be creative, so they do not fall into the *I'm not creative* self-fulfilling prophecy. Simply requiring creativity as an assessable component of, for example, project work will allow students to see that they can be creative and that their creativity is an asset. This requires faculty to understand creativity and how it is manifest in engineering products, and to encourage students to build this in to the work they do.
- **Find what excites them**—Help students explore a broad range of areas of their chosen discipline so that they have a better chance of finding the part that really turns them on. A wide variety of *broadening subjects*, as well as the opportunity for diverse, real-world projects, sponsored by real-world organizations, gives students the best chance of finding their chosen field before they graduate. I do not mean only finding a particular disciplinary specialization—e.g., power electronics versus telecommunications—but also a question of activities such as *design* versus *testing*.
- **Understand the importance of delaying gratification**—Foster a sense that sometimes you need to work a little longer and harder to get the reward. Pushing students to the full extent of their abilities is necessary. Both in regular coursework and in project work, we must ensure that students have the opportunity to push boundaries. This may require more flexibility in assessment so that each student can be pushed to his or her limit, without always being assessed in a norm-based fashion. In every case, however, as faculty, we should have the option of pushing students beyond their comfort zones. This does not, however, mean doing 20 convergent homework problems instead of 10, but pushing students further across *all aspects* of their program.
- **Provide a favorable environment**—Engineering educators need to role model creativity. We need to demonstrate our own flexibility, openness, tolerance for ambiguity, and resilience—all 12 of the items mentioned. More simply, we need to demonstrate that we understand what creativity is, why it is valuable, and why it is in the curriculum.

The concepts in earlier chapters provide the background to these 12 strategies. If, as programs are updated and reaccredited, we ensure that students are given the opportunity to develop the creativity habit by embedding these 12 strategies across the program, we will go a long way towards reconnecting creativity to engineering.

DESIGNING A CURRICULUM FOR ENGINEERING CREATIVITY

Notwithstanding my arguments about the importance of a holistic, program-level approach to creativity in engineering education, and the 12 strategies for shaping the design of an engineering program with embedded creativity, it is also helpful to discuss the development of specific courses that tackle this topic. I believe that one way to overcome some of the barriers that I have described is the development of an *exemplar*—a model course design that can be used by faculty to kick-start the process of embedding creativity in engineering education.

Curriculum Objectives

The development of a credible, effective, and relevant curriculum is founded on solid pedagogy. My own preference is Biggs' (1999) (and updated in Biggs & Tang, 2011) approach to *constructive alignment*. Under his framework, a curriculum is stated in the form of clear objectives that specify the level of understanding required. Teaching and learning activities (TLAs) directly address those objectives. Finally, assessment tasks are chosen to give students the opportunity to demonstrate that they have achieved the level of understanding specified in the objectives (Figure 10.2).

The first step in the process of developing an aligned curriculum is to define the kinds of knowledge that are relevant to the subject and level—in this case, an introductory course in engineering creativity and innovation.

FIGURE 10.2 Aligning curriculum objectives, teaching and learning activities, and assessment tasks. *(Adapted from Biggs & Tang, 2011).*

Biggs (1999) distinguishes among four *kinds of knowledge:*

- Knowledge of things or facts (*declarative* or *propositional* knowledge—"what?");
- Knowledge in the form of competencies or skills (*procedural* knowledge—"how?);
- Knowledge of the applicability of facts and skills (*conditional* knowledge—"when?" and "why?");
- Knowledge as an ability to apply facts and skills in an appropriate manner (*functioning* knowledge—"application" and "performance").

It is particularly important in a course on engineering creativity that our focus is on functioning knowledge, characterized by Biggs (1999) as the ability to "... put declarative knowledge to work by solving problems..." (p. 40). He goes on to describe how "functioning knowledge requires a solid foundation of declarative knowledge, at relational level at least, but it also involves ... procedural ... and conditional knowledge" (p. 40). Functioning knowledge in this course, and in the context of engineering creativity, is demonstrated by the ability to develop novel and effective solutions to practical, realistic technological problems.

An introductory course on engineering creativity must impart some declarative knowledge to the student. The primary purpose of the course is to teach engineering students to *be* creative and to embed creativity in the work they do as engineers. However, achieving this in a deep sense, whereby the student is able not only to execute simple procedures but also to understand why those procedures work and to apply them to different situations requires a foundation of factual (declarative) knowledge.

Required *declarative knowledge* for an introductory course in engineering creativity will include

- What is creativity?
- What contribution does creativity make to engineering and society?
- What are the stages in the development of a creative engineering solution?
- What factors affect the role of creativity in the engineering process?
- What role does creativity play in innovation?

Required *procedural knowledge* builds on this declarative base and includes

- How do engineers solve problems?
- How is creativity measured?
- How are creative ideas generated?
- How is creativity fostered in people?
- How is creativity managed?

Conditional knowledge extends the declarative/procedural base by adding further richness:

- When and why do engineers use creativity to solve problems?
- When and why do different thinking styles play a role in creative problem solving?
- Why is creativity valuable in products?
- When and why are different tools used to support engineering creativity?
- When and why are different factors active in fostering/inhibiting creativity?

This foundation then facilitates the practical application of the knowledge—functioning knowledge—for the purpose of solving real engineering problems in a creative manner.

The question of which topics are required to achieve the development of the kinds of knowledge outlined here is preempted by the subject matter of this book. The titles of Chapters 1–8 tell us that in order to develop the package of knowledge specified, we need to cover these topics: The Definition of Creativity; The Importance of Creativity in Engineering; The Phases of Creativity and the Design Process; Product Creativity; Creative Processes; Personal Factors and Creativity; Creativity and Climate; Innovation.

Levels of Understanding

The kinds of knowledge and the topics specified come together with the specification of the *level of understanding* required for each topic. Biggs (1999) emphasizes the relationship between the four kinds of knowledge and the range of levels of understanding that are possible for that knowledge. In simple terms, we can speak of the difference between "surface" and "deep" understanding of any given body of knowledge. Biggs describes five levels of understanding, each of which can be characterized by "learning verbs." Any given topic, directed toward addressing any given kind of knowledge, can be understood at a level ranging from *no understanding* (prestructural) through superficial, *surface understanding* (e.g., unistructural) up to *deep understanding* (extended abstract). Readers will notice that I used these terms earlier in this chapter to discuss pseudo-expertise.

Take, for example, the topic *operational amplifiers* in electronic engineering. I may know facts about this topic (*declarative* knowledge), or I may know how to calculate the gain of an op amp (*procedural* knowledge), or I may know when and why op amps are used in electronic circuits (*conditional* knowledge). Collectively, I may also know how to tackle design problems using operational amplifiers (*functioning* knowledge). However, for each of these kinds of knowledge, my understanding may be very shallow—for example, my declarative knowledge may not extend beyond being able to identify an op amp in a circuit diagram because I have memorized the symbol used to represent it—or my understanding may be very deep. In the latter case, my conditional knowledge of op amps may extend to being able to compare and contrast different types of op amps in order to select the one most appropriate to a given situation. Even my functioning knowledge of operational amplifiers may be shallow—I can solve simple, well-defined, familiar design problems—or deep, allowing me to handle unusual, unfamiliar problems and still develop an appropriate operational amplifier solution to the problem.

Recent thinking in higher education pedagogy makes a compelling case for qualitative assessment (for example, Biggs, 1999). This examines your depth of understanding using a set of criteria called the SOLO Taxonomy (Structure of the Observed Learning Outcome). These criteria hold that the best learning outcomes are associated with an ability to demonstrate a deep understanding of content. "Deep" is characterized by verbs such as *compare, contrast, analyze, apply*. Surface learning, where students fail fully to engage with the material, is characterized by verbs such as *identify, name, list*, and an ability to *do simple procedures*.

Deep learning results in an ability to integrate knowledge and to apply it to new situations, whereas surface learning generally results in an ability to regurgitate facts without real understanding or application.

The SOLO Taxonomy is a hierarchy, ranging from *Prestructural* to *Extended Abstract* levels of understanding, and is summarized in Table 10.2.

The combination of particular kinds of knowledge, specific topics, and the required level of understanding allows us now to state the *curriculum in the form of a set of objectives*. To this is added the specific teaching and learning activities that realize the objectives, and the assessment tasks that evaluate how well the learning verbs are deployed by students. I bring this information together in the form of a model syllabus statement.

TABLE 10.2 Levels of Understanding in Education

Level of understanding	Learning verbs
Prestructural	Misses point
Unistructural	Identify, Name, Do simple procedures
Multistructural	Enumerate, Describe, List, Combine
Relational	Compare, Contrast, Explain, Analyze
Extended Abstract	Theorize, Generalize, Reflect

Model Syllabus

An excellent way to summarize the package of topics, kinds of knowledge, and levels of understanding is in the form of a model syllabus statement. I have used a format typical of those used at my own university, and one that I believe will be broadly similar to other institutions. This is intended not only to summarize the information in this chapter, but also to be used by readers to create their own syllabus. With minor modifications, for example, where semester lengths are slightly different, readers can adapt this syllabus to their own needs.

INTRODUCTION TO CREATIVITY IN ENGINEERING

Course Code: ABCD 0001
School: XY
Study Period: BBB
Credit Hours: AAA

Aim

The aim of this course is to provide the student with a foundation of factual and practical knowledge of creativity and creative problem solving in the engineering domain.

Curriculum Objectives

On completion of this course, students will be able to

- **Analyze** the role of teamwork in successful engineering problem solving;
- **Characterize** the real need in engineering problem situations;
- **Apply** general problem-solving skills to the successful solution of practical problems;
- **Apply** rapid prototyping to the solution of engineering problems;
- **Analyze** the creativity of engineering solutions;
- **Integrate** the characteristics of product creativity into the solution of engineering problems;
- **Apply** convergent thinking at the appropriate stage in the process of solving engineering problems;
- **Apply** divergent thinking at the appropriate stage in the process of solving engineering problems;
- **Explain** the impact of personal psychological factors on the process of solving engineering problems;
- **Apply** knowledge of creativity and innovation to the successful solution of general technological problems;

- **Explain** the impact of the *press* on the solution of engineering problems.

Syllabus (topics)

The Definition of Creativity; The Importance of Creativity in Engineering; The Phases of Creativity and the Design Process; Product Creativity; Creative Processes; Personal Factors and Creativity; Creativity and Climate; Innovation.

Teaching and Learning Activities

This course will be delivered through a combination of lectures, tutorials, and practical exercises (see suggested teaching plan).

Assessment

Assignment A—*Plank* Exercise & Written Report (10%)
Assignment B—*Egg* Exercise & Written Report (15%)
Assignment C—*Spaghetti* Exercise & Written Report (15%)
Assignment D—*Project* Exercise & Written Report (60%)

Textbooks

Cropley, D. H. (2014) *Creativity in Engineering*. San Diego: Academic Press.
Kaufman, J. C. (2009) *Creativity 101*. New York: Springer.

References

Sternberg, R. & Kaufman, J. C. (2010) *The Cambridge Handbook of Creativity*. New York: Cambridge University Press.

Prerequisites: None

Curriculum Implementation

This section develops the information given in the preceding syllabus statement. It is intended as a guide for faculty wishing to implement an introductory course in engineering creativity based on the model syllabus. The lecture, tutorial, and practical exercise components are intended to achieve the curriculum objectives described in the syllabus statement. Lecture content is based on the material presented in the earlier chapters of this book, and will not be summarized except to refer to the general subject matter of each chapter. Tutorial and practical exercises are designed to supplement the lecture material and, consistent with Biggs (1999), to support the curriculum objectives by *presenting opportunities for students to develop the desired level of understanding of each curriculum objective*. The assessment activities will be described separately in a subsequent section. Table 10.3 shows my recommendation for a teaching plan that will implement the model syllabus. I have assumed a general plan of 15 teaching weeks, with one hour of lecture, one hour of tutorial, and two hours of practical activity each week.

Teaching Plan

Although this plan was developed as an exemplar, it is not possible to give full details of lectures, tutorials, and practical exercises here. In the following sections, I will therefore briefly summarize the content of key activities, highlighting useful sources and resources. In addition, you can obtain details of specific activities, such as the Tower Build exercise, by contacting me (david.cropley@unisa.edu.au) or by visiting my profile on

TABLE 10.3 Teaching Plan

Week	Lecture (1)	Tutorial (1)	Practical (2)
1	Introduction	Individual Creativity Test (TCT-DP)	Practical problem-solving exercise (tower build)
2	Creativity in Engineering	Convergent problem solving (Logic problems, numerical problems)	Divergent problem solving (open-ended problems, ambiguous problems)
3	Creativity in Engineering	Problem definition (Solving the *right* problem)	*Plank* Exercise: Teamwork and problem solving
4	Creativity and the Design Process	Rapid prototyping (Using prototyping to understand the problem)	Rapid prototyping exercise
5	Creativity and Products	Measuring the creativity of products (CAT and CSDS ratings)	Using measures to enhance product creativity (Design task + CSDS)
6	Creativity and Products	Evaluating product creativity (I/O technique, Screening matrix)	*Egg* Exercise: Building creativity into solutions
7	Cognitive Processes for Creativity	Creative cognition—Divergent production (Torrance Tests)	Creative cognition—Priming the pump

(Continued)

TABLE 10.3 (Continued)

Week	Lecture (1)	Tutorial (1)	Practical (2)
8	Cognitive Processes for Creativity	Creative cognition—Tools (mind-mapping)	Creative cognition—Brainstorming + Enhancements
9	Cognitive Processes/Person	Creative cognition—Locks and blocks (the slime task)	*Spaghetti* Exercise: *Being* more creative
10	Creativity and the Person	Personality and creativity—The Big 5	Open-ended, team-based design problem (Mousetrap Vehicle, Berlin Airlift)
11	Creativity and the Person	Engineering creativity—Preference for complexity (BWAS) and technical knowledge	
12	Creativity and Press	*Climate* scales	
13	Creativity and Press	*Climate* scales	
14	Innovation	Assessing organizational innovation (IPAI)	
15	Creativity Training	Individual creativity test (TCT-DP)	

Academia.edu (https://unisa-au.academia.edu/DavidCropley), where readers will find full descriptions of the exercises and tests in the *Resources* folder. A further, more general discussion of designing activities to foster creativity is given by D. H. Cropley (2014b).

Week 1

- Lecture—Introduction to Creativity (Chapter 1);
- Tutorial—Test of Creative Thinking-Drawing Production (TCT-DP):
  - Students complete TCT-DP (Form A) in class. Instructor to discuss the scoring categories and show examples of different responses (available in the test manual). No formal scoring required. The purpose of this exercise is to stimulate thinking about creativity and creative thinking in the student;

- Practical—Tower Build:
  - Students complete a two-stage problem-solving exercise in teams. The purpose of this exercise is to develop problem-solving skills, including understanding the need, and to begin to see the role that creativity plays in a technological problem domain.

Week 2

- Lecture—Creativity in Engineering (Chapter 2);
- Tutorial—Convergent Problem Solving (Logic problems, numerical problems):
  - Students complete a set of logical and numerical problems that are characterized by the fact that they all have a single right answer. Instructors emphasize that these analytical problems, while important in many stages of the engineering process, are not the only kind of problem the students may have to solve in engineering. As each problem is completed, the instructor should ask for feedback and discussion: How do the students characterize these problems? Are there particular skills or strategies employed in solving them? Where do they encounter this type of problem in engineering? Do they encounter other problems in engineering that are fundamentally different from these? If so, what is that difference?
- Practical—Divergent Problem Solving (Open-ended problems, ambiguous problems):
  - Students complete a set of open-ended and/or ambiguous problems, the fundamental characteristic of which is that there is more than one possible answer to the problem. Here the instructor should emphasize the contrast to convergent problems, and each problem should be used as a basis for discussion along similar lines to the discussion of convergent problems. Emphasize the fundamental differences between the two types of problems and the fact that both figure in engineering problem solving.

Week 3

- Lecture—Creativity in Engineering (Chapter 2, continued);
- Tutorial—Problem Definition (Solving the Right Problem):
  - This tutorial is an exercise in problem definition and needs analysis. Options for this exercise are described on my website.
- Practical—*Plank* Exercise (Assignment A):
  - Teamwork & Problem Solving. This exercise is the first assessed activity in the course. It is complemented with a written report, and a full description is given in a later section.

Week 4

- Lecture—Creativity and the Design Process (Chapter 3);
- Tutorial—Rapid Prototyping (Using prototyping to understand the problem):
    - This exercise is focused on the value of prototyping as a means for understanding the problem, and as a mechanism for aiding idea generation and development. A key concept here is that, very often, *the solution defines the need*. It is only when a tangible solution is available that a better understanding of the real need is possible. This tutorial activity will introduce the concept of rapid prototyping in preparation for the practical exercise on the same subject.
- Practical—Rapid Prototyping Exercise:
    - Students will employ rapid prototyping in the solution of a practical problem. Students should be given a problem statement in the "Voice of the Customer" that is *reasonably ambiguous*. For example, "I need something that can make it easier for my grandmother to open tin cans that have a ring pull on them." The problem chosen should be amenable to rapid prototyping using everyday materials such as paper, cardboard, clay, plasticine, string, wire, straws, tape, paperclips, and so forth. The instructor should ask students to reflect on their prototypes. Did the process of rapid prototyping help them to understand better the user need? Did their understanding of the need change as a result of the prototyping? Did the process of prototyping lead to any solution ideas that may not have been obvious without it?

Week 5

- Lecture—Creativity & Products (Chapter 4);
- Tutorial—Measuring the creativity of products (CAT & CSDS ratings):
    - Students will use the Consensual Assessment Technique (CAT; see Chapter 4) and the Creative Solution Diagnosis Scales (CSDS; see Chapter 4) to rate the creativity of a selection of products. Use of the CAT will help to demonstrate to students that there is a reasonable degree of consistency among the group in terms of their intrinsic beliefs about creativity. In other words, engineers have a certain common understanding of what makes a product creative. The CSDS is used to illustrate that product creativity can be broken down into more specific categories, and that these, too, have a degree of commonality across the group. The CSDS is

also useful for illustrating how an instruction like "make it more creative" can be made more explicit.

- Practical—Using measures to enhance product creativity (Design task + CSDS):
  - Student teams are given a simple design task, amenable to rapid prototyping. They are given a fixed period of time to create a prototype solution. The groups are then invited to rate the creativity of another group's design, using the CSDS. The instructor then leads a discussion of the ratings. Each group should be asked to give the overall rating for the product it assessed and then the scores for each major category (effectiveness, novelty, etc.). The discussion should then turn to how the creativity of each design could be improved. This is achieved by examining scores on individual indicators. What could a group do to improve scores on individual indicators?

Week 6

- Lecture—Creativity & Products (Chapter 4, continued);
- Tutorial—Evaluating product creativity (I/O technique, Screening matrix):
  - Students will be taught the use of typical evaluation techniques (I/O; Screening Matrix) and apply these to sample products.
- Practical—*Egg* Exercise: Building Creativity into Solutions (Assignment B):
  - The second assessed activity builds on the well-known egg-drop design task. A particular focus, however, is building creativity (i.e., effectiveness, novelty, elegance) into the solution. This exercise in described in more detail in a later section.

Week 7

- Lecture—Cognitive Processes for Creativity (Chapter 5);
- Tutorial—Creative Cognition—Divergent Production (Torrance Tests):
  - Students will undertake some of the TTCT exercises, without formal scoring, to illustrate concepts of divergent thinking, including fluency, flexibility, and originality.
- Practical—Creative Cognition (Priming the Pump):
  - The purpose of this exercise is to present students with an opportunity to experience typical idea-generation techniques, e.g., brainstorming. This should be done with a discussion following each technique to reflect on strengths and weaknesses. If time permits, students should be given an opportunity to experience differences such as brainstorming in *real* and *nominal*

groups. Students should also practice describing the function of common objects as an abstract verb–noun pair (e.g., screwdriver = torque applier) and then generating other solutions that satisfy the same verb–noun functional description.

Week 8

- Lecture—Cognitive Processes for Creativity (Chapter 5 continued);
- Tutorial—Creative Cognition—Tools (Mind-Mapping):
  - This exercise gives students an opportunity to experience other tools that support creativity and idea generation. This exercise will involve describing a problem situation (e.g., see the Public Transport example and mind-mapping in Chapter 9) using a mind map, and then using the mind map to help generate solutions (e.g., How can we increase patronage on public transport?).
- Practical—Creative Cognition (Brainstorming + Enhancements):
  - This exercise will present students with a problem to be addressed through brainstorming. The exercise will be conducted as normal, and when the flow of ideas begins to slow, the instructor will then explain the concept of hierarchies (see the discussion of remote associates and categories in Chapter 5 and the KJ Method in Chapter 9).

Week 9

- Lecture—Cognitive Processes for Creativity (Chapter 5, continued)/ Creativity & the Person (Chapter 6);
- Tutorial—Creative Cognition (Locks & Blocks, the Slime Task):
  - In this exercise, the instructor will present students with the *Slime Task*. This is a partially constrained problem-solving exercise involving extracting *slime* from a test tube, and moving it through some obstacles and then back into the tube, using only materials provided (full details on the website). The purpose of this is to illustrate the various ways that creativity is blocked or inhibited.
- Practical—*Spaghetti* Exercise: *Being* More Creative (Assignment C):
  - This exercise—the third assessment item—is based on the well-known Marshmallow Challenge, but with a focus on applying the students' knowledge of creativity (including *Process, Person,* and *Product*). Student groups must therefore be encouraged to apply their knowledge of idea generation, and use their knowledge of product criteria, to maximize the creativity of their design. They should also reflect on *Person* factors that may have

influenced their solutions (e.g., willing to take risks) and discuss how these factors affected their final design.

Week 10

- Lecture—Creativity & the Person (Chapter 6, continued);
- Tutorial—Personality and Creativity (The Big 5):
  - Students will complete one of the widely available Big 5 personality factor questionnaires. This does not need to be formally scored, but students should be encouraged to reflect on their personality profile in comparison to those factors known to be associated with creativity.
- Practical—Open-ended, team-based design problem (Assignment D):
  - The major assignment spans the final six weeks of the course and involves a practical design problem and report. This activity is intended as an opportunity for students to demonstrate their understanding of *all* the curriculum objectives.

Week 11

- Lecture—Creativity & the Person (Chapter 6 cont.);
- Tutorial—Engineering Creativity (Preference for Complexity (BWAS) and Technical Knowledge):
  - This exercise will explore other Person factors associated with creativity, including preference for complexity (measured by the Barron–Welsh Art Scale) and the role of technical knowledge in creativity.
- Practical—Open-ended, team-based design problem (continued).

Week 12

- Lecture—Creativity & Press (Chapter 7);
- Tutorial—*Climate* (Press) scales:
  - This exercise will explore a variety of scales designed to assess the organizational/social climate and creativity. This will include scales with empirical support, such as those described in Chapter 7, as well as other informal scales. Students will reflect on the institutional climate at their college, as well as the climate in their project teams, and discuss how these impact on their creativity.
- Practical—Open-ended, team-based design problem (continued).

Week 13

- Lecture—Creativity & Press (Chapter 7, continued);
- Tutorial—Climate scales:
  - This session will be used for further exploration of organizational and team climate. Details of recommended scales are available on my website.
- Practical—Open-ended, team-based design problem (continued).

Week 14

- Lecture—Innovation (Chapter 8);
- Tutorial—Assessing organizational innovation (IPAI):
  - This exercise will use the Innovation Phase Assessment Instrument to diagnose institutional or team alignment to the ideal conditions for creativity. Students will discuss results of the assessment and reflect on strategies to maintain strengths and improve weaknesses.
- Practical—Open-ended, team-based design problem (continued).

Week 15

- Lecture—Creativity Training (Chapter 9);
- Tutorial—Individual Creativity Test (TCT-DP):
  - Students complete TCT-DP (Form B) in class. While no formal scoring is required, students should reflect on any changes compared to their score on Form A in Week 1.
- Practical—Open-ended, team-based design problem (continued).

Assessment Guide

According to Biggs (1999), *assessment tasks* are the means by which students demonstrate their level of understanding of the specified curriculum objectives. Table 10.4 shows the relationship between the curriculum objectives specified in the syllabus and the assessment tasks of the course.

The learning verb identified in the syllabus for each curriculum objective (e.g., Characterize the real need in engineering problem situations) represents the highest level of understanding that students are expected to demonstrate. In other words, a student showing that she can characterize the real need in engineering problem situations (a relational level of understanding), as distinct, for example, from only showing that she can identify the real need (a unistructural level of understanding), will receive the highest grade for that part of the course.

TABLE 10.4 Objectives and Assessment Activities

Assessment Curriculum objective	A (10%)	B (15%)	C (15%)	Project (60%)
1	✓			✓
2	✓			✓
3	✓			✓
4		✓		✓
5		✓		✓
6		✓		✓
7			✓	✓
8			✓	✓
9			✓	✓
10				✓
11				✓

Each assignment is therefore assessed qualitatively by looking for the highest level learning verbs that are evident in the student's work. Table 10.5 shows the relationship between grade levels, curriculum objectives, and learning verbs for Assignment A (Plank Exercise). I use the Australian grading system; however, this can be translated directly into comparable letter grades (A–E, and Fail, for example).

In the remaining sections, I give a brief description of each of the four assignments. Similar grading criteria can be defined for each of these and are available from my website. A more general discussion of the use of assessment as a mechanism for fostering creativity is given by A. J. Cropley and Cropley (2007).

Assignment A—*Plank* Exercise (10%)

The *Plank* Exercise assignment consists of two parts: a practical exercise and a written summary.

The practical exercise is based on the following and can be modified to suit particular classes and circumstances. In teams of 8–10, students use the available equipment to move the team from the start area to the finish area (see the exercise plan available on my website) as quickly as possible, without any members *falling off* the equipment (wooden or cardboard planks are used flat on the ground, so there is no danger to participants). The instructor will set up the exercise in a suitable area (lab or outdoors) according to the specification in exercise plan.

TABLE 10.5 Grading Criteria, Assignment A

Grade level	Curriculum objectives and learning verbs (To get this grade, you need to demonstrate that you can...)
High Distinction (Relational understanding)	**Analyze** the role of teamwork in successful problem solving; **Characterize** the real need in problem situations; **Apply** general problem-solving skills to the successful solution of practical problems.
Distinction (mix)	A grade of D will be a mix of elements of HD and C.
Credit (Multistructural understanding)	**Describe** the role of teamwork in problem solving; **Discuss** the real need in problem situations; **Use** general problem-solving skills to the solution of a practical problem.
P1 (mix)	A grade of P1 will be a mix of elements of P2 and Credit.
P2 (Unistructural understanding)	**Identify** some aspects of teamwork in problem solving; **Identify** the real need in a problem situation; **Identify** problem-solving skills used in the solution of a practical problem.
F1 (Partial Unistructural understanding)	**Identify** some aspects of teamwork; **Identify** a need in a problem situation; **Identify** a problem-solving skill used in the solution of a practical problem.
F2 (Prestructural understanding)	**Fail to identify** some aspects of teamwork; **Fail to identify** a need in a problem situation; **Fail to identify** problem-solving skills used in the solution of a practical problem.

Key points:

- Students to participate in teams of 8–10 (minimum 8);
- Instructor to set up exercise area (squares and ropes) as shown in exercise plan;
- Planks (1 × 2.0 m, 2 × 1.7 m) placed in start zone;
- Team members all begin in start zone;
- Team must move entire team, including planks, from start zone to end zone as quickly as possible;
- Nobody is allowed to step off planks or squares. If a person steps on ground, the team must return to start zone and begin again;
- Ropes and squares must be placed so that gaps are as indicated in the plan. This means that only certain planks and combinations can be used to move across the gaps. It is important that gaps are the width indicated so that the short planks cannot bridge the

long gaps, and also so that the long planks cannot bridge a diagonal gap (see plan);
- More than one person can stand on a square at a given time.

The task requires teamwork and coordination, as well as problem-solving skills.

With several groups in a tutorial class, acknowledge that the first group has a bit of a disadvantage, but encourage subsequent groups to try to improve, both on time and on teamwork. If time permits and a team wants to try again, it may do so after all teams have had a turn.

Initially, give minimal instructions. Simply tell teams the basic rules. Do not suggest any solutions, and discourage other teams from making suggestions. Remind them that teamwork will help.

On completion of the exercise, the instructor should ask students for comments on how they think they performed. Did they achieve the goal? Did they work well as a team? Did teamwork help them? Did lack of teamwork hinder them? Was communication important? Did any team member(s) take a leadership role? Did this make a difference? If someone acted as leader, was this deliberate, or did it just work out that way? Did the leader explain what was needed and encourage teamwork, or did the leader just tell people what to do? Was there any conflict? Did it get resolved? Did it affect performance? If the team tried again, did it improve? What do team members think would make the biggest difference to the team's performance?

After completion of the exercise, students will submit a written assignment (typically 500 words) in which they attempt to demonstrate their understanding of the relevant curriculum objectives:

- the role of teamwork in successful problem solving;
- the identification of the real need in problem situations;
- the use of general problem-solving skills to the successful solution of practical problems.

Each student's grade for this assignment is formed by assessing highest level learning verbs, for the curriculum objectives in question, that were manifest in the student's written assignment and practical exercise performance.

Assignment B—*Egg* Exercise (15%)

The *Egg* Exercise consists of two components: a practical activity and a written summary.

The practical activity requires students to move a raw egg from a defined starting point—typically a platform at a set height above the ground—to ground level without breaking the egg. A variety of constraints can be placed on this task including the material permitted, the degree of human intervention, the height dropped, and so on. The focus in this course should be firmly on developing a solution that can be assessed as creative, as per the definition described in this book (i.e., as a minimum effective and novel). Instructors should stress that assessment of this activity (see curriculum objectives) is based on demonstrating an ability to design a creative solution. Students are also graded on their ability to use rapid prototyping, and a variety of materials will need to be provided. This also acts as a challenging constraint—students do not have access to any possible material but must demonstrate creativity under conditions of constraint.

Students also submit a written assignment in which they reflect on and discuss the exercise. The emphasis in this written activity is on further demonstrating their level of understanding of the particular curriculum objectives in question.

Assignment C—*Spaghetti* Exercise (15%)

The *Spaghetti* Exercise consists of two components: a practical exercise and a written summary.

Like the *Egg* Exercise, this exercise is based on a popular task (the Marshmallow Challenge); however, it is important to note that there are differences in the focus and purpose of the exercise in this course. The practical exercise is typically run over 18 minutes, with teams given 20 sticks of spaghetti, one yard of masking tape, one yard of string, and one marshmallow. Their task is to build the tallest possible freestanding structure, with the marshmallow on top. The exercise is modified somewhat in this course to focus on particular curriculum objectives. For this reason, the times and conditions/constraints may be modified. Further details are available on my website.

Students also submit a written assignment in which they reflect on and discuss this exercise. The emphasis in this written activity is on further demonstrating their level of understanding of the particular curriculum objectives in question.

Assignment D—*Project* Exercise (60%)

The *Project* Exercise also consists of two components: a practical exercise and a written summary.

The purpose of this exercise is to give students an opportunity to demonstrate understanding of all curriculum objectives in a single, holistic exercise. The exercise is presented to students with a minimal set of instructions, and they are given a period of up to six weeks to work on the exercise (e.g., Week 9–Week 15). Students are informed that they will demonstrate their solutions in the final practical session (i.e., Week 15). I have found two problems particularly effective and suitable for engineering students. One is the design of a mousetrap-powered wheeled vehicle, and one is a laboratory recreation of the Berlin Airlift. In both cases, the problem is described in an open-ended and somewhat ambiguous manner, and students are able to use any reasonable materials that they can find. Emphasis is again placed on all characteristics of creativity (novelty, effectiveness, etc.). Full details are available on my website.

Students also submit a written assignment in which they reflect on all aspects of the curriculum objectives: what they did, why it was creative, the methods they used, the impact of the team climate, factors that may have inhibited their creativity, and so on. This written report will be considerably longer than the previous reports (e.g., 2,500 words). Students' final grades will be determined from both the written report and also the practical demonstration, both of which will show their level of understanding of the relevant objectives.

SUMMARY

There are many engineering faculties, around the world, who understand the importance of creativity and innovation in engineering. There are also many examples of attempts to embed creativity in engineering programs and curricula. While this is to be applauded and encouraged, one feature of much of this work is that it is guilty of reinventing the wheel. I have seen many examples of papers from engineering faculties, and even government-funded workshops, which begin from first principles. The result is that these efforts to embed creativity in engineering rarely progress beyond recognizing and defining the problem. If we are to see real progress in changing engineering education, then it is vital to build on the foundations that already exist. We have to recognize that just because these questions may be new to us, they are not new to everyone. Engineering education is ready to move to the next level—having embedded evidence-based approaches to creativity in programs and curricula, we need to start gathering evidence that these are delivering the desired results.

CONCLUDING REMARKS

"And it ought to be remembered that there is nothing more difficult to take in hand, more perilous to conduct, or more uncertain in its success, than to take the lead in the introduction of a new order of things. Because the innovator has for enemies all those who have done well under the old conditions, and lukewarm defenders in those who may do well under the new." ***Niccolo Machiavelli, 1469–1527, Diplomat, Philosopher***

I hope that I have made a strong case for the relationship between creativity and engineering. Creativity is concerned with finding novel, effective solutions to problems. Engineering is a special case of creative problem solving—one where the needs, problems, and solutions revolve around technology. Finding novel and effective solutions to problems is not unique to engineering—we see it in every facet of human endeavor—but there seems to be a special relationship between engineering creativity and the continuing advance of humankind. Throughout history, our most basic leaps forward have depended on technology. Taking nothing away from the great advances in medicine, education, art, music, psychology, farming, etc., it seems that none of these would have been possible without the basic tools, structures, and products that have been developed by engineers.

If we look at the most fundamental of Abraham Maslow's *Hierarchy of Needs* (Maslow, 1943)—*physiological* and *safety*, for example—it is not hard to make the case that many if not all of these (breathing, food, water, sleep, excretion, shelter, health, etc.) are only possible, or at least greatly aided, by the products of engineering.

Critics might also point to the damage that engineering has inflicted on humankind. Careless exploitation of resources and undesirable emergent properties of many products and systems have caused pollution and contamination, and have greatly increased our ability to harm each other. While this cannot be denied, it is futile to cry over spilled milk. Creative engineering problem solving has almost unlimited power not only to solve new problems as they arise, but also to undo past mistakes. I like to tell my students that it is not environmentalists who will save the planet (although their role in drawing attention to the problem and acting as agents of change has been vital), it is engineers!

This book cannot hope to cover all of the material that is available on creativity, innovation, and engineering. What I hope I have achieved is to give readers a firm grounding in the basic and core material so that you now talk the talk. New material and new research are added, of course, to the body of knowledge all the time. I hope now that readers coming to this material for the first time, or who have dabbled in creativity in engineering, now know where to look to develop their own ideas. Engineers, of all people, should *not* be guilty of reinventing the wheel. To make progress in creativity and innovation, the engineering domain must accept the extant body of knowledge and move forward with new questions and problems, and not waste energy revisiting questions that have already been answered.

As you do that, remember: psychologists are your friends!

"All we know about the new economic world tells us that nations which train engineers will prevail over those which train lawyers. No nation has ever sued its way to greatness." ***Richard Lamm, 1935–, Author, Politician***

References

ABET (2011). *Criteria for accrediting engineering programs*. Baltimore, MD: ABET Engineering Accreditation Commission.

Abra, J. (1994). Collaboration in creative work: An initiative for investigation. *Creativity Research Journal, 7*(1), 1–20.

Acar, B. S. (1998). Releasing creativity in an interdisciplinary systems engineering course. *European Journal of Engineering Education, 23*(2), 133–140.

Ahern, A., O'Connor, T., McRuairc, G., McNamara, M., & O'Donnell, D. (2012). Critical thinking in the university curriculum–the impact on engineering education. *European Journal of Engineering Education, 37*(2), 125–132.

Ajzen, I. (1991). The theory of planned behavior. *Organizational Behavior and Human Decision Processes, 50*(2), 197–211.

Albert, R. S., & Runco, M. A. (1988). Independence and the creative potential of gifted and exceptionally gifted boys. *Journal of Youth and Adolescence, 18*(3), 221–230.

Alliance of Artists' Communities (1996). *American creativity at risk: Restoring creativity as a priority in public policy, cultural philanthropy, and education*. Portland, OR: Alliance of Artists' Communities.

Altshuller, G. S. (1988). *Creativity as an exact science*. New York: Gordon and Breach.

Amabile, T. M. (1983). *The social psychology of creativity*. New York: Springer.

Amabile, T. M. (1996). *Creativity in context*. Boulder: Westview Press.

Amabile, T. M., Goldfarb, P., & Brackfleld, S. C. (1990). Social influences on creativity: Evaluation, coaction, and surveillance. *Creativity Research Journal, 3*(1), 6–21.

Amabile, T. M., & Gryskiewicz, N. D. (1989). The creative environment scales: Work environment inventory. *Creativity Research Journal, 2*(4), 231–253.

Amabile, T. M., & Tighe, E. (1993). Questions of creativity. In J. Brockman (Ed.), *Creativity. The reality club* (Vol. 4, pp. 7–27). New York: Simon and Schuster.

Amoussou, G. A., Porter, M., & Steinberg, S. J. (2011). *Assessing creativity practices in design*. Paper presented at the Frontiers in Education Conference, Rapid City, SD.

Anderson, J. R. (1976). *Language, memory and thought*. Hillsdale, NJ: Erlbaum.

Anderson, N. R., & West, M. A. (1994). *The team climate inventory: Manual and users' guide*. Windsor: NFER-Nelson.

Andreasen, N. C. (1987). Creativity and mental illness: Prevalence rates in writers and their first degree relatives. *American Journal of Psychiatry, 144*, 1288–1292.

Anthony, E. J. (1987). Risk, vulnerability and resilience: An overview. In E. J. Anthony, & B. J. Cohen (Eds.), *The invulnerable child* (pp. 3–48). New York: Guilford Press.

Austin, J. H. (1978). *Chase, chance, and creativity*. New York: Columbia University Press.

Ausubel, D. P., & Paul, D. (2000). *The acquisition and retention of knowledge: A cognitive view*. Dordrecht: Kluwer Academic.

Bacon, F. (1909). *Essays, civil and moral and the new Atlantis [1627]*. New York: Collier.

Baer, J. M. (1996). The effects of task-specific divergent-thinking training. *The Journal of Creative Behavior, 30*(3), 183–187.

Baer, J. M. (1998). The case for domain specificity of creativity. *Creativity Research Journal, 11*(2), 173–177.

Baer, J. M. (2010). Is creativity domain specific? In J. C. Kaufman, & R. J. Sternberg (Eds.), *The Cambridge handbook of creativity* (pp. 321–341). New York: Cambridge University Press.

Baer, J. M. (2012). Domain specificity and the limits of creativity theory. *The Journal of Creative Behavior, 46*(1), 16–29.
Baer, J. M., & Kaufman, J. C. (2005). Bridging generality and specificity: The amusement park theoretical (APT) model of creativity. *Roeper Review, 27*(3), 158–163.
Baer, J. M., Kaufman, J. C., & Gentile, C. A. (2004). Extension of the consensual assessment technique to nonparallel creative products. *Creativity Research Journal, 16*(1), 113–117.
Baer, M., Oldham, G. R., Jacobsohn, G. C., & Hollingshead, A. B. (2008). The personality composition of teams and creativity: The moderating role of team creative confidence. *The Journal of Creative Behavior, 42*(4), 255–282.
Bailin, S. (1988). *Achieving extraordinary ends: An essay on creativity*. Dordrecht, Holland: Kluwer.
Baillie, C. (2002). Enhancing creativity in engineering students. *Engineering Science, & Education Journal, 11*(5), 185–192.
Baillie, C., & Walker, P. (1998). Fostering creative thinking in student engineers. *European Journal of Engineering Education, 23*(1), 35–44.
Bain, A. (1868). *The sense and the intellect* (3rd ed.). London: Longman Green.
Bandura, A. (1962). Social learning through imitation. In M. R. E. Jones (Ed.), *Nebraska symposium on motivation* (pp. 211–269). Lincoln, NE: University of Nebraska Press.
Barron, F. X. (1969). *Creative person and creative process*. New York: Holt, Rinehart, & Winston.
Barron, F. X. (1972). *Artists in the making*. New York: Seminar Press.
Barron, F. X., & Harrington, D. M. (1981). Creativity, intelligence, and personality. *Annual Review of Psychology, 32*(1), 439–476.
Basadur, M., & Hausdorf, P. A. (1996). Measuring divergent thinking attitudes related to creative problem solving and innovation management. *Creativity Research Journal, 9*(1), 21–32.
Bateman, K. (2013, April 18, 2013). IT students miss out on roles due to lack of creativity. *ComputerWeekly.com*.
Batey, M., Chamorro-Premuzic, T., & Furnham, A. (2010). Individual differences in ideational behavior: Can the big five and psychometric intelligence predict creativity scores? *Creativity Research Journal, 22*(1), 90–97.
Batey, M., & Furnham, A. (2006). Creativity, intelligence, and personality: A critical review of the scattered literature. *Genetic, Social, and General Psychology Monographs, 132*(4), 355–429.
Beghetto, R. A. (2006). Creative self-efficacy: Correlates in middle and secondary students. *Creativity Research Journal, 18*(4), 447–457.
Benner, M. J., & Tushman, M. L. (2003). Exploitation, exploration, and process management: The productivity dilemma revisited. *The Academy of Management Review*, 238–256.
Benson, C. (2004). Professor John Eggleston Memorial Lecture 2004–Creativity: Caught or taught? *Journal of Design, & Technology Education, 9*(3), 138–144.
Berlyne, D. E. (1962). *Conflict, arousal and curiosity*. New York: McGraw Hill.
Besemer, S. P. (1998). Creative Product Analysis Matrix: Testing the model structure and a comparison among products—Three novel chairs. *Creativity Research Journal, 11*(4), 333–346.
Besemer, S. P. (2006). *Creating products in the age of design: How to improve your new product ideas!* Stillwater, OK: New Forums Press.
Besemer, S. P., & O'Quin, K. (1987). Creative product analysis: Testing a model by developing a judging instrument. In S. G. Isaksen (Ed.), *Frontiers of creativity research: Beyond the basics* (pp. 367–389). Buffalo: Brady.
Besemer, S. P., & O'Quin, K. (1999). Confirming the three-factor creative product analysis matrix model in an American sample. *Creativity Research Journal, 12*(4), 287–296.

Biermann, K.-R. (1985). Über Stigmata der Kreativität bei Mathematikern des 17. bis 19. Jahrhunderts [On indicators of creativity in mathematicians of the seventeenth to nineteenth centuries]. *Rostocker Mathematik Kolloquium [Rostock Mathematics Colloquium], 27*, 5–22.

Biggs, J. (1999). *Teaching for quality learning at university: What the student does*. Buckingham, UK: SRHE and Open University Press.

Biggs, J., & Tang, C. (2011). *Teaching for quality learning at university*. Maidenhead, UK: McGraw-Hill International.

Blanchard, B. S., & Fabrycky, W. J. (2006). *Systems engineering and analysis* (4th ed.). Upper Saddle River, NJ: Pearson Prentice Hall.

Bledow, R., Frese, M., Anderson, N., Erez, M., & Farr, J. (2009). A dialectic perspective on innovation: Conflicting demands, multiple pathways, and ambidexterity. *Industrial and Organizational Psychology, 2*(3), 305–337.

Bloom, B. S., & Sosniak, L. A. (1985). *Developing talent in young people*. New York: Ballantine Books.

Blumrich, J. F. (1970). Design. *Science, 168*, 1551–1554.

Boden, M. (1995). Creativity and unpredictability. *Constructions of the Mind: Artificial Intelligence and the Humanities. Stanford Electronic Humanities Review, 4*(2).

Boden, M. A. (1994a). Introduction. In M. A. Boden (Ed.), *Dimensions of creativity* (pp. 1–8). Cambridge, MA: MIT Press.

Boden, M. A. (1994b). What is creativity? In M. A. Boden (Ed.), *Dimensions of creativity* (pp. 75–118). Cambridge, MA: MIT Press.

Bransford, J. D., Brown, A. L., & Cocking, R. R. (Eds.), (2000). *How people learn* Washington, DC: National Academy Press.

Brockman, J. B. (2009). *Introduction to engineering: Modeling and problem solving*. Hoboken, NJ: John Wiley, & Sons Inc.

Brophy, D. R. (1998). Understanding, measuring, and enhancing individual creative problem-solving efforts. *Creativity Research Journal, 11*(2), 123–150.

Bruner, J. S. (1962). The conditions of creativity. In H. Gruber, G. Terrell, & M. Wertheimer (Eds.), *Contemporary approaches to cognition* (pp. 1–30). New York: Atherton Press.

Bruner, J. S. (1964). The course of cognitive growth. *American Psychologist, 19*(1), 1.

Bruner, J. S. (1975). Child development: Play is serious business. *Psychology Today, 8*, 80–83.

Bryan, W. L., & Harter, N. (1899). Studies on the telegraphic language: The acquisition of a hierarchy of habits. *Psychological Review, 4*, 345–375.

Buel, W. D. (1960). The validity of behavioral rating scale items for the assessment of individual creativity. *Journal of Applied Psychology, 44*(6), 407–412.

Buel, W. D. (1965). Biographical data and the identification of creative research personnel. *Journal of Applied Psychology, 49*(5), 318–321.

Buel, W. D., Albright, L. E., & Glennon, J. R. (1966). A note on the generality and cross-validity of personal history for identifying creative research scientists. *Journal of Applied Psychology, 50*(3), 217–219.

Buel, W. D., & Bachner, V. M. (1961). The assessment of creativity in a research setting. *Journal of Applied Psychology, 45*(6), 353–358.

Buhl, H. R. (1960). *Creative engineering design*. Ames, IA: Iowa State University Press.

Burghardt, M. D. (1995). *Introduction to the engineering profession* (2nd ed.). New York: HarperCollins College Publishers.

Burkhardt, H. (1985). *Gleichheitswahn, Parteienwahn: Massenpsychosen der Gegenwart*. Düsseldorf: Hohenrain-Verlag.

Butler, D. L., & Kline, M. A. (1998). Good versus creative solutions: A comparison of brainstorming, hierarchical, and perspective-changing heuristics. *Creativity Research Journal, 11*(4), 325–331.

Buzan, A. (2003). *The mind map book*. New York: Dutton.
Byrd, R. (1986). *Creativity and risk-taking*. San Diego, CA: Pfeiffer International Publishers.
Campbell, D. T. (1960). Blind variation and selective survival as a general strategy in knowledge processes. In M. C. Yovits, & S. Cameron (Eds.), *Self-organizing systems* (pp. 205–231). New York: Pergamon Press.
Carson, S. H., Peterson, J. B., & Higgins, D. M. (2005). Reliability, validity, and factor structure of the Creative Achievement Questionnaire. *Creativity Research Journal, 17*(1), 37–50.
Cattell, J., Glascock, J., & Washburn, M. (1918). Experiments on a possible test of aesthetic judgment of pictures. *The American Journal of Psychology, 29*, 333–336.
Cattell, R. B., & Butcher, H. J. (1968). *The prediction of achievement and creativity*. New York: Bobbs-Merrill.
Cattell, R. B., & Drevdahl, J. E. (1955). A comparison of personality profile (16PF) of eminent researchers with that of eminent teachers and administrators and of general population. *British Journal of Psychology, 46*(4), 248–261.
Chang, C. P., Hsu, C. T., & Chen, I. J. (2013). The relationship between the playfulness climate in the classroom and student creativity. *Quality & Quantity, 47*(3), 1493–1510.
Charyton, C., & Snelbecker, G. E. (2007). General, artistic and scientific creativity attributes of engineering and music students. *Creativity Research Journal, 19*(2–3), 213–225.
Child, I. L., & Iwao, S. (1968). Personality and esthetic sensitivity: Extension of findings to younger age and to different culture. *Journal of Personality and Social Psychology, 8*, 308–312.
Christensen, C. M. (1997). *The innovator's dilemma*. Boston: Harvard Business School Press.
Christensen, C. M. (1999). *Innovation and the general manager*. Boston: Irwin/McGraw-Hill.
Clapham, M. M. (2003). The development of innovative ideas through creativity training. *The International Handbook on Innovation*, 366–376.
Colangelo, N., Kerr, B., Hallowell, K., Huesman, R., & Gaeth, J. (1992). The Iowa Inventiveness Inventory: Toward a measure of mechanical inventiveness. *Creativity Research Journal, 5*(2), 157–163.
Cooper, C., Altman, W., & Garner, A. (2002). *Inventing for business success*. New York: Texere.
Cooperrider, D. L., & Srivastva, S. (1987). Appreciative inquiry in organizational life. *Research in Organizational Change and Development, 1*(1), 129–169.
Costa, P. T., Jr, & McCrae, R. R. (1992). Four ways five factors are basic. *Personality and Individual Differences, 13*(6), 653–665.
Cox, C. M., & Terman, L. M. (1926). *Genetic studies of genius. Vol. 2, The early mental traits of three hundred geniuses*. Stanford, CA: Stanford University Press.
Crawford, V. M., & Brophy, S. (2006). *Adaptive expertise: Methods, findings, and emerging issues*. Paper presented at the SRI International.
Cropley, A. J. (1967). *Creativity*. London: Longman.
Cropley, A. J. (1972). Creativity test scores under timed and untimed conditions. *Australian Journal of Psychology, 24*, 31–36.
Cropley, A. J. (1990). Creativity and mental health in everyday life. *Creativity Research Journal, 3*(3), 167–178.
Cropley, A. J. (1992a). Glück und Kreativität: Förderung von Aufgeschlossenheit für den zündenden Gedanken [Luck and creativity: Fostering openness for the spark of inspiration]. In K. Urban (Ed.), *Begabungen Entwickeln, Erkennen und Fördern [Developing, Recognizing and Fostering Gifts]* (pp. 216–221). Hannover, Germany: University of Hannover, Faculty of Education.
Cropley, A. J. (1992b). *More ways than one: Fostering creativity*. New York: Ablex Publishing.
Cropley, A. J. (1994). Creative intelligence: A concept of "true" giftedness. *European Journal for High Ability, 5*(1), 6–23.

Cropley, A. J. (1997a). Creativity: A bundle of paradoxes. *Gifted and Talented International, 12*(1), 8–14.

Cropley, A. J. (1997b). Fostering creativity in the classroom. In M. A. Runco (Ed.), *The creativity research handbook* (pp. 83–114). Cresskill, NJ: Hampton Press.

Cropley, A. J. (1999). Creativity and cognition: Producing effective novelty. *Roeper Review, 21*(4), 253–260.

Cropley, A. J. (2001). *Creativity in education and learning: A guide for teachers and educators*. London: Kogan Page.

Cropley, A. J. (2002). Creativity and innovation: Men's business or women's work? *Baltic Journal of Psychology, 3*, 77–88.

Cropley, A. J. (2006). In praise of convergent thinking. *Creativity Research Journal, 18*(3), 391–404.

Cropley, A. J., & Cropley, D. H. (2007). Using assessment to foster creativity. In A.-G. Tan (Ed.), *Creativity: A handbook for teachers* (pp. 209–230). Singapore: World Scientific.

Cropley, A. J., & Cropley, D. H. (2008). Resolving the paradoxes of creativity: An extended phase model. *Cambridge Journal of Education, 38*(3), 355–373.

Cropley, A. J., & Cropley, D. H. (2009). *Fostering creativity: A diagnostic approach for education and organizations*. Cresskill, NJ: Hampton Press.

Cropley, A. J., & Sikand, J. S. (1973). Creativity and schizophrenia. *Journal of Consulting and Clinical Psychology, 40*(3), 462.

Cropley, A. J., & Urban, K. K. (2000). Programs and strategies for nurturing creativity. In F. J. Monks, K. A. Heller, R. J. Sternberg, & R. F. Subotnik (Eds.), *International handbook research and development of giftedness and talent* (pp. 481–484). Oxford, UK: Pergamon.

Cropley, D. H. (2005). *Eleven principles of creativity and terrorism*. Paper presented at the Fourth Homeland Security Summit and Exposition, Canberra, Australia.

Cropley, D. H. (2006). *The role of creativity as a driver of innovation*. Paper presented at the Management of Innovation and Technology, 2006 IEEE International Conference.

Cropley, D. H. (2014a). Engineering, ethics and creativity: N'er the twain shall meet? In S. Moran, D. H. Cropley, & J. C. Kaufman (Eds.), *The ethics of creativity* (pp. 152–169). Basingstoke, UK: Palgrave MacMillan Ltd.

Cropley, D. H. (2014b). From rhetoric to reality: Designing activities to foster creativity. *Knowledge Quest, 42*(5), 24–27.

Cropley, D. H. (2014c). Teaching engineers to think creatively: Barriers and challenges in STEM disciplines. In R. Wegerif, L. Li, & J. C. Kaufman (Eds.), *Handbook of research on teaching thinking*. London: Routledge.

Cropley, D. H., & Cropley, A. J. (2000a). *Creativity and innovation in the systems engineering process*. Paper presented at the Proceedings of the Tenth Annual International Symposium on Systems Engineering.

Cropley, D. H., & Cropley, A. J. (2000b). Fostering creativity in engineering undergraduates. *High Ability Studies, 11*(2), 207–219.

Cropley, D. H., & Cropley, A. J. (2005). Engineering creativity: A systems concept of functional creativity. In J. C. Kaufman, & J. Baer (Eds.), *Faces of the muse: How people think, work and act creatively in diverse domains* (pp. 169–185). Hillsdale: NJ: Lawrence Erlbaum.

Cropley, D. H., & Cropley, A. J. (2008). Elements of a universal aesthetic of creativity. *Psychology of Aesthetics, Creativity, and the Arts, 2*(3), 155–161.

Cropley, D. H., & Cropley, A. J. (2010a). Functional creativity: Products and the generation of effective novelty. In J. C. Kaufman, & R. J. Sternberg (Eds.), *The Cambridge handbook of creativity* (pp. 301–317). New York: Cambridge University Press.

Cropley, D. H., & Cropley, A. J. (2010b). Understanding the innovation-friendly institutional environment: A psychological framework. *Baltic Journal of Psychology, 11*(1&2), 73–87.

Cropley, D. H., & Cropley, A. J. (2011). Understanding value innovation in organizations: A psychological framework. *International Journal of Creativity and Problem Solving, 21*(1), 17–36.

Cropley, D. H., & Cropley, A. J. (2012). A psychological taxonomy of organizational innovation: Resolving the paradoxes. *Creativity Research Journal, 24*(1), 29–40.

Cropley, D. H., & Cropley, A. J. (2013). *Creativity and crime: A psychological approach.* Cambridge, UK: Cambridge University Press.

Cropley, D. H., & Cropley, A. J. (2014). Managing entrepreneurship for innovation: A psychological analysis. In R. Sternberg, & G. Krauss (Eds.), *Handbook of research on entrepreneurship and creativity* (pp. 21–59). Cheltenham: Edward Elgar Publishing.

Cropley, D. H., Cropley, A. J., Chiera, B. A., & Kaufman, J. C. (2013). Diagnosing organizational innovation: Measuring the capacity for innovation. *Creativity Research Journal, 25* (4), 388–396.

Cropley, D. H., & Kaufman, J. C. (2012). Measuring functional creativity: Non-expert raters and the creative solution diagnosis scale. *The Journal of Creative Behavior, 46*(2), 119–137.

Cropley, D. H., & Kaufman, J. C. (2013). Rating the creativity of products. In K. Thomas, & J. Chan (Eds.), *Handbook of research on creativity* (pp. 196–211). Cheltenham, UK and Northampton, MA, USA: Edward Elgar Publishing.

Cropley, D. H., Kaufman, J. C., & Cropley, A. J. (2008). Malevolent creativity: A functional model of creativity in terrorism and crime. *Creativity Research Journal, 20*(2), 105–115.

Cropley, D. H., Kaufman, J. C., & Cropley, A. J. (2011). Measuring creativity for innovation management. *Journal of Technology Management, & Innovation, 6*(3), 13–30.

Crowe, A., Dirks, C., & Wenderoth, M. P. (2008). Biology in bloom: Implementing Bloom's taxonomy to enhance student learning in biology. *CBE-Life Sciences Education, 7*(4), 368–381.

Crozier, W. R. (1999). Age and individual differences in artistic productivity: Trends within a sample of British novelists. *Creativity Research Journal, 12*(3), 197–204.

Csikszentmihalyi, M. (1988). Society, culture, and person: A systems view of creativity. In R. J. Sternberg (Ed.), *The nature of creativity* (pp. 325–339). New York: Cambridge University Press.

Csikszentmihalyi, M. (1996). *Creativity: Flow and the psychology of discovery and invention.* New York: Harper Collins.

Csikszentmihalyi, M. (1999). Implications of a systems perspective for the study of creativity. In R. J. Sternberg (Ed.), *Handbook of creativity* (pp. 313–335). Cambridge, UK: Cambridge University Press.

Csikszentmihalyi, M., Rathunde, K. R., & Whalen, S. (1993). *Talented teenagers: The roots of success and failure.* New York: Cambridge University Press.

Dacey, J. S. (1989). *Fundamentals of creative thinking.* Lexington, MA: Lexington Press.

Damasio, A. R. (1994). *Descartes' error: Emotion, reason, and the human brain.* New York: Groset/Putnam.

Dasgupta, S. (2004). Is creativity a Darwinian process? *Creativity Research Journal, 16*(4), 403–413.

Davis, M. A. (2009). Understanding the relationship between mood and creativity: A meta-analysis. *Organizational Behavior and Human Decision Processes, 108*(1), 25–38.

de Bono, E. (1978). *Teaching thinking.* London: Pelican.

de Bono, E. (1993). *Water logic.* New York: Viking Penguin.

DeHaan, R. L. (2009). Teaching creativity and inventive problem solving in science. *CBE-Life Sciences Education, 8*(3), 172–181.

Dellas, M., & Gaier, E. L. (1970). Identification of creativity: The individual. *Psychological Bulletin, 73*(1), 55–73.

Dennis, W. (1973). *Children of the crèche.* New York: Appleton-Century-Crofts.

Descartes, R. (1991 [1644]). *Principles of Philosophy.* (Trans. V. R. Miller , & R. P. Miller). Boston: Kluwer.
DeVellis, R. F. (2012). *Scale development: Theory and applications.* Newbury Park, CA: Sage Publications.
Diamond, A., Barnett, W. S., Thomas, J., & Munro, S. (2007). Preschool program improves cognitive control. *Science, 318*, 1387–1388.
Diaz de Chumaceiro, C. L. (1999). Research on career paths: Serendipity and its analog. *Creativity Research Journal, 12*(3), 227–229.
Dickson, P. (2001). *Sputnik: The shock of the century.* New York: Walker Publishing Company.
Dieter, G. E., & Schmidt, L. C. (2012). *Engineering design* (5th ed.). New York: McGraw-Hill Higher Education.
Dillon, J. T. (1982). Problem finding and solving. *Journal of Creative Behavior, 16*, 97–111.
Dillon, T. A., Lee, R. K., & David, M. (2005). Value innovation: Passport to wealth creation. *Research-Technology Management, 48*(2), 22–36.
Dollinger, S. J., Urban, K. K., & James, T. A. (2004). Creativity and openness: Further validation of two creative product measures. *Creativity Research Journal, 16*(1), 35–47.
Doolittle, J. H. (1990). *Creative reasoning test.* Pacific Grove, CA: Midwest Publications/Critical Thinking Press.
Dow, G. T., & Mayer, R. E. (2004). Teaching students to solve insight problems: Evidence for domain specificity in creativity training. *Creativity Research Journal, 16*(4), 389–398.
Drevdahl, J. E., & Cattell, R. B. (1958). Personality and creativity—Artists and writers. *Journal of Clinical Psychology, 14*, 107–111.
Dudek, S. Z., & Hall, W. B. (1991). Personality consistency: Eminent architects 25 years later. *Creativity Research Journal, 4*(3), 213–231.
Dul, J., Ceylan, C., & Jaspers, F. (2011). Knowledge workers' creativity and the role of the physical work environment. *Human Resource Management, 50*(6), 715–734.
Edwards, S. M. (2001). The technology paradox: Efficiency versus creativity. *Creativity Research Journal, 13*(2), 221–228.
Eiduson, B. T. (1958). Artist and nonartist: a comparative study. *Journal of Personality, 26* (1), 13–28.
Eisenberger, R., & Armeli, S. (1997). Can salient reward increase creative performance without reducing intrinsic creative interest? *Journal of Personality and Social Psychology, 72*(3), 652–663.
Eisenberger, R., & Byron, K. (2011). Rewards and creativity. In M. A. Runco, & S. R. Pritzker (Eds.), *Encyclopedia of creativity* (pp. 313–318). New York: Elsevier.
Eisenberger, R., & Rhoades, L. (2001). Incremental effects of reward on creativity. *Journal of Personality and Social Psychology, 81*(4), 728.
Eisenman, R. (1999). Creative prisoners: Do they exist? *Creativity Research Journal, 12*(3), 205–210.
Ekvall, G. (1996). Organizational climate for creativity and innovation. *European Journal of Work and Organizational Psychology, 5*(1), 105–123.
Elliott, C., & Deasley, P. (Eds.), (2007). *Creating systems that work: Principles of engineering systems for the 21st century.* London: The Royal Academy of Engineering.
Ericsson, K. A., & Lehmann, A. C. (1999). Expertise. In M. A. Runco, & S. R. Pritzker (Eds.), *Encyclopedia of creativity* (Vol. 1, pp. 695–707). San Diego: Academic Press.
Ericsson, K. A., & Smith, J. (1991). *Toward a general theory of expertise: Prospects and limits.* Cambridge: Cambridge University Press.
Evans, F. T. (1991). The creative engineer. In R. A. Smith (Ed.), *Innovative teaching in engineering* (pp. 497–502). London: Ellis Horwood.
Eysenck, H. J. (1940). The general factor in aesthetic judgements. *British Journal of Psychology, 31*(1), 94–102.

Eysenck, H. J. (1997). Creativity and personality. In M. A. Runco (Ed.), *The creativity research handbook* (Vol. 1, pp. 41–66). Cresskill, NJ: Hampton Press.

Facaoaru, C. (1985). *Kreativität in Wissenschaft und Technik [Creativity in science and technology]*. Bern: Huber.

Fagot, B. I., & Leinbach, M. D. (1993). Gender-role development in young children: From discrimination to labeling. *Developmental Review, 13*(2), 205–224.

Farisha, B. (1978). Mental imagery and creativity. Review and speculation. *Journal of Mental Imagery, 2*, 209–238.

Fasko, D. (2001). Education and creativity. *Creativity Research Journal, 13*(3–4), 317–327.

Feist, G. J. (1998). A meta-analysis of personality in scientific and artistic creativity. *Personality and Social Psychology Review, 2*(4), 290–309.

Feist, G. J. (2010). The function of personality in creativity: The nature and nurture of the creative personality. In J. C. Kaufman, & R. J. Sternberg (Eds.), *The Cambridge handbook of creativity* (pp. 113–130). New York: Cambridge University Press.

Feldhusen, J. F. (1988). *The Purdue creative thinking program* (3rd ed.). Lafayette, IN: Gifted Education Resource Institute.

Feldhusen, J. F. (1995). Creativity: A knowledge base, metacognitive skills, and personality factors. *The Journal of Creative Behavior, 29*(4), 255–268.

Finke, R. A., Ward, T. B., & Smith, S. M. (1992). *Creative cognition*. Boston: MIT Press.

Flavell, J. H. (1976). Metacognitive aspects of problem solving. In L. B. Resnick (Ed.), *The nature of intelligence* (pp. 231–236). Hillsdale, NJ: Erlbaum.

Florida, R. (2002). *The rise of the creative class*. New York: Basic Books.

Forsberg, K., Mooz, H., & Cotterman, H. (2000). *Visualizing project management: A model for business and technical success*. Hoboken, NJ: John Wiley and Sons.

Frängsmyr, T. (1997). *Les Prix Nobel. The Nobel Prizes 1996*. Stockholm: Nobel Foundation.

Freeman, S., O'Connor, E., Parks, J. W., Cunningham, M., Hurley, D., Haak, D., et al. (2007). Prescribed active learning increases performance in introductory biology. *CBE-Life Sciences Education, 6*(2), 132–139.

Fromm, E. (1980). *Greatness and limitations of Freud's thought*. New York: New American Library.

Gardner, H. (1993). *Creating minds*. New York: Basic Books.

Genco, N., Hölttä-Otto, K., & Seepersad, C. C. (2012). An experimental investigation of the innovation capabilities of undergraduate engineering students. *Journal of Engineering Education, 101*(1), 60–81.

Gertner, J. (2012). *The idea factory: Bell Labs and the great age of American innovation*. London: The Penguin Press.

Getzels, J. A., & Jackson, P. W. (1962). *Creativity and intelligence*. New York: Wiley.

Getzels, J. W., & Csikszentmihalyi, M. (1976). *The creative vision: A longitudinal study of problem finding in art*. New York: Wiley.

Glover, J. A., Ronning, R. R., & Reynolds, C. R. (Eds.), (1989). *Handbook of creativity*. New York: Plenum Press.

Glück, J., Ernst, R., & Unger, F. (2002). How creatives define creativity: Definitions reflect different types of creativity. *Communication Research Journal, 14*(1), 55–67.

Goertzel, M. C., Goertzel, V., & Goertzel, T. C. (1978). *300 eminent personalities*. San Francisco: Jossey-Bass.

Gordon, W. J. (1961). *Synectics*. New York: Harper.

Götz, K. O. (1985). *Visual Aesthetic Sensitivity Test (VAST)* (4th ed.). Düsseldorf: Concept Verlag.

Götz, K. O., & Götz, K. (1979). Personality characteristics of professional artists. *Perceptual and Motor Skills, 49*(1), 327–334.

Gough, H. G. (1979). A creative personality scale for the Adjective Check List. *Journal of Personality and Social Psychology, 37*(8), 1398–1405.

Graham, B. C., Sawyers, J. K., & DeBord, K. B. (1989). Teachers' creativity, playfulness, and style of interaction with children. *Creativity Research Journal*, *2*(1–2), 41–50.

Gribov, I. A. (1989). Psychological and educational conditions of development of creative self-expression of students and teachers. *Voprosy–Psikhologii*, *2*, 75–82.

Gruber, H. E. (1993). Creativity in the moral domain. *Creativity Research Journal*, *6*(1–2), 3–15.

Gruber, H. E., & Barrett, P. H. (1974). *Darwin on man: A psychological study of scientific creativity*. Boston: EP Dutton.

Grudin, R. (1990). *The grace of great things: Creativity and innovation*. New York: Ticknor and Fields.

Guilford, J. P. (1950). Creativity. *American Psychologist*, *5*, 444–454.

Guilford, J. P. (1959). Traits of creativity. In H. H. Anderson (Ed.), *Creativity and its cultivation* (pp. 142–161). New York: Harper.

Guilford, J. P. (1967). *The nature of human intelligence*. New York: McGraw-Hill.

Guilford, J. P. (1976). *Creativity tests for children*. Orange, CA: Sheridan Psychological Services.

Gupta, A. K., Smith, K. G., & Shalley, C. E. (2006). The interplay between exploration and exploitation. *Academy of Management Journal*, *49*(4), 693–706.

Hadamard, J. (1945). *The psychology of invention in the mathematical field*. New York: Dover.

Han, S. H., Hwan Yun, M., Kim, K.-J., & Kwahk, J. (2000). Evaluation of product usability: Development and validation of usability dimensions and design elements based on empirical models. *International Journal of Industrial Ergonomics*, *26*(4), 477–488.

Haner, U.-E. (2005). Spaces for creativity and innovation in two established organizations. *Creativity and Innovation Management*, *14*, 288–298.

Harman, J. (2013). *The Shark's paintbrush: Biomimicry and how nature is inspiring innovation*. London: Nicholas Brealey Publishing.

Harrington, D. M. (1999). Conditions and settings/Environment. In M. A. Runco, & S. R. Pritzker (Eds.), *Encyclopedia of creativity* (Vol. 1, pp. 323–340). San Diego: Academic Press.

Hassenstein, M. (1988). *Bausteine zu einer Naturgeschichte der Intelligenz [Building blocks for a natural history of creativity]*. Stuttgart: Deutsche Verlags-Anstalt.

Hatano, G., & Oura, Y. (2003). Commentary: Reconceptualizing school learning using insight from expertise research. *Educational Researcher*, *32*(8), 26–29.

Hausman, C. R. (1984). *A discourse on novelty and creation*. Albany: State University of New York Press.

Heinelt, G. (1974). *Kreative Lehrer/kreative Schüler [Creative Teachers/Creative Students]*. Freiburg: Herder.

Heller, K. A. (2007). Scientific ability and creativity. *High ability studies*, *18*(2), 209–234.

Helson, R. (1983). Creative mathematicians. In R. S. Albert (Ed.), *Genius and eminence: The social psychology of creativity and exceptional achievement* (pp. 311–330). Elmsford, NY: Pergamon.

Helson, R. (1996). In search of the creative personality. *Creativity Research Journal*, *9*(4), 295–306.

Helson, R. (1999). A longitudinal study of creative personality in women. *Creativity Research Journal*, *12*(2), 89–101.

Henle, M. (1974). The cognitive approach: The snail beneath the shell. In S. Rosner, & L. E. Aber (Eds.), *Essays in creativity* (pp. 23–44). Croton on Hudson, NY: North River Press.

Hennessey, B. A. (1994). The consensual assessment technique: An examination of the relationship between ratings of product and process creativity. *Creativity Research Journal*, *7*(2), 193–208.

Hennessey, B. A. (2010). The creativity-motivation connection. In J. C. Kaufman, & R. J. Sternberg (Eds.), *The Cambridge handbook of creativity* (pp. 342–365). New York: Cambridge University Press.

Hennessey, B. A., & Amabile, T. (1999). Consensual assessment. In M. A. Runco, & S. R. Pritzker (Eds.), *Encyclopedia of creativity* (pp. 347–359). San Diego: Academic Press.

Heron, W. (1957). The pathology of boredom. *Scientific American*, 52–56.

Higgins, J. M. (1994). *101 Creative problem solving techniques: The handbook of new ideas for business*. Winter Park, FL: The New Management Publishing Company.

Hoffman, M. L. (1971). Identification and conscience development. *Child Development, 42*, 1071–1082.

Hoffmann, O., Cropley, D., Cropley, A., Nguyen, L., & Swatman, P. (2005). Creativity, requirements and perspectives. *Australasian Journal of Information Systems, 13*(1), 159–175.

Horenstein, M. N. (2002). *Design concepts for engineers* (2nd ed.). Upper Saddle River, NJ: Prentice-Hall, Inc.

Horn, D., & Salvendy, G. (2006). Consumer-based assessment of product creativity: A review and reappraisal. *Human Factors and Ergonomics in Manufacturing, & Service Industries, 16*, 155–175.

Hruby, F. M. (1998). *TechnoLeverage: Using the power of technology to outperform the competition*. New York: Amacom.

Hruby, G. G. (1999). Review of Jensen, E (1998). Teaching with the brain in mind. *Roeper Review, 21*, 326–327.

Huczynski, W. (1983). *Encyclopedia of management development methods*. Aldershot: Gower.

Hudson, L. (1968). *Frames of mind*. London: Methuen.

Hulsheger, U. R., Anderson, N., & Salgado, J. F. (2009). *Selecting for innovation: What is good for job performance is not necessarily good for innovative performance*. Paper presented at the EAWOP Conference, Santiago de Compostela.

Hunsaker, S. L. (2005). Outcomes of creativity training programs. *Gifted Child Quarterly, 49* (4), 292–299.

Ihsen, S., & Brandt, D. (1998). Editorial: Creativity: How to educate and train innovative engineers. *European Journal of Engineering Education, 23*(1), 3–4.

Isaksen, S. G., Lauer, K. J., Ekvall, G., & Britz, A. (2001). Perceptions of the best and worst climates for creativity: Preliminary validation evidence for the situational outlook questionnaire. *Creativity Research Journal, 13*(2), 171–184.

Isen, A. M., Daubman, K. A., & Nowicki, G. P. (1987). Positive affect facilitates creative problem solving. *Journal of personality and social psychology, 52*(6), 1122–1131.

Jackson, P. W., & Messick, S. (1965). The person, the product, and the response: Conceptual problems in the assessment of creativity. *Journal of Personality, 33*, 309–329.

Jamison, K. R. (1993). *Touched with fire: Depressive illness and the artistic temperament*. New York: Free Press.

Jay, E. S., & Perkins, D. N. (1997). Problem finding: The search for mechanisms. In M. A. Runco (Ed.), *The creativity research handbook* (Vol. 1, pp. 257–294). Cresskill, NJ: Hampton Press.

Jensen, J. N. (2006). *A user's guide to engineering*. Upper Saddle River, NJ: Pearson: Prentice Hall.

Johnson, D. L. (1979). *The creativity checklist*. Wood Dale, IL: Stoelting.

Kasof, J. (1997). Creativity and breadth of attention. *Creativity Research Journal, 10*(4), 303–315.

Katz, D., & Kahn, R. L. (1978). *The social psychology of organizations* (2nd ed.). New York: Wiley.

Kaufman, J. C. (2009). *Creativity 101*. New York: Springer Publishing Company.

Kaufman, J. C., & Baer, J. (2002). I bask in dreams of suicide: Mental illness, poetry, and women. *Review of General Psychology, 6*(3), 271.

Kaufman, J. C., Baer, J., Cropley, D. H., Reiter-Palmon, R., & Sinnett, S. (2013). Furious activity vs. understanding: How much expertise is needed to evaluate creative work? *Psychology of Aesthetics, Creativity, and the Arts, 7*(4), 332–340.

Kaufman, J. C., & Beghetto, R. A. (2009). Beyond big and little: The four c model of creativity. *Review of General Psychology, 13*, 1–12.

Kawenski, M. (1991). Encouraging creativity in design. *The Journal of Creative Behavior, 25* (3), 263–266.

Kazerounian, K., & Foley, S. (2007). Barriers to creativity in engineering education: A study of instructors' and students' perceptions. *Journal of Mechanical Design, 129*, 761.

Kim, K. H. (2006). Can we trust creativity tests? A review of the Torrance Tests of Creative Thinking (TTCT). *Creativity Research Journal, 18*(1), 3–14.

Kim, K. H. (2011). The creativity crisis: The decrease in creative thinking scores on the Torrance Tests of Creative Thinking. *Creativity Research Journal, 23*(4), 285–295.

Kim, K. H., & Coxon, S. V. (2013). The creativity crisis, possible causes, and what schools can do. In J. B. Jones, & L. J. Flint (Eds.), *The creative imperative* (pp. 53–68). Santa Barbara, CA: ABC-CLIO.

Kim, W. C., & Mauborgne, R. (2004). Value innovation: The strategic logic of high growth. *Harvard Business Review, 82*(7/8), 172–180.

King, L. A., Walker, L. M., & Broyles, S. J. (1996). Creativity and the five-factor model. *Journal of Research in Personality, 30*(2), 189–203.

Kinney, D. K., Richards, R., Lowing, P. A., LeBlanc, D., Zimbalist, M. E., & Harlan, P. (2001). Creativity in offspring of schizophrenic and control parents: An adoption study. *Creativity Research Journal, 13*, 17–26.

Kirton, M. (1989). *Adaptors and innovators: Styles of creativity and problem solving*. London: Routledge.

Kitto, J., Lok, D., & Rudowicz, E. (1994). Measuring creative thinking: An activity-based approach. *Creativity Research Journal, 7*(1), 59–69.

Kogan, N. (1983). Stylistic variation in childhood and adolescence: Creativity, metaphor, and cognitive styles. In P. Mussen (Ed.), *Handbook of child psychology* (Vol. 3, pp. 631–706). New York: Wiley.

Kohlberg, L. A. (1966). A cognitive-developmental analysis of childhood sex role concepts and attitudes. In E. E. Maccoby (Ed.), *The development of sex differences* (pp. 179–204). Palo Alto, CA: Stanford University Press.

Köstler, A. (1964). *The act of creation*. London: Hutchinson.

Kottler, J. A. (2005). *Divine madness*. San Francisco: Jossey-Bass.

Kozbelt, A., Beghetto, R. A., & Runco, M. A. (2010). Theories of creativity. In J. C. Kaufman, & R. J. Sternberg (Eds.), *The Cambridge handbook of creativity* (pp. 20–47). New York: Cambridge University Press.

Krystal, H. (1988). On some roots of creativity. *Psychiatric Clinics of North America, 11*, 475–491.

Larey, T. S., & Paulus, P. B. (1999). Group preference and convergent tendencies in small groups: A content analysis of group brainstorming performance. *Creativity Research Journal, 12*(3), 175–184.

Lehman, H. C. (1953). *Age and achievement*. Princeton, NJ: Princeton University Press.

Lewis, M. W., Welsh, M. A., Dehler, G. E., & Green, S. G. (2002). Product development tensions: Exploring contrasting styles of project management. *Academy of Management Journal, 45*(3), 546–564.

Lindauer, M. S. (1993). The span of creativity among long-lived historical artists. *Creativity Research Journal, 6*, 221–240.

Lipman-Blumen, J. (1991). *Individual and organizational achieving styles: A handbook for researchers and human resource professionals* (4th ed.). Claremont, CA: Achieving Styles Institute.

Lipman-Blumen, J. (1996). *Women in corporate leadership: Reviewing a decade's research*. Wellesley, MA: Wellesley College Center for Research on Women.

Litwin, G., & Stringer, R. (1968). *Motivation and organizational climate*. Boston: Harvard University Press.
Liu, Z., & Schonwetter, D. J. (2004). Teaching creativity in engineering. *International Journal of Engineering Education, 20*(5), 801–808.
Lonergan, D. C., Scott, G. M., & Mumford, M. D. (2004). Evaluative aspects of creative thought: Effects of appraisal and revision standards. *Creativity Research Journal, 16* (2–3), 231–246.
Lubart, T. I. (2001). Models of the creative process: Past, present and future. *Creativity Research Journal, 13*, 295–308.
Ludwig, A. M. (1998). Method and madness in the arts and sciences. *Creativity Research Journal, 11*(2), 93–101.
Luecke, R., & Katz, R. (2003). *Managing creativity and innovation*. Boston: Harvard Business School Press.
Ma, H. H. (2006). A synthetic analysis of the effectiveness of single components and packages in creativity training programs. *Creativity Research Journal, 18*(4), 435–446.
Maccoby, E. E., & Jacklin, C. N. (1974). *The psychology of sex differences*. Stanford, CA: Stanford University Press.
Mach, E. (1896). On the part played by accident in invention and discovery. *The Monist, VI* (2), 161–175.
Mack, R. W. (1987). Are methods of enhancing creativity being taught in teacher education programs as perceived by teacher educators and student teachers? *The Journal of Creative Behavior, 21*(1), 22–33.
Mackay, A. L. (1991). *A dictionary of scientific quotations*. Bristol: Adam Hilger.
MacKinnon, D. W. (1978). *In search of human effectiveness: Identifying and developing creativity*. Buffalo, NY: Creative Education Foundation.
MacKinnon, D. W. (1983). Creative architects. In R. S. Albert (Ed.), *Genius and eminence: The social psychology of creativity and exceptional achievement* (pp. 291–301). Elmsford, NY: Pergamon.
Mann, E. L. (2009). The search for mathematical creativity: Identifying creative potential in middle school students. *Creativity Research Journal, 21*(4), 338–348.
Mansfield, R. S., Busse, T. V., & Krepelka, E. J. (1978). The effectiveness of creativity training. *Review of Educational Research, 48*, 517–536.
Martindale, C. (1990). *The clockwork muse*. New York: Basic Books.
Martinsen, Ø. (1995). Cognitive styles and experience in solving insight problems: Replication and extension. *Creativity Research Journal, 8*(3), 291–298.
Maslow, A. H. (1943). A theory of human motivation. *Psychological Review, 50*(4), 370–396.
Maslow, A. H. (1973). Creativity in self-actualizing people. In A. Rothenberg, & C. R. Hausman (Eds.), *The creative question* (pp. 86–92). Durham, NC: Duke University Press.
Mathisen, G. E., & Einarsen, S. (2004). A review of instruments assessing creative and innovative environments within organizations. *Creativity Research Journal, 16*(1), 119–140.
May, R. (1976). *The courage to create*. New York: Bantam.
McCrae, R. R. (1987). Creativity, divergent thinking and openness to experience. *Journal of Personality and Social Psychology, 52*, 1258–1265.
McFadzean, E. (2002). Developing and supporting creative problem-solving teams: Part 1—A conceptual model. *Management Decision, 40*(5), 463–475.
McGraw-Hill (2003). *McGraw-Hill dictionary of engineering* (2nd ed.). New York: McGraw-Hill.
McGregor, G. D. (2001). *Creative thinking instruction for a college study skills program: A case study*. Ph.D. dissertation, Baylor University.
McLaren, R. B. (1993). The dark side of creativity. *Creativity Research Journal, 6*, 137–144.

McMullan, W. E. (1978). Creative individuals: Paradoxical personages. *Journal of Creative Behavior, 10*, 265–275.

McWilliam, E., Dawson, S., & Tan, J. P.-L. (2011). Less elusive, more explicit. The challenge of 'seeing' creativity in action. In P. Thomson, & J. Sefton-Green (Eds.), *Researching creative learning: Methods and issues* (pp. 113–125). London: Routledge.

Mednick, S. A. (1962). The associative basis of creativity. *Psychological Review, 69*, 220–232.

Meeker, M. (1985). *Structure of Intellect Learning Abilities Test*. Los Angeles, CA: Western Psychological Services.

Michael, W. B., & Colson, K. R. (1979). The development and validation of a life experience inventory for the identification of creative electrical engineers. *Educational and Psychological Measurement, 39*(2), 463–470.

Michalewicz, Z., & Michalewicz, M. (2008). *Puzzle-based learning*. Melbourne, Australia: Hybrid Publishers.

Michalko, M. (1996). *Thinkertoys*. Berkeley, CA: Ten Speed Press.

Michalko, M. (2001). *Cracking creativity*. Berkeley, CA: Ten Speed Press.

Milgram, R. M., & Hong, E. (1999). Creative out-of-school activities in intellectually gifted adolescents as predictors of their life accomplishments in young adults: A longitudinal study. *Creativity Research Journal, 12*, 77–88.

Miller, A. I. (1992). Scientific creativity: A comparative study of Henri Poincare and Albert Einstein. *Creativity Research Journal, 5*(4), 385–414.

Millward, L. J., & Freeman, H. (2002). Role expectations as constraints to innovation: The case of female managers. *Creativity Research Journal, 14*, 93–110.

Miron, E., Erez, M., & Naveh, E. (2004). Do personal characteristics and cultural values that promote innovation, quality, and efficiency compete or complement each other? *Journal of Organizational Behavior, 25*(2), 175–199.

Mishra, P., & Henriksen, D. (2013). A NEW approach to defining and measuring creativity: Rethinking technology, & creativity in the 21st century. *TechTrends, 57*(5), 10–13.

Mohan, M. (1973). Is there a need for a course in creativity in teacher education? *The Journal of Creative Behavior, 7*(3), 175–186.

Mokyr, J. (1990). *The lever of riches: Technological creativity and economic progress*. New York: Oxford University Press.

Moran, S., Cropley, D. H., & Kaufman, J. C. (Eds.), (2014). *The ethics of creativity* Basingstoke: Palgrave MacMillan.

Morgan, D. N. (1953). Creativity today: A constructive analytic review of certain philosophical and psychological work. *The Journal of Aesthetics and Art Criticism, 12*(1), 1–24.

Motamedi, K. (1982). Extending the concept of creativity. *Journal of Creative Behavior, 16*, 75–88.

Moustakis, C. E. (1977). *Creative life*. New York: Van Nostrand.

Mumford, M. D. (2011). *Handbook of organizational creativity*. Amsterdam: Academic Press.

Mumford, M. D., Baughman, W. A., Maher, M. A., Costanza, D. P., & Supinski, E. P. (1997). Process-based measures of creative problem-solving skills: IV. Category combination. *Creativity Research Journal, 10*(1), 59–71.

Mumford, M. D., Baughman, W. A., Threlfall, K. V., Supinski, E. P., & Costanza, D. P. (1996). Process-based measures of creative problem-solving skills: I. Problem construction. *Creativity Research Journal, 9*(1), 63–76.

Mumford, M. D., Marks, M. A., Connelly, M. S., Zaccaro, S. J., & Johnson, J. F. (1998). Domain-based scoring of divergent-thinking tests: Validation evidence in an occupational sample. *Creativity Research Journal, 11*, 151–163.

Mumford, M. D., & Moertl, P. (2003). Cases of social innovation: Lessons from two innovations in the 20th century. *Creativity Research Journal, 15*(2–3), 261–266.

Myers-Briggs, I., & McCaulley, M. H. (1992). *Manual: A guide to the development and use of the Myers-Briggs Type Indicator*. Palo Alto, CA: Consulting Psychologists Press.

Nardi, K., & Martindale, C. (1981). *Creativity and preference for tones varying in dissonance and intensity*. Paper presented at the Eastern Psychological Association Convention, New York.

Neff, G. (1975). Kreativität und Gruppe [Creativity and the group]. In G. Neff (Ed.), *Kreativität in Schule und Gesellschaft [Creativity in school and society]* (pp. 17–29). Ravensburg: Otto Maier.

Nettle, D. (2002). *Strong imagination*. Oxford: Oxford University Press.

Newell, A., Shaw, J. C., & Simon, H. A. (1962). The processes of creative thinking. In H. E. Gruber, G. Terrell, & M. Wertheimer (Eds.), *Contemporary approaches to creative thinking* (pp. 63–119). New York: Atherton.

Newell, A., & Simon, H. A. (1972). *Human problem solving*. Englewood Cliffs, NJ: Prentice-Hall.

Nicholls, J. (1972). Creativity in the person who will never produce anything original or useful. The concept of creativity as a normally distributed trait. *American Psychologist, 27*, 717–727.

Nickerson, R. S. (1999). Enhancing creativity. In R. J. Sternberg (Ed.), *Handbook of creativity* (pp. 392–430). Cambridge: Cambridge University Press.

Oldham, G. R., & Cummings, A. (1996). Employee creativity: Personal and contextual factors at work. *Academy of Management Journal, 39*, 607–634.

Olken, H. (1964). Creativity training for engineers—Its past, present and future. *International Association for Engineering Education Transactions in Education*, 149–161.

Olson, M. (1982). *The rise and decline of nations*. New Haven, CT: Yale University Press.

Osborn, A. F. (1953). *Applied imagination*. New York: Scribner's.

Park, J., & Jang, K. (2005). Analysis of the actual scientific inquiries of physicists. Accessed September 17, 2006, from <www.arxiv.org/abs/physics/0506191>.

Parker, W. N. (1984). *Europe, America and the wider world: Essays on the economic history of western capitalism. Vol. 1, Europe and the world economy*. Cambridge University Press.

Parloff, M. B., Datta, L., Kleman, M., & Handlon, J. H. (1968). Personality characteristics which differentiate creative male adolescents and adults. *Journal of Personality, 36*, 530–552.

Parnes, S. J. (1981). *Magic of your mind*. Buffalo, NY: Creative Education Foundation.

Parnes, S. J., & Noller, R. B. (1972). Applied creativity: The creative studies project. *The Journal of Creative Behavior, 6*(3), 164–186.

Paulus, P. B. (1999). Group creativity. In M. A. Runco, & S. R. Pritzker (Eds.), *Encyclopedia of creativity* (pp. 779–784). San Diego: Academic Press.

Paulus, P. B., & Nijstad, B. A. (2003). *Group creativity: Innovation through collaboration*. Oxford: Oxford University Press.

Perkins, D. N. (1981). *The mind's best work*. Cambridge, MA: Harvard University Press.

Petersen, S. (1989). *Motivation von Laienautoren [Motivation of hobby authors]*. Unpublished Master's thesis, University of Hamburg.

Peterson, H. (Ed.), (1954). *A treasury of the world's great speeches* Danbury, CT: Grolier.

Pfeiffer, S. I., & Thompson, T. L. (2013). Creativity from a talent development perspective. In K. H. Kim, J. C. Kaufman, & J. Baer (Eds.), *Creatively gifted students are not like other gifted students* (pp. 231–255). New York: Springer.

Pilzer, P. Z. (1990). *Unlimited wealth: The theory and practice of economic alchemy*. New York: Crown Publishers.

Plucker, J. A. (1998). Beware of simple conclusions: The case for content generality of creativity. *Creativity Research Journal, 11*, 179–182.

Plucker, J. A. (1999). Is the proof in the pudding? Reanalyses of Torrance's (1958 to present) longitudinal data. *Creativity Research Journal, 12*(2), 103–114.

Plucker, J. A., Beghetto, R. A., & Dow, G. T. (2004). Why isn't creativity more important to educational psychologists? Potentials, pitfalls, and future directions in creativity research. *Educational Psychologist, 39*(2), 83–96.

Plucker, J. A., & Makel, M. C. (2010). Assessment of creativity. In J. C. Kaufman, & R. J. Sternberg (Eds.), *The Cambridge handbook of creativity* (pp. 48–73). New York: Cambridge University Press.

Plucker, J. A., & Renzulli, J. S. (1999). Psychometric approaches to the study of human creativity. In R. J. Sternberg (Ed.), *Handbook of creativity* (pp. 35–61). New York: Cambridge University Press.

Poincaré, J. H. (2003). *Science and method*. Mineola, NY: Courier Dover.

Porter, M. E. (1996). What is strategy? *Harvard Business Review, November–December*, 61–78.

Powell, G. N. (1993). *Women and men in management* (2nd ed.). Newbury Park, CA: Sage Publications.

Prindle, E. J. (1906). The art of inventing. *Transactions of the American Institute for Engineering Education, 25*, 519–547.

Proctor, R. M. J., & Burnett, P. C. (2004). Measuring cognitive and dispositional characteristics of creativity in elementary students. *Creativity Research Journal, 16*(4), 421–429.

Puccio, G. J. (1999). Teams. In M. A. Runco, & S. R. Pritzker (Eds.), *Encyclopedia of creativity* (pp. 640–649). San Diego: Academic Press.

Puccio, G. J., & Cabra, J. F. (2010). Organizational creativity: A systems approach. In J. C. Kaufman, & R. J. Sternberg (Eds.), *The Cambridge handbook of creativity* (pp. 145–173). New York: Cambridge University Press.

Puccio, G. J., Treffinger, D. J., & Talbot, R. J. (1995). Exploratory examination of the relationship between creativity styles and creative products. *Creativity Research Journal, 8*, 152–157.

Pugh, S. (1981). *Concept selection: A method that works*. Paper presented at the Proceedings from the International Conference on Engineering Design, Rome.

Pugh, S. (1991). *Total design: Integrated methods for successful product engineering*. Wokingham, UK: Addison-Wesley.

Rechtin, E., & Maier, M. W. (2000). *The art of systems architecting*. Boca Raton, FL: CRC Press.

Reis, S. M., & Renzulli, J. S. (1991). The assessment of creative products in programs for gifted and talented students. *Gifted Child Quarterly, 35*(3), 128–134.

Reiter-Palmon, R., Illies, M. Y., Cross, L. K., Buboltz, C., & Nimps, T. (2009). Creativity and domain specificity: The effect of task type on multiple indexes of creative problem-solving. *Psychology of Aesthetics, Creativity, and the Arts, 3*(2), 73–80.

Renzulli, J. S. (1986). The three-ring conception of giftedness: A developmental model for creative productivity. In R. J. Sternberg, & J. E. Davidson (Eds.), *Conceptions of giftedness* (pp. 53–92). Cambridge: Cambridge University Press.

Rhodes, M. (1961). An analysis of creativity. *The Phi Delta Kappan, 42*(7), 305–310.

Richards, R. (2010). Everyday creativity: Process and way of life—Four key issues. In J. Kaufman, & R. Sternberg (Eds.), *The Cambridge handbook of creativity* (pp. 189–215). New York: Cambridge University Press.

Richards, R., Kinney, D. K., Bennet, M., & Merzel, A. P. C. (1988). Assessing everyday creativity: Characteristics of the Lifetime Creativity Scales and validation with three large samples. *Journal of Personality and Social Psychology, 54*, 476–485.

Rickards, T. J. (1993). Creativity from a business school perspective: Past, present and future. In S. G. Isaksen, M. C. Murdock, R. L. Firestien, & D. J. Treffinger (Eds.), *Nurturing and developing creativity: The emergence of a discipline* (pp. 155–176). Norwood, NJ: Ablex.

Rickards, T. J. (1999). Brainstorming. In M. A. Runco, & S. R. Pritzker (Eds.), *Encyclopedia of creativity* (Vol. 1, pp. 219–227). San Diego: Academic Press.

Ritchey, T. (2006). Problem structuring using computer-aided morphological analysis. *Journal of the Operational Research Society, 57*(7), 792–801.

Roberson, W., & Johns, J. A. (2007). Fuel conservation strategies: Takeoff and climb. *Boeing Aero Quarterly, 2*(2007), 25–29.

Roe, A. (1953). A psychological study of eminent psychologists and anthropologists, and a comparison with biological and physical scientists. *Psychological Monographs: General and Applied, 67/352.*

Rogers, C. R. (1961). *On becoming a person.* Boston: Houghton.

Root-Bernstein, R. S. (1989). *Discovery.* New York: Cambridge University Press.

Root-Bernstein, R. S., Bernstein, M., & Garnier, H. (1993). Identification of scientists making long-term high-impact contributions, with notes on their methods of working. *Creativity Research Journal, 6*, 329–343.

Rose, L. H., & Lin, H.-T. (1984). A meta-analysis of long-term creativity training programs. *The Journal of Creative Behavior, 18*(1), 11–22.

Ross, L., & Nisbett, R. (1991). *The person and the situation: Perspectives of social psychology.* New York: McGraw Hill.

Rossman, J. (1931). *The psychology of the inventor: A study of the patentee.* Washington: Inventors' Publishing Company.

Rothenberg, A. (1983). Psychopathology and creative cognition: A comparison of hospitalized patients, Nobel laureates and controls. *Archives of General Psychiatry, 40*, 937–942.

Runco, M. A. (1993). Creative morality: Intentional and unconventional. *Creativity Research Journal, 6*, 17–28.

Runco, M. A. (Ed.), (2003). *Critical creative processes* Cresskill, NJ: Hampton Press.

Runco, M. A. (2010). Divergent thinking, creativity, and ideation. In J. C. Kaufman, & R. J. Sternberg (Eds.), *The Cambridge handbook of creativity* (pp. 413–446). New York: Cambridge University Press.

Runco, M. A., & Charles, R. E. (1997). Developmental trends in creative potential and creative performance. In M. A. Runco (Ed.), *The creativity research handbook* (Vol. 1, pp. 115–152). Creskill, NJ: Hampton Press.

Runco, M. A., & Nemiro, J. (2003). Creativity in the moral domain: Integration and implications. *Creativity Research Journal, 15*, 91–105.

Runco, M. A., Plucker, J. A., & Lim, W. (2001). Development and psychometric integrity of a measure of ideational behavior. *Creativity Research Journal, 13*(3–4), 393–400.

Runco, M. A., & Richards, R. (Eds.), (1997). *Eminent creativity, everyday creativity, and health* Greenwich, CT: Ablex.

Russ, S. W., & Fiorelli, J. A. (2010). Developmental approaches to creativity. In J. C. Kaufman, & R. J. Sternberg (Eds.), *The Cambridge handbook of creativity* (pp. 233–249). New York: Cambridge University Press.

Savransky, S. D. (2000). *Engineering of creativity: Introduction to TRIZ methodology of inventive problem solving.* Boca Raton, FL: CRC Press.

Sawyer, R. K. (2006). Educating for innovation. *Thinking Skills and Creativity, 1*(1), 41–48.

Sawyer, R. K. (2010). Individual and group creativity. In J. C. Kaufman, & R. J. Sternberg (Eds.), *The Cambridge handbook of creativity* (pp. 366–380). New York: Cambridge University Press.

Schein, V. E. (1994). Managerial sex typing: A persistent and pervasive barrier to women's opportunities. In M. J. Davidson, & R. J. Burke (Eds.), *Women in management: Current research issues* (pp. 65–84). London: Chapman.

Schrage, M. (2001). Playing around with brainstorming. *Harvard Business Review, 79*(3), 149–154.

Schuldberg, D. (2001). Six subclinical spectrum traits in normal creativity. *Creativity Research Journal, 13*(1), 5–16.

Schumpeter, J. A. (1942). *The theory of economic development.* Cambridge, MA: Harvard University Press.

Schwartz, D. L., Bransford, J. D., & Sears, D. (2005). Efficiency and innovation in transfer. In J. P. Mestre (Ed.), *Transfer of learning from a modern multidisciplinary perspective* (pp. 1–51). Greenwich, CT: Infrmation Age Publishing.

Schwebel, M. (1993). Moral creativity as artistic transformation. *Creativity Research Journal, 6*(1–2), 65–81.

Scott, G., Leritz, L. E., & Mumford, M. D. (2004a). The effectiveness of creativity training: A quantitative review. *Creativity Research Journal, 16*(4), 361–388.

Scott, G., Leritz, L. E., & Mumford, M. D. (2004b). Types of creativity training: Approaches and their effectiveness. *The Journal of Creative Behavior, 38*(3), 149–179.

Scott, S. G., & Bruce, R. A. (1994). Determinants of innovative behaviour: A path model of individual innovation in the workplace. *Academy of Management Journal, 37*, 580–607.

Scott, T. E. (1999). Knowledge. In M. A. Runco, & S. R. Pritzker (Eds.), *Encyclopedia of creativity* (Vol. 2, pp. 119–129). San Diego: Academic Press.

Seger, C. A. (1994). Implicit learning. *Psychological Bulletin, 115*(2), 163–196.

Semmer, E. (1870). Resultate der Injektion von Pilzsporen und Pilzhefen in's Bluth der Thiere [Effects of injecting fungus spores into the blood of animals]. *Virchows Archiv, 50*, 158–160.

Shaughnessy, M. F., & Manz, A. F. (1991). Personological research on creativity in the performing and fine arts. *European Journal for High Ability, 2*(1), 91–101.

Shaw, M. P. (1989). The Eureka process: A structure for the creative experience in science and engineering. *Creativity Research Journal, 2*, 286–298.

Siegel, S. M., & Kaemmerer, W. F. (1978). Measuring the perceived support for innovation in organizations. *Journal of Applied Psychology, 63*(5), 553.

Silvia, P. J., & Kaufman, J. C. (2010). Creativity and mental illness. In J. C. Kaufman, & R. J. Sternberg (Eds.), *The Cambridge handbook of creativity* (pp. 381–394). New York: Cambridge University Press.

Simon, H. A. (1989). The scientist as problem solver. In D. Klahr, & K. Kotovsky (Eds.), *Complex information processing: The impact of Herbert A. Simon* (pp. 375–398). Hillsdale, NJ: Erlbaum.

Simon, H. A. (1996). *The sciences of the artificial*. Cambridge, MA: MIT Press.

Simon, H. A., & Chase, W. (1973). Skill in chess. *American Scientist, 61*, 394–403.

Simonton, D. K. (1988a). Age and outstanding achievement: What do we know after a century of research? *Psychological Bulletin, 104*(2), 251–267.

Simonton, D. K. (1988b). *Scientific genius: A psychology of science*. New York: Cambridge University Press.

Simonton, D. K. (1994). *Greatness: Who makes history and why?* New York: Guilford.

Simonton, D. K. (1997). Historiometric studies of creative genius. In M. A. Runco (Ed.), *The creativity research handbook* (Vol. 1, pp. 3–28). Creskill, NJ: Hampton Press.

Simonton, D. K. (1998). Masterpieces in music and literature: Historiometric inquires. *Creativity Research Journal, 11*(2), 103–110.

Simonton, D. K. (1999). *Origins of genius: Darwinian perspectives of creativity*. New York: Oxford University Press.

Simonton, D. K. (2004). *Creativity in science: Chance, logic, genius, and zeitgeist*. Cambridge: Cambridge University Press.

Simonton, D. K. (2009). Varieties of (scientific) creativity: A hierarchical model of domain-specific disposition, development, and achievement. *Perspectives on Psychological Science, 4*(5), 441–452.

Snyder, A., Mitchell, J., Bossomaier, T., & Pallier, G. (2004). The creativity quotient: An objective scoring of ideational fluency. *Creativity Research Journal, 16*(4), 415–419.

Sosa, R., & Gero, J. S. (2003). Design and change: A model of situated creativity. In C. Bento, A. Cardosa, & J. S. Gero (Eds.), *Approaches to creativity in artificial intelligence and cognitive science* (pp. 25–34). Acapulco: IJCAI03.

Souriau, P. (1881). *Theorie de l'invention [A theory of invention]*. Paris: Hachette.
Spencer, S. J., Steele, C. M., & Quinn, D. M. (1999). Stereotype threat and women's math performance. *Journal of Experimental Social Psychology, 35*(1), 4–28.
Sprecher, T. B. (1959). A study of engineers' criteria for creativity. *Journal of Applied Psychology, 43*(2), 141–148.
Stein, M. I. (1953). Creativity and culture. *Journal of Psychology: Interdisciplinary and Applied, 36*, 311–322.
Sternberg, R. J. (1985). *Beyond IQ: A triarchic theory of human intelligence*. New York: Cambridge University Press.
Sternberg, R. J. (1999). A propulsion model of types of creative contributions. *Review of General Psychology, 3*(2), 83–100.
Sternberg, R. J. (2003a). What is an "expert student?" *Educational Researcher, 32*(8), 5–9.
Sternberg, R. J. (2003b). *Wisdom, intelligence, and creativity synthesized*. New York: Cambridge University Press.
Sternberg, R. J. (2006). The nature of creativity. *Creativity Research Journal, 18*(1), 87–98.
Sternberg, R. J. (2007). Creativity as a habit. In A.-G. Tan (Ed.), *Creativity: A handbook for teachers* (pp. 3–25). Singapore: World Scientific.
Sternberg, R. J., & Davidson, J. E. (1999). Intuition. In M. A. Runco, & S. R. Pritzker (Eds.), *Encyclopedia of creativity* (Vol. 2, pp. 57–69). San Diego: Academic Press.
Sternberg, R. J., & Kaufman, J. C. (2010). *The Cambridge handbook of creativity*. New York: Cambridge University Press.
Sternberg, R. J., & Kaufman, J. C. (2012). When your race is almost run, but you feel you're not yet done: Application of the propulsion theory of creative contributions to late-career challenges. *The Journal of Creative Behavior, 46*(1), 66–76.
Sternberg, R. J., Kaufman, J. C., & Pretz, J. E. (2002). *The creativity conundrum: A propulsion model of kinds of creative contributions*. New York: Psychology Press.
Sternberg, R. J., Kaufman, J. C., & Pretz, J. E. (2003). A propulsion model of creative leadership. *The Leadership Quarterly, 14*(4), 455–473.
Sternberg, R. J., & Lubart, T. I. (1995). *Defying the crowd: Cultivating creativity in a culture of conformity*. New York: Free Press.
Sternberg, R. J., & Lubart, T. I. (1999). The concept of creativity: Prospects and paradigms. In R. J. Sternberg (Ed.), *Handbook of creativity* (pp. 3–15). Cambridge: Cambridge University Press.
Sternberg, R. J., & Williams, W. M. (1996). *How to develop student creativity*. Alexandria, VA: Association for Supervision and Curriculum Development.
Stevens, R., Brook, P., Jackson, K., & Arnold, S. (1998). *Systems engineering: Coping with complexity*. London: Pearson Prentice Hall.
Stokes, D. E. (1997). *Pasteur's quadrant: Basic science and technological innovation*. Washington, DC: Brookings Institution Press.
Sullivan, L. H. (1896). The tall office building artistically considered. *Lippincott's Magazine, 57*(3), 403–409.
Sweetland, R. C., & Keyser, D. J. (1991). *A comprehensive reference for assessment in psychology, education and business*. Austin, TX: Pro-Ed.
Tardif, T. Z., & Sternberg, R. J. (1988). What do we know about creativity? In R. J. Sternberg (Ed.), *The nature of creativity* (pp. 429–440). New York: Cambridge University Press.
Taylor, I. A. (1975). An emerging view of creative actions. In I. A. Taylor, & J. W. Getzels (Eds.), *Perspectives in creativity* (pp. 297–325). Chicago: Aldine.
Thanksgiving for innovation (2002). *Economist Technology Quarterly*, 13–14.
Thomson, W. (1889). Electrical units of measurement. In W. Thomson (Ed.), *Nature series: Popular lectures and addresses* (Vol. 1, pp. 73–136). London: MacMillan and Co. Constitution of Matter.

Tierney, P., & Farmer, S. M. (2011). Creative self-efficacy development and creative performance over time. *Journal of Applied Psychology, 96*(2), 277–293.

Tilbury, D., Reid, A., & Podger, D. (2003). *Action research for university staff: Changing curricula and graduate skills towards sustainability, Stage 1 Report*. Canberra: Environment Australia.

Törnkvist, S. (1998). Creativity: Can it be taught? The case of engineering education. *European Journal of Engineering Education, 23*(1), 5–12.

Torrance, E. P. (1963). *Education and the Creative Potential*. Minneapolis, MN: University of Minnesota Press.

Torrance, E. P. (1965). *The Minnesota studies of creative thinking: Widening horizons in creativity*. New York: Wiley.

Torrance, E. P. (1966). *Torrance tests of creative thinking: Technical norms manual*. Lexington, MA: Personnel Press.

Torrance, E. P. (1972). Can we teach children to think creatively? *The Journal of Creative Behavior, 6*(2), 114–143.

Torrance, E. P. (1992). A national climate for creativity and invention. *Gifted Child Today, 15*(1), 10–14.

Torrance, E. P. (1998). *Torrance Test of Creative Thinking: Norms and technical manual*. Bensenville, IL: Scholastic Testing Services.

Torrance, E. P., & Safter, H. T. (1999). *Making the creative leap beyond*. Buffalo, NY: Creative Education Foundation Press.

Treffinger, D. J. (1985). Review of Torrance Tests of Creative Thinking. In J. V. Mitchell (Ed.), *Ninth mental measurements yearbook* (pp. 1632–1634). Lincoln, NB: University of Nebraska Press.

Treffinger, D. J. (1995). Creative problem solving: Overview and educational implications. *Educational Psychology Review, 7*(3), 301–312.

Treffinger, D. J., Isaksen, S. G., & Dorval, K. B. (1995). *Creative problem solving: An introduction*. Sarasota, FL: Center for Creative Learning.

Treffinger, D. J., Isaksen, S. G., & Firestien, R. L. (1983). Theoretical perspectives on creative learning and its facilitation: An overview. *The Journal of Creative Behavior, 17*(1), 9–17.

Treffinger, D. J., Sortore, M. R., & Cross, J. A. (1993). Programs and strategies for nurturing creativity. In K. Heller, F. J. Monks, & A. H. Passow (Eds.), *International handbook for research on giftedness and talent* (pp. 555–567). Oxford, UK: Pergamon.

Urban, K. K. (1997). Modeling creativity: The convergence of divergence or the art of balancing. In J. Chan, R. Li, & J. Spinks (Eds.), *Maximizing potential: Lengthening and strengthening our stride* (pp. 39–50). Hong Kong: University of Hong Kong Social Sciences Research Centre.

Urban, K. K., & Jellen, H. G. (1996). *Test for Creative Thinking–Drawing Production (TCT-DP)*. Lisse, Netherlands: Swets and Zeitlinger.

van Dam, C. P., Holmes, B. J., & Pitts, C. (1981). Effect of winglets on performance and handling qualities of general aviation aircraft. *Journal of Aircraft, 18*(7), 587–591.

van der Heijden, B. I. J. M. (2000). The development and psychometric evaluation of a multidimensional measurement instrument of professional expertise. *High Ability Studies, 11*(1), 9–39.

VanGundy, A. B. (1984). *Managing group creativity: A modular approach to problem solving*. New York: American Management Association.

Vosburg, S. K. (1998). Mood and the quantity and quality of ideas. *Creativity Research Journal, 11*(4), 315–324.

Walberg, H. J., & Stariha, W. E. (1992). Productive human capital: Learning, creativity and eminence. *Creativity Research Journal, 5*, 323–340.

Walk, C. L. (1996). *Management and leadership. MBTI applications: A decade of research on the Myers-Briggs Type Indicator*. Palo Alto, CA: Consulting Psychology Press.

Wallach, M. A. (1985). Creativity testing and giftedness. In F. D. Horowitz, & M. O'Brien (Eds.), *The gifted and talented: Developmental perspectives* (pp. 99–123). Washington, DC: American Psychological Association.

Wallach, M. A., & Kogan, N. (1965). *Modes of thinking in young children*. New York: Holt: Rinehart and Winston.

Wallas, G. (1926). *The Art of Thought*. New York: Harcourt Brace.

Walther, J., Kellam, N., Sochacka, N., & Radcliffe, D. F. (2011). Engineering competence? An interpretive investigation of engineering students' professional formation. *Journal of Engineering Education, 100*(4), 703–740.

Walther, J., & Radcliffe, D. F. (2007). The competence dilemma in engineering education: Moving beyond simple graduate attribute mapping. *Australasian Journal of Engineering Education, 13*(1), 41–51.

Ward, T. B., Saunders, K. N., & Dodds, R. A. (1999). Creative cognition in gifted adolescents. *Roeper Review, 21*, 260–266.

Weeks, D. J., & Ward, K. (1988). *Eccentrics: The scientific investigation*. Stirling: Stirling University Press.

Welsh, G. S. (1975). *Creativity and intelligence: A personality approach*. Chapel Hill, NC: Institute for Research in Social Science.

West, M. A., & Rickards, T. (1999). Innovation. In M. A. Runco, & S. R. Pritzker (Eds.), *Encyclopedia of creativity* (pp. 45–55). San Diego: Academic Press.

Wilbur, R. (2013). *Boxed in: The lack of creative thinking in engineering students*. Undergraduate Research Posters. Virginia Commonwealth University.

Yeomans, J. (1968). *The other Taj Mahal*. London: Longmans.

Zuckerman, M. (1969). Theoretical formulations. In J. Zubek (Ed.), *Sensory deprivation: Fifteen years of research* (pp. 407–432). New York: Appleton-Century-Crofts.

Index

Note: Page numbers followed by "*f*" and "*t*" refer to figures and tables, respectively.